D0577493

CHEMICAL PHYSICS
OF FREE MOLECULES

CHEMICAL PHYSICS OF FREE MOLECULES

Norman H. March

Oxford University
Oxford, England

and

Joseph F. Mucci

Vassar College
Poughkeepsie, New York

PLENUM PRESS • NEW YORK AND LONDON

Library of Congress Cataloging-in-Publication Data

March, Norman H. (Norman Henry), 1927-
    Chemical physics of free molecules / Norman H. March and Joseph F.
  Mucci.
        p.   cm.
    Includes bibliographical references and index.
    ISBN 0-306-44270-1
    1. Chemical bonds. 2. Intermolecular forces. 3. Molecules.
  I. Mucci, Joseph F.   II. Title.
  QD461.M26 1994
  541.2'24--dc20   93-24193
                                                                   CIP

ISBN 0-306-44270-1

© 1993 Plenum Press, New York
A Division of Plenum Publishing Corporation
233 Spring Street, New York, N.Y. 10013

All rights reserved

No part of this book may be reproduced, stored in a retrieval system, or transmitted
in any form or by any means, electronic, mechanical, photocopying, microfilming,
recording, or otherwise, without written permission from the Publisher

Printed in the United States of America

QD
461
m26
1993

PREFACE

The aim of this book is to provide a basic treatment of both electronic and nuclear motions in molecules, which is suitable for chemical physicists and for chemists. The book may also be of use, as an introduction to molecules, to materials scientists who wish to employ the methods of quantum chemistry to study condensed molecular matter, though the present work is almost exclusively about the gas phase.* The level is for advanced undergraduate students and postgraduates.

From the outset, attention is focused on the molecule, with its attendant multinuclear character. Thus, the simplest problem treated in the body of the text is the hydrogen molecule ion. However, some background is presented in the Appendixes, regarding the electronic structure of atoms and their binding energies. The main tool employed in treating electronic behavior is the variation method. While this is widely used in current undergraduate texts for dealing with wave functions, a unique feature of the present account is that it is also used, within the framework of the "inhomogeneous electron gas,"† to treat the ground state electron density. Both variational approaches, wave function and electron density, are brought to full fruition when used to establish a self-consistent field. This concept, going back to Hartree, is a recurring theme. We therefore elaborate a little on its meaning immediately below.

Only in the one problem of the H_2^+ ion referred to above do we have a situation of chemical interest where electron repulsion does not occur. It is our view in this work that the existence of localized chemical bonding is a rather direct manifestation of the importance of the correlated motion of electrons, both because the electrons have spin, and parallel spin electrons avoid one another as a consequence of the Pauli Exclusion Principle; and also because all electrons, whether with parallel or antiparallel spins, avoid one another by virtue of Coulomb repulsion. Nevertheless, certain properties of molecules, most importantly the ground state electron density (the pioneers having been Thomas, Fermi, and Dirac), can be calculated by one-electron (orbital) methods, provided the electron correlation, both Pauli principle and Coulombic repulsion, is built into a suitable self-consistent field in which the electrons move independently. This recognition means that the "conflict" which sometimes appeared historically to exist between orbital methods on the one hand and chemical, valence bond, methods on the

---

*A few examples taken from solids and liquids have been used to illustrate points concerned with chemical bonding in free molecules.

†For the reader who becomes deeply interested in the electron density description, advanced accounts can be found in *The Theory of the Inhomogeneous Electron Gas*, S. Lundqvist and N. H. March, eds., Plenum Press, New York (1983); *Density Functional Theory of Atoms and Molecules* by R. G. Parr and W. Yang, Oxford University Press (1989); and *Electron Density Theory of Atoms and Molecules* by N. H. March, Academic Press, New York (1992).

v

other is only apparent. For some purposes, orbital methods are very convenient and can be formulated, at least in principle, exactly.

For specified (i.e., fixed) nuclear positions, knowledge of the ground state electron density suffices (via the so-called Hellmann–Feynman theorem described in the Appendix) to calculate the forces on all nuclei, which are evidently needed to discuss vibrational degrees of freedom. Though this approach is not explicitly utilized in the present work (see Chapter 3 for an alternative approach to molecular force fields), it serves to highlight further the long-term importance for chemical physics of the ground state electron density.*

While we are both practicing theorists, we have also sought in our presentation here to emphasize the crucial interplay between experiment and theory in this subject. This, we believe, is made especially clear in Chapters 1, 5, and 7, the last constituting a brief introduction to the now vast field of chemical reactions.

Finally we note (i) that much of the material presented here has been used in undergraduate courses at Vassar College and at Oxford University, and (ii) that if our book should find favor with university teachers of such courses, we shall be most grateful for constructive criticism, to which we shall do our best to respond if the opportunity arises in the future.

We gratefully acknowledge the encouragement and help of members of the Editorial Staff at Plenum; namely, Amelia McNamara, Laura Troup, and Barbara Sonnenschein. We especially want to thank our copyeditor, Evelyn Grossberg, for her understanding and expertise. In addition, we thank Marilyn Bontempo for the production of the art work.

Norman H. March
Joseph F. Mucci

---

*In principle a directly observable property, unlike the many-electron wave function, via X-ray or electron diffraction experiments.

ACKNOWLEDGMENTS

We gratefully acknowledge permission to reproduce figures and tables from other sources as follows:

Chapter 1

Figure 1.2. *Chemical Bonding Clarified Through Quantum Mechanics* by G. C. Pimentel and R. D. Spratley, Holden-Day Inc., San Francisco (1969). Reprinted by permission of the authors, the copyright owners.

Figure 1.4. *High Resolution Spectroscopy* by J. M. Hollas, Butterworths Publishers (1982). Reprinted by permission of Butterworth–Heinemann Publishers.

Figure 1.5. *ECSA Applied to Free Molecules* by K. Siegbahn, C. Nordling, G. Johansson, P. F. Heden, K. Hamerin, U. Gelius, T. Bergmark, L. O. Worme, R. Manne, and Y. Baer, copyright (c) 1969 by North-Holland Publishing Co., Amsterdam. Reprinted by permission of Elsevier Science Publishers.

Figures 1.7 and 1.8. *Quanta* by P. W. Atkins, 2nd Ed., Oxford University Press, Oxford (1991). Reprinted by permission of Oxford University Press.

Figures 1.9 and 1.10. H. L. Chen and C. B. Moore, *J. Chem. Phys.* **54**, 4072 (1971). Reprinted by permission of the American Institute of Physics and the authors.

Figure 1.11. *Molecular Structure and Dynamics* by W. H. Flygare, Prentice-Hall Inc., Englewood Cliffs, N.J. (1978). Reprinted by permission of Ruth M. Flygare, the current copyright owner.

Figure 1.12. *Modern Spectroscopy* by J. M. Hollas, copyright (c) 1987 by John Wiley and Sons Ltd., Chichester. By permission of copyright owner.

Tables 1.1 and 1.4. Reprinted from *The Nature of the Chemical Bond and the Structure of Molecules and Crystals: An Introduction to Modern Structural Chemistry* 3rd. Ed. by Linus Pauling. 3rd edition copyright (c) 1960 by Cornell University Press. Used by permission of the publisher, Cornell University Press.

Tables 1.2 and 1.3. *Chemical Structure and Bonding* by R. L. DeKock and H. B. Gray, Benjamin/Cummings Publishing Co., Menlo Park, CA. (1980). Reprinted by permission of the current copyright holder, Roger DeKock.

Table 1.5. *Chemical Constitution*, 2nd. Ed. by J. A. A. Ketelaar, Elsevier Science Publishers, Amsterdam, 1958. Reprinted by permission of Elsevier Science Publishers and the author.

Table 1.6. J. L. Hollenberg, *J. Chem. Educ.* **47**, 2 (1970). Reprinted by permission of the Division of Chemical Education of the American Chemical Society.

Chapter 2

Figure 2.4.   *Physical Chemistry* by R. S. Berry, S. A. Rice, and J. Ross. Copyright (c) 1980 by John Wiley and Sons, Inc., New York. Reprinted by permission of John Wiley and Sons, Inc.

Figures 2.5, 2.7, 2.9, 2.10, 2.11, and 2.13.   *Coulson's Valence*, 3rd Ed. by R. McWeeny, Oxford University Press, Oxford, 1979. Reprinted by permission of Oxford University Press.

Figure 2.14.   R. F. W. Bader, I. Keaveney, and P. E. Cade, *J. Chem. Phys.* **47**, 3381 (1967). Reprinted by permission of the American Institute of Physics and the authors.

Table 2.2.   *Wave Mechanics for Chemists* by C. W. N. Cumper, Academic Press Inc., 1966. Reprinted by permission of the current copyright owner Butterworth–Heinemann Ltd. and the author.

Table 2.5.   *The Structure of Small Molecules* by W. J. Orville-Thomas, Elsevier Science Publishers, 1966. Reprinted by permission of the author and the current copyright owner.

Chapter 3

Figures 3.1, 3.9, 3.10, 3.15, 3.17, and Table 3.4.   *The Forces Between Molecules* by M. Rigby, E. B. Smith, W. A. Wakeham, and G. C. Maitland, Oxford University Press, Oxford, 1986. Reprinted by permission of Oxford University Press.

Figure 3.2.   *Introduction to Physical Chemistry* by Arthur M. Lesk, (c) 1982. Reprinted by permission of Prentice-Hall, Inc., Englewood Cliffs, New Jersey.

Figure 3.8.   M. Ross, H. K. Mao, P. M. Bell, and J. A. Xu, *J. Chem. Phys.* **85**, 1028 (1986). Reprinted by permission of the American Institute of Physics and the authors.

Figure 3.11.   *Bonding Theory* by D. J. Royer, McGraw-Hill, New York, 1986. Copyright (c) 1968 by McGraw-Hill Inc. Reprinted by permission by McGraw-Hill, Inc.

Figure 3.12.   *Advanced Physical Chemistry* by J. C. Davis, Ronald Press Co., N.Y., 1965. Reprinted by permission of John Wiley and Sons, Inc., the current copyright owners.

Figure 3.13.   *Kinetic Theory of Gases* by W. Kauzmann, W. A. Benjamin Inc., New York, 1966. Reprinted by permission of Benjamin/Cummings Publishing Co.

Figures 3.14 and 3.16.   *Physical Chemistry*, 2nd Ed. by P. W. Atkins. Copyright (c) 1982 by P. W. Atkins. Reprinted by permission of W. H. Freeman and Company, New York.

Figure 3.18.   P. A. Egelstaff and F. Barocchi, *Phys. Lett.* **3**, 313 (1985). Reprinted by permission of Elsevier Science Publishers and the authors.

Table 3.5.   *The Chemical Bond*, 2nd Ed. by J. N. Murrell, S. F. A. Kettle, and J. M. Tedder. Copyright (c) 1985 by John Wiley and Sons Ltd. Reprinted by permission of John Wiley and Sons Ltd.

Chapter 4

Figure 4.2.   N. L. Allan, D. L. Cooper, C. G. West, P. J. Grout, and N. H. March, *J. Chem. Phys.* **83**, 239 (1983). Reprinted by permission of the American Institute of Physics and the authors.

Figures 4.3 and 4.4.   J. F. Mucci and N. H. March, *J. Chem. Phys.* **71**, 5270 (1979). Reprinted by permission of the American Institute of Physics.

Chapter 5

Figure 5.1.   *Fundamentals of Molecular Spectroscopy* by C. M. Banwell, McGraw-Hill Book Co. Reprinted by permission of McGraw-Hill, Inc.

Figure 5.5.   *Modern Spectroscopy* by J. M. Hollas, 1987. Copyright (c) by John Wiley and Sons Ltd., Chichester. By permission of the copyright owner.

Figures 5.6, 5.7, and 5.10.   *High-Resolution Spectroscopy* by J. M. Hollas, Butterworths Publishers (1982). Reprinted by permission of Butterworth–Heinemann Publishers.

Figure 5.11.   E. A. Ballik and D. A. Ramsay, *Astrophys. J.* **137**, 84 (1963). Reprinted by permission of the publisher and the authors.

Figure 5.13.   L. Karlsson, K. Mattsson, R. Jadrny, T. Bergmark, and K. Siegbahn, *Physica Scripta* **14**, 230 (1976). Reprinted by permission of the Royal Swedish Academy of Sciences.

Figure 5.14.   *Molecular Structure and Dynamics* by W. H. Flygare, Prentice-Hall Inc., Englewood Cliffs, New Jersey, 1978. Reprinted by permission of Ruth M. Flygare the current copyright owner.

Figure 5.17.   *Spectroscopy and Structure* by R. N. Dixon, John Wiley and Sons Inc., 1965. Reprinted by permission of Chapman and Hall, London the current copyright owner.

Figure 5.18.   A. A. Bothner-By and C. Naar-Colin, *Ann. N.Y. Acad. Sci.* **70**, 833 (1958). Reprinted by permission of the New York Academy of Science and the authors.

Figure 5.22.   *Quantum Chemistry* by J. P. Lowe, copyright (c) Academic Press Inc., New York, 1978. Reprinted by permission of Academic Press and the author.

Chapter 6

Figure 6.3.   *Molecular Photoelectron Spectroscopy* by D. W. Turner, C. Baker, A. D. Baker, and C. R. Brundle, Copyright (c) 1970 by John Wiley and Sons Ltd. Reprinted by permission of John Wiley and Sons Ltd.

Figures 6.4, 6.15, and 6.30.   *High Resolution Spectroscopy* by J. M. Hollas, Butterworths Publishers, 1982. Reprinted by permission of Butterworth–Heinemann Publishers.

Figure 6.10.   *Valence*, 2nd Ed, by C. A. Coulson, Oxford University Press, Oxford, 1961. Reprinted by permission of Oxford University Press.

Figure 6.27.   L. Karlsson, L. Mattson, R. Jadrny, T. Bergmark, and K. Siegbahn, *Physica Scripta*, **230** (1976). Reprinted by permission of the Royal Swedish Academy of Sciences.

Figure 6.29.   M. A. Coplan, J. H. Moore, and J. A. Tossell, *J. Chem. Phys.* **68**, 329 (1978). Reprinted by permission of the American Institute of Physics and the authors.

Chapter 7

Figures 7.5, 7.6, 7.7, 7.8, 7.9, and Tables 7.2, 7.3, and 7.4.   *Physical Organic Chemistry* by K. B. Wiberg. John Wiley and Sons Inc., New York 1964. Reprinted by permission of the author the current copyright owner.

Figures 7.10, 7.11, 7.12, 7.13, 7.14, 7.15, 7.16, and 7.17, 7.19, 7.20.   *Quantum Chemistry* by J. P. Lowe, Academic Press 1978. Reprinted by permission of the publisher and author.

Figure 7.18.   *Molecular Orbital Theory* by A. Streitwieser Jr., John Wiley and Sons Inc., N.Y., 1961. Reprinted by permission of the author the current copyright owner.

Figure 7.21.   R. N. Zare and R. B. Bernstein, *Physics Today*, Vol. 33, No. 11 (1980). Reprinted by permission of the American Institute of Physics and the authors.

Figure 7.22 and Table 7.6.   *Chemical Kinetics* 3rd Ed. by Keith J. Laidler. Copyright (c) 1987 by Harper and Row Publishers, Inc., N.Y. Reprinted by permission of Harper Collins Publishers.

Table 7.5.   *Quanta*, 2nd Ed. by P. W. Atkins, Oxford University Press, Oxford, 1991. Reprinted by permission of Oxford University Press.

Appendix

Figure A.5.   C. A. Coulson and I. Fischer, *Philosophical Magazine* (*London*) **40**, 386 (1949). Reproduced with permission of the publisher, Taylor and Francis, Inc.

Figure A.9.   L. S. Bartell and L. O. Brockway, *Phys. Rev.* **90**, 833 (1953). Reprinted by permission of the American Institute of Physics and the authors.

Figures A.11, A.12, and A.13.   J. A. Alonso and N. H. March, *Chem. Phys.* **76**, 121 (1983). Reprinted by permission of Elsevier Science Publishers BV and the authors.

Figures A.14.   K. E. Banyard and N. H. March, *Acta. Cryst.* **9**, 385 (1956). Reprinted by permission of International Union of Crystallography.

Figures A.20, A.21, and A.22 and Table A.1.   *Quantum Chemistry* by J. P. Lowe. Copyright (c) Academic Press, Inc., N.Y. 1978. Reprinted by permission of the publishers and the author.

Figures A.23 and A.24 and Tables A.3 through A.8.   M. Raimondi, M. Simonetta, and J. Gerratt, *Chem. Phys. Letts.* **77**, 12 (1981). Reprinted by permission of Elsevier Science Publishers BV and the authors.

While the above makes abundantly clear our indebtedness to many authors, we want to acknowledge the extensive use we have made of the fine books *High Resolution Spectroscopy* (1982) by J. M. Hollas, *Quantum Chemistry* (1978) by J. P. Lowe, and *Physical Organic Chemistry* (1964) by K. B. Wiberg. Finally, one of us (NHM) chaired sets of splendid pedagogical lectures by Professor N. C. Handy at consecutive Coulson Summer Schools in Theoretical Chemistry in Oxford. Three of the Advanced General Appendixes have their origins directly in these lectures; however, it must be emphasized that the sole responsibility for the final presentation of this and, of course, all other material rests with the present authors.

CONTENTS

Chapter One

CHEMICAL CONCEPTS AND EXPERIMENTAL TECHNIQUES

1.1. Chemical Bonds: Types and Strengths	1
1.2. Single, Double, and Triple Bond Formation via Octet Rule	1
1.2.1. Homonuclear Examples	2
1.2.2. Heteronuclear Examples	3
1.3. Bond Dissociation	4
1.4. Ionization Potential and Electron Affinity	5
1.5. The Concept of Electronegativity	9
1.6. Dipole Moments and Electronegativity	13
1.7. Bond Type and Length: Associated Force Constants	13
1.8. Some Important Experimental Methods	15
1.8.1. Background	15
1.8.2. Mainly Thermodynamics: The Born–Haber Cycle	15
1.8.3. Techniques for Probing Electronic Structure	18
1.8.4. Molecular Spectroscopy and Molecular Properties	22
1.8.5. Coherent Radiation: Laser Studies	22
Problems	34
References	35

Chapter Two

THE NATURE OF BONDING IN DIATOMS

2.1. Valence Bond and Molecular Orbital Methods	37
2.2. Nature of Wave Functions for H_2^+ and H_2	37
2.3. The LCAO Approximation to Ground States of H_2^+ and H_2	39
2.3.1. LCAO Treatment of the H_2 Molecule	41
2.3.2. The Molecular Energy of the H_2 Molecule in Its Ground State	42

2.3.3. The United Atom: He 44
2.3.4. Coulomb and Multicenter Integrals in H_2 45
2.4. Valence Bond Theory of the H_2 Molecule 46
2.5. Comparison of Valence Bond and Molecular Orbital Theories
for H_2 49
2.6. Chemical Bonding, Molecular Orbital Energy Levels, and
Correlation Diagrams for Homonuclear Diatoms 52
2.6.1. Symmetry Classification 52
2.6.2. Correlation Diagrams 54
2.6.3. Electron Configurations 57
2.7. Heteronuclear Diatomic Molecules 60
Problems 66
References 67

Chapter Three

MOLECULAR INTERACTIONS

3.1. Introduction 69
3.2. Long-Range Forces 69
3.2.1. The London Dispersion Force 70
3.2.2. The Drude Model 71
3.2.3. Introduction of Molecular Polarizability into the
Dispersion Interaction 73
3.2.4. Van der Waals Equation of State Related to the London
Dispersion Force 74
3.3. Short-Range Forces 76
3.3.1. The Hard-Sphere Model 76
3.3.2. Point Centers of Inverse Power Law Repulsion 77
3.3.3. The Square-Well Model 77
3.3.4. The Lennard-Jones Potential 78
3.3.5. The Exponential -6 Potential 79
3.4. Short-Range Energies and Quantum-Chemical Studies 80
3.5. The Hydrogen Bond 82
3.6. The Relation to Experimental Studies 85
3.6.1. The Virial Equation: Imperfect Gases 85
3.6.2. Transport Properties of Gases 88
3.6.3. Pair Potentials and Transport Coefficients 90
3.6.4. Molecular Beam Studies 93
3.6.5. Spectra of Molecular Pairs 95
3.6.6. The Inversion of Measured Fluid Structure to Extract
Pair Potentials 96
Problems 98
References 98

Chapter Four

ELECTRON DENSITY DESCRIPTION OF MOLECULES

4.1. Approximate Electron Density–Potential Relation 101
4.2. Energy Relations in Molecules at Equilibrium 103
4.3. Can the Total Energy of a Molecule Be Related to the Sum of
     Orbital Energies? 106
4.4. Foundations of Walsh's Rules for Molecular Shape 108
4.5. Self-Consistent Field Treatment for Diatomic Molecules and the
     Roothaan–Hall Formulation 110
4.6. Roothaan's Approach Compared with the Electron Density
     Description... 112
     4.6.1. Kinetic Energy 112
     4.6.2. Test of Energy Scaling Relations 114
4.7. Chemical Potential in Relation to Electronegativity 115
Problems 119
References 121

Chapter Five

MOLECULAR PARAMETERS DETERMINED BY SPECTROSCOPIC METHODS

5.1. Rotational Spectroscopy 123
     5.1.1. Principal Moments of Inertia of Molecules 125
     5.1.2. The Symmetric Rotor or Symmetric Top 127
     5.1.3. Spherical Rotors 127
     5.1.4. Rotational Spectra of Diatomic and Linear Polyatomic
            Molecules 128
     5.1.5. Rotational Selection Rules 128
5.2. Centrifugal Distortion 132
5.3. Vibrational Spectroscopy of Diatomic Molecules: Harmonic
     Approximation 133
5.4. Anharmonicity 134
5.5. Vibrational Selection Rules 135
5.6. Vibration–Rotation Spectrum.................................. 136
5.7. Electronic Spectroscopy 138
     5.7.1. Rydberg Orbitals 138
     5.7.2. Classification of Electronic States 138
     5.7.3. Electric Dipole Selection Rules for Transitions with $\Delta S = 0$ .. 140
     5.7.4. Transitions with $\Delta S \neq 0$ 141
5.8. Excited-State Potential Energy Curves 142
     5.8.1. Progressions and Sequences 144
     5.8.2. Vibronic Transitions 144

5.9. Determination of Ionization Potentials of Polyatomic Molecules 145
5.10. Nuclear Magnetic Resonance Spectra 147
    5.10.1. Chemical Shifts 149
    5.10.2. Chemical Applications 150
    5.10.3. Interaction Constants: Spin–Spin Coupling 151
    5.10.4. Comments on Analyses of NMR Spectra 151
    5.10.5. Rules for Interpreting First-Order Spectra 152
    5.10.6. The Relation between Chemical Shift and Electronegativity 153
5.11. Electron Spin Resonance and π-Electron Densities: Hyperfine
    Structure 154
Problems 158
References 159

Chapter Six

MOLECULAR ORBITAL METHODS AND POLYATOMIC MOLECULES

6.1. Directed Bonds: Conformation of H_2O and NH_3 161
6.2. Bent Molecules: Walsh Diagrams and Rules 162
    6.2.1. HeI UPS Spectrum of the H_2O Molecule 163
    6.2.2. HeI UPS Spectrum of H_2S 165
6.3. UPS of AH_3 Molecules 166
6.4. The Need for Hybrid Orbitals 168
6.5. Some Carbon Compounds with sp^n Hybridization ($n = 1, 2, 3$) 173
6.6. Molecular Orbital Theory of π-Electron Systems 175
    6.6.1. The Hückel Molecular Orbital Method 177
    6.6.2. An Example of π-Level Spectra: Ethylene 180
    6.6.3. The HeI UPS Spectrum of Ethylene 181
    6.6.4. Other Examples of π-Level Spectra 183
    6.6.5. The UPS Spectrum of Benzene 192
6.7. Determination of Molecular Orbitals and LCAO Coefficients 195
6.8. Molecular Indexes 197
    6.8.1. Electron Density on Atoms 198
    6.8.2. Charge Density on Atoms 199
    6.8.3. Bond Order 201
6.9. Some Quantum-Mechanical and Semiempirical Methods Applied
    to Polyatomic Molecules 202
    6.9.1. The Pariser–Parr–Pople Method 205
    6.9.2. The *Ab Initio* Method 207
6.10. Molecular Orbitals in Acetylene and the ($e, 2e$) Experiment 207
    6.10.1. The Technique and Theory of ($e, 2e$) Studies 207
    6.10.2. Results of Measurements on Acetylene 209
6.11. Slater–Kohn–Sham One-Electron Equations 211
6.12. Renner–Teller Effect 211
Problems 213
References 216

Chapter Seven

**CHEMICAL REACTIONS, DYNAMICS,
AND LASER SPECTROSCOPY**

7.1. Introduction and Background 219
    7.1.1. Rates of Chemical Reactions: Reaction Rates and Rate Laws 219
    7.1.2. Largely Qualitative Considerations 224
    7.1.3. The Transition State: Reaction Coordinate and Potential
           Energy Diagrams 224
7.2. Rates of Chemical Reactions: Absolute Rate Theory 227
    7.2.1. The Methane–Chlorine-Atom Reaction 228
    7.2.2. A Reaction Coordinate Representation 232
    7.2.3. Determination of the Equilibrium Constant 233
7.3. The Woodward–Hoffmann Rules 236
    7.3.1. Introduction 236
    7.3.2. Cycloaddition Reactions 241
    7.3.3. Other Types of Chemical Reactions 242
7.4. Localization Energy and the Rate of Reaction for Aromatic
    Hydrocarbons 246
7.5. Oxidation–Reduction and Orbital Energies of Hydrocarbons 247
7.6. Laser Spectroscopy and Chemical Dynamics 248
    7.6.1. Rate Coefficient Analysis 248
    7.6.2. State-to-State Chemical Reaction Dynamics: Hydrogen
           Exchange Reaction 251
7.7. Electron Density Theory and Chemical Reactivity 252
    7.7.1. Hardness and Softness 252
    7.7.2. Acids and Bases 253
7.8. Other Studies in Chemical Kinetics 254
    7.8.1. Electron Impact Studies 254
    7.8.2. Photons and Chemical Reactions 255
    7.8.3. Rate Studies for Reactions of the Hydrated Electron 256
    7.8.4. Topics for Further Study Involving Chemical Reactivity 257
Problems 257
References 259

APPENDIX

A1.1. Wave Functions for the Hydrogen Atom 261
A1.2. The Periodic Table and Atomic Ground States 271
A1.3. Total Energies of Heavy Atomic Ions—Coulomb Field Model .. 273
A1.4. Orthogonality of Solutions of the Schrödinger Equation 275
A2.1. Variation Principle 276
A2.2. Integrals Involved in LCAO Treatment of the H_2^+ Ground State 277

A2.3.   Born–Oppenheimer Approximation   278
A2.4.   Slater- and Gaussian-Type Orbitals   280
A2.5.   The Coulson–Fischer Wave Function for the Ground State
          of the H_2 Molecule   281
A2.6.   Virial and Hellmann–Feynman Theorems   282
A3.1.   Topics Relevant to the Treatment of Intermolecular Forces   289
A3.2.   Bibliography for Further Study of Intermolecular Forces   290
A4.1.   The Correspondence between Cells in Phase Space and
          Quantum-Mechanical Energy Levels   290
A4.2.   The Kinetic Energy Density of an Inhomogeneous Electron Gas   294
A4.3.   The Chemical Potential, Teller's Theorem, and Scaling of
          Energies of Homonuclear Diatoms   295
A4.4.   The Self-Consistent Field in the Helium Atom   296
A4.5.   The Self-Consistent Field Treatment of Binding Energies of
          Heavy Positive Atomic Ions   297
A4.6.   The Hartree–Fock Self-Consistent Field Method   299
A4.7.   The Dirac–Slater Exchange Energy and Existence of a
          One-Body Potential Including Both Exchange and Correlation . .   308
A4.8.   Proof that the Ground-State Energy of a Molecule Is
          Uniquely Determined by the Electron Density   312
A4.9.   Modeling of the Chemical Potential in Hydrogen Halides
          and Mixed Halides   314
A4.10.   X-Ray Scattering by Neon-Like Molecules   318
A4.11.   Two-Center Calculations from the Thomas–Fermi Theory   322
A5.1.   Rotational Energy Levels of Some Simple Classes
          of Molecules: The Symmetric Rotor   326
A5.2.   The Rotational Partition Function in Relation to
          Spectroscopic Intensities   328
A5.3.   Normal Modes of Vibration of Molecules   330
A5.4.   The Franck–Condon Principle   335
A5.5.   Time-Dependent Perturbation Theory and Selection Rules for
          Electric Dipole Transitions in Atoms   338
A5.6.   Spin–Orbit Coupling   342
A5.7.   Dirac's Relativistic Wave Equation for One Electron   345
A5.8.   Relativistic Electron Density Theory for Molecules Composed
          of Heavy Atoms   349
A6.1.   The Jahn–Teller Effect   350
A6.2.   Koopmans' Theorem and Its Use in Interpreting
          Photoelectron Spectra   351
A7.1.   Symmetry Arguments for Electrocyclic Reactions   353
A7.2.   Chemical Reactions: Arrhenius' Empirical Work, the
          Collision Theory, and the Absolute Rate or Transition
          State Theory   356
Advanced Problems   360
AI.   Some Advanced Aspects of Quantum-Mechanical Perturbation
          Theory   362
AII.   The Formation of Acetylene from Two CH Fragments   367
AIII.   MacDonald's Theorem   374

AIV. Spectroscopic Nomenclature 375
AV. The Wentzel–Kramers–Brillouin Semiclassical Method for
Calculating Eigenvalues for Central Fields 377
AVI. Electron Correlation in the Helium Atom and the Slow
Convergence of Configuration Interaction 379
Further Problems 380
References 386

INDEX   389

CHEMICAL CONCEPTS AND EXPERIMENTAL TECHNIQUES

This introductory chapter contains: a summary of some basic chemical concepts and a brief survey of techniques used to probe the electronic structure of molecules and other molecular properties. These two aspects then lead naturally into a discussion of the way in which the wave mechanics of electrons in molecules can be employed to deepen insight into the nature of the chemical bond.

1.1. CHEMICAL BONDS: TYPES AND STRENGTHS

We start from a pictorial description of the different types of chemical bonds in free molecules, restricting the discussion in this and the following chapter to diatomic cases. In the first part of the present chapter, homonuclear diatomics will first be considered as specific examples of the important types of chemical bonds that can arise. This will be followed by a treatment of the relationship between the bonding of heteronuclear diatoms and the properties of the constituent atoms in their own homonuclear diatomic molecules. This will afford an opportunity, at the outset, to motivate the concept of electronegativity, which is of the first rank in importance for understanding the chemical physics of free molecules. Later, in Chapter 4, we shall discuss how this concept can be made quantitative, within one particular description of the electronic structure of the ground state of molecules, namely electron density (rather than wave function) theory. Though it is hardly necessary to emphasize the interest of excited states for chemistry, it nevertheless remains of great importance to have a full, and preferably compact, picture of the nature of chemical bonding in the ground state of molecules. Electron density theory provides just this.

1.2. SINGLE, DOUBLE, AND TRIPLE BOND FORMATION VIA OCTET RULE

The study of the electronic configuration of atoms, plus the Pauli exclusion principle, which in elementary terms, states that two electrons with parallel spins are not allowed to occupy the same quantum state, yields the periodic behavior of the elements. For completeness, the reader can refer to Appendix A1.1 for the treatment of the quantum numbers and the electronic wave functions of the hydrogen (like) atom.

All we need do at this stage is recall that atoms with the same outer electronic configuration are in the same group (column) of the Periodic Table (see Appendix A1.2). It is this outer electronic configuration that is all-important for understanding the formation of chemical bonds and for the existence of stable molecules. In the broadest sense, chemical bonds form because the total energy of the system decreases thereby, that is, the equilibrium molecule has a lower energy than its separated constituent atoms. This point will be taken up in more detail in Chapter 2.

Valency is the term used to describe the ability of atoms to combine with each other to form molecules. The very stable nature of an outer octet of electrons in the noble gases, with the exception of helium, has proved extremely useful in the understanding of molecular formation and stability. The outer electronic configuration of the noble gases is ns^2np^6. Hence the atoms of other elements with atomic numbers close to those of the noble gases tend to achieve the outer octet configuration in molecular bonding. Elements immediately following the noble gases in atomic number may achieve a stable octet electronic configuration by losing electrons and forming thereby positively charged ions. Elements such as these are said to have positive valence numbers. Those elements immediately preceding the noble gases could, in contrast, achieve the stable octet structure by gaining electrons and would consequently become negatively charged ions, these elements would be assigned negative valence numbers. Consequently, it may be said that such elements can enter into chemical combination by losing or gaining electrons in order to take up a noble gas type of electronic configuration. The chemical compounds that are formed by such loss or gain of electrons are plainly ionic in character, with the ions held together by electrostatic attraction. The valence number is therefore equal to the ionic charge (i.e., the number of electrons lost or gained). Many compounds can be classified using this idea of ionic bond formation. However, we must emphasize here that there are many molecules which do not fit into such a classification. Those that fail to fit into this conceptual framework are not electrolytes, and it is also often true that the atoms comprising the molecule are the same (homonuclear molecules), in which case electron transfer is plainly not possible.

To overcome the apparent difficulty of molecules that were formed without the loss or gain of electrons by their constituents, Lewis (1916) introduced a further important concept. He asserted that atoms might attain noble gas electronic structure by sharing electrons. Such shared electrons would then be considered as belonging to both atoms and attain the noble gas structure for both of them. Each pair of such shared electrons make up what is known as a covalent bond. With this idea of Lewis's, it is immediately possible to consider the formation of so-called single, double, and triple bonds in some homonuclear diatomic molecules. We turn now to some important examples.

1.2.1. Homonuclear Examples

(a) Hydrogen Molecule H_2

A single bond is formed by sharing electrons to yield the electronic configuration of the rare gas He. This is, as will be elaborated on later, the so-called united atom formed when the H_2 molecule is compressed (thought experiment!) to zero bond length or internuclear distance. This process, envisaged by Lewis, may be represented as

$$H(1s^1) + H(1s^1) \rightarrow H:H \text{ or } H—H \tag{1.1}$$

(b) Oxygen Molecule O_2

Here a double bond is formed, electrons being shared to yield the stable outer octet structure of Ne. One can depict this double bond formation by writing

$$O(1s^2 2s^2 2p^4) + O(1s^2 2s^2 2p^4) \rightarrow \overset{\times\times}{\underset{\times}{\times}O}\overset{\circ\circ}{\underset{\circ}{O}}\text{ or } O{=}O$$

A cautionary remark in this example is to note that the ground state of the O_2 molecule is found to have characteristic magnetic properties. This point will come up in Chapter 2, when a simple explanation will be offered that requires a new ingredient beyond the Lewis concept exploited above.

(c) Nitrogen Molecule N_2

In this case, a triple bond is formed and the shared electrons yield the stable octet electronic configuration of Ne. One can represent this type of bond formation by writing

$$N(1s^2 2s^2 2p^3) + N(1s^2 2s^2 2p^3) \rightarrow \overset{\times}{\underset{\times}{\times}N}\overset{\circ}{\underset{\circ}{N}}\text{ or } N{\equiv}N$$

1.2.2. Heteronuclear Examples

Pressing the Lewis concept of electronic octet formation, one can represent the formation of some heteronuclear covalent diatoms; we take the following two simple examples.

(a) Hydrogen Chloride Molecule HCl

This time we may write

$$H(1s^1) + Cl(1s^2 2s^2 2p^6 3s^2 3p^5) \rightarrow H\ \overset{\circ\circ}{\underset{\circ\circ}{\times}Cl}\overset{}{\underset{}{}}\text{ or } H—Cl$$

Again it should be noted that the outer He-type electronic configuration satisfies the H atom, while the rare gas octet structure satisfies the Cl constituent. Similar considerations can be applied to the HF, HBr, and HI molecules.

(b) *Bromine Chloride Molecule BrCl*

The molecular formation can be depicted thus:

$$Br(1s^2 2s^2 2p^6 3s^2 3p^6 3d^{10} 4s^2 4p^5) + Cl(1s^2 2s^2 2p^6 3s^2 3p^5)$$

$$\rightarrow \overset{\times\times}{\underset{\times\times}{\times Br^{\times}_{\times}}} \overset{oo}{\underset{oo}{Cl^o_o}} \text{ or } Br—Cl$$

Once again one may note the attainment of the outer stable electronic configuration of the rare gases. The same procedure can be used in the formation of other heteronuclear covalent compounds.

It must be stressed at this point, however, that while the Lewis octet formation by sharing electrons to attain the outer electronic configuration of the rare gases is very useful in the study of chemical bond formation for many compounds, it does not have universal validity. Indeed the octet rule has some serious limitations, but the concept that electrons like to pair with one another has almost universal applicability. In later chapters, we shall develop a fully wave-mechanical view of the covalent bond, but that does not in the least deny the importance of considering some very useful concepts related to the formation of the chemical bond in the above simple manner.

1.3. BOND DISSOCIATION

After these qualitative concepts to help understand bond formation have been set out, it is natural to appeal to experiment to gain further insight into the nature of the various bonds.

An immediate and important deduction that follows from experimental data is that heats of formation associated with particular bonds are additive. This means, in practice, that a quantity known as bond energy can be associated with a particular bond. This bond energy is equivalent to the energy of dissociation of diatomic molecules into atoms. In the case of polyatomic molecules (which will be discussed in some detail in Chapter 6), when all the bonds are alike, the bond energy is the average of the energies required to break all the bonds in the molecule. This average energy is often termed the bond strength.

In order to obtain quantitative values for the bond energies, one must have data for the heats of formation of the monatomic gas relative to the standard states and the heats of formation of the molecules.

Let us take as a specific example the bond energy of the O—H linkage. The heats of dissociation of molecular hydrogen H_2 and oxygen O_2 are 104.2 and 118.3 kcal/mole, respectively, which can be represented as (g indicating the gas);

$$H_2(g) \rightarrow 2H(g), \qquad \Delta H = 104.2 \text{ kcal}$$

$$O_2(g) \rightarrow 2O(g), \qquad \Delta H = 118.3 \text{ kcal}$$

The heat of formation of gaseous H_2O from its elements is 57.8 kcal/mole or

$$H_2(g) + \tfrac{1}{2}O_2(g) \rightarrow H_2O(g), \qquad \Delta H = -57.8 \text{ kcal}$$

The heat of formation of gaseous H_2O from gaseous hydrogen and oxygen atoms is found to be 221.1 kcal/mole as follows:

$$2H(g) \rightarrow H_2(g), \qquad \Delta H = -104.2 \text{ kcal}$$

$$O(g) \rightarrow \tfrac{1}{2}O_2(g), \qquad \Delta H = -59.2 \text{ kcal}$$

$$H_2(g) + \tfrac{1}{2}O_2(g) \rightarrow H_2O(g), \qquad \Delta H = -57.8 \text{ kcal}$$

or

$$2H(g) + O(g) \rightarrow H_2O(g), \qquad \Delta H = -221.1 \text{ kcal}$$

The H_2O molecule contains two O—H bonds; hence the average bond energy for this linkage is $\tfrac{1}{2}(221.1) = 110.6$ kcal/mole. Single-bond energies can be calculated for several bonds by the method just described. Data thus obtained are shown in Table 1.1. Some of the bond energies in the table were determined spectroscopically (see Chapter 5). In any case, the most important conclusion of this section is that it is possible to assign quantitative values of bond energies to various linkages.

In the case of molecules containing double and triple bonds, the bond energies can be calculated by a similar procedure involving thermochemical data and consequently we need not go into more detail here.

1.4. IONIZATION POTENTIAL AND ELECTRON AFFINITY

The ionization potential of an atom is the energy required to remove an electron from an atom in the gaseous phase. The first ionization potential I_1 is the

TABLE 1.1
Energy Values for Single Bonds (in kcal/mole)[a]

Bond	Bond energy	Bond	Bond energy	Bond	Bond energy	Bond	Bond energy
H—H	104.2	C—N	69.7	Sn—Sn	34.2	O—F	44.2
H—C	98.8	C—O	84.0	N—N	38.4	O—Cl	48.5
H—Si	70.4	C—S	62.0	N—F	64.5	S—S	50.9
H—N	93.4	C—F	105.4	N—C	47.7	S—Cl	59.7
H—P	76.4	C—Cl	78.5	P—P	51.3	S—Br	50.7
H—As	58.6	C—Br	65.9	P—Cl	79.1	Se—Se	44.0
H—O	110.6	C—I	57.4	P—Br	65.4	Te—Te	33.0
H—S	81.1	Si—Si	42.2	P—I	51.4	F—F	36.6
H—Se	66.1	Si—O	88.2	As—As	32.1	Cl—F	60.6
H—Te	57.5	Si—S	54.2	As—F	111.3	Cl—Cl	58.0
H—F	134.6	Si—F	129.3	As—Cl	68.9	Br—Cl	52.3
H—Cl	103.2	Si—Cl	85.7	As—Br	56.5	Br—Br	46.1
H—Br	87.5	Si—Br	69.1	As—I	41.6	I—Cl	50.3
H—I	71.4	Si—I	50.9	Sb—Sb	30.2	I—Br	42.5
C—C	33.1	Ge—Ge	37.6	Bi—Bi	25.0	I—I	36.1
C—Si	69.3	Ge—Cl	97.5	O—O	33.2		

[a]Data from Pauling (1960).

energy that must be supplied to remove one electron from the neutral atom, which results in the production of a gaseous ion which is unipositive, i.e.,

$$\text{atom(g)} + I_1 \rightarrow \text{ion}^+(\text{g}) + e^-, \qquad \Delta E = I_1$$

For example, in the case of the Na atom we have

$$\text{Na(g)} + 5.14\,\text{eV} \rightarrow \text{Na}^+(\text{g}) + e^-, \qquad \Delta E = I_1 = 5.14\,\text{eV}$$

We can proceed to take a further electron away from $\text{Na}^+(\text{g})$ to form $\text{Na}^{2+}(\text{g})$; the removal of a second electron would require energy I_2, and we can write

$$\text{Na}^+(\text{g}) + I_2 \rightarrow \text{Na}^{2+}(\text{g}) + e^-, \qquad \Delta E = I_2 = 47.3\,\text{eV}$$

If yet a further electron is stripped off we have

$$\text{Na}^{2+}(\text{g}) + I_3 \rightarrow \text{Na}^{3+}(\text{g}) + e^-, \qquad \Delta E = I_3 = 71.6\,\text{eV}$$

I_1, I_2, and I_n are referred to as the first, second, and nth ionization potentials, respectively. We note immediately that

$$I_1 < I_2 < I_3 < \cdots < I_n$$

Examination of Table 1.2 shows that this set of inequalities is true for all atoms. H, having only one electron, has only one ionization potential.

Figure 1.1 illustrates the variation of the first ionization potential of atoms through the Periodic Table. It will be noted that the first ionization potential increases along a period and decreases in a group (for numerical values, see Table 1.2). There are reasons for the departures from this trend but it is not our purpose to go into detail of atomic properties in the main text, unless these relate directly to molecular properties. Later we shall see that there are indeed series of molecules in which the molecular ionization potentials relate, rather directly, to atomic ionization potentials. The theory underlying some of the regularities in atomic ionization potentials is set out in Appendix A1.3, where the so-called $1/Z$ expansion for the total energy $E(Z, N)$ of atomic ions with N electrons is reviewed briefly (see also Problems 1.2 and 1.3).

Another important property of an atom is the electron affinity A, which is defined as the energy released when an electron is added to a neutral atom in the gaseous phase. Thus we may write

$$\text{atom(g)} + e^- \rightarrow \text{ion}^-(\text{g}), \qquad \Delta E = -A \qquad \textbf{(1.2)}$$

The alternative definition used on occasion is the reverse of the above process, i.e., the energy required to remove an electron from the negative ion. This means when using this second definition, that we simply change the sign of A. Electron affinities are rather difficult to measure experimentally; those that have been determined are listed in Table 1.3.

TABLE 1.2
Ionization Energies of the Elements (in eV)^a

Z	Element	I_1	I_2	I_3	I_4	I_5	I_6	I_7	I_8
1	H	13.598							
2	He	24.587	54.416						
3	Li	5.392	75.638	122.451					
4	Be	9.322	18.211	153.893	217.713				
5	B	8.298	25.154	37.930	259.368	340.217			
6	C	11.260	24.383	47.887	64.492	392.077	489.981		
7	N	14.534	29.601	47.448	77.472	97.888	552.057	667.029	
8	O	13.618	35.116	54.934	77.412	113.896	138.116	739.315	871.387
9	F	17.422	34.970	62.707	87.138	114.240	157.161	185.182	953.886
10	Ne	21.564	40.962	63.45	97.11	126.21	157.93	207.27	239.09
11	Na	5.139	47.286	71.64	98.91	138.39	172.15	208.47	264.18
12	Mg	7.646	15.035	80.143	109.24	141.26	186.50	224.94	265.90
13	Al	5.986	18.828	28.447	119.99	153.71	190.47	241.43	284.59
14	Si	8.151	16.345	33.492	45.141	166.77	205.05	246.52	303.17
15	P	10.486	19.725	30.18	51.37	65.023	220.43	263.22	309.41
16	S	10.360	23.33	34.83	47.30	72.68	88.049	280.93	328.23
17	Cl	12.967	23.81	39.61	53.46	67.8	97.03	114.193	348.28
18	Ar	15.759	27.629	40.74	59.81	75.02	91.007	124.319	143.456
19	K	4.341	31.625	45.72	60.91	82.66	100.0	117.56	154.86
20	Ca	6.113	11.871	50.908	67.10	84.41	108.78	127.7	147.24
21	Sc	6.54	12.80	24.76	73.47	91.66	111.1	138.0	158.7
22	Ti	6.82	13.58	27.491	43.266	99.22	119.36	140.8	168.5
23	V	6.74	14.65	29.310	46.707	65.23	128.12	150.17	173.7
24	Cr	6.766	16.50	30.96	49.1	69.3	90.56	161.1	184.7
25	Mn	7.435	15.640	33.667	51.2	72.4	95	119.27	196.46
26	Fe	7.870	16.18	30.651	54.8	75.0	99	125	151.06
27	Co	7.86	17.06	33.50	51.3	79.5	102	129	157
28	Ni	7.635	18.168	35.17	54.9	75.5	108	133	162
29	Cu	7.726	20.292	36.83	55.2	79.9	103	139	166
30	Zn	9.394	17.964	39.722	59.4	82.6	108	134	174
31	Ga	5.999	20.51	30.71	64				
32	Ge	7.899	15.934	34.22	45.71	93.5			
33	As	9.81	18.633	28.351	50.13	62.63	127.6		
34	Se	9.752	21.19	30.820	42.944	68.3	81.70	155.4	
35	Br	11.814	21.8	36	47.3	59.7	88.6	103.0	192.8
36	Kr	13.999	24.359	36.95	52.5	64.7	78.5	111.0	126
37	Rb	4.177	27.28	40	52.6	71.0	84.4	99.2	136
38	Sr	5.695	11.030	43.6	57	71.6	90.8	106	122.3
39	Y	6.38	12.24	20.52	61.8	77.0	93.0	116	129
40	Zr	6.84	13.13	22.99	34.34	81.5			
41	Nb	6.88	14.32	25.04	38.3	50.55	102.6	125	
42	Mo	7.099	16.15	27.16	46.4	61.2	68	126.8	153
43	Tc	7.28	15.26	29.54					
44	Ru	7.37	16.76	28.47					
45	Rh	7.46	18.08	31.06					
46	Pd	8.34	19.43	32.93					
47	Ag	7.576	21.49	34.83					
48	Cd	8.993	16.908	37.48					
49	In	5.786	18.869	28.03	54				
50	Sn	7.344	14.632	30.502	40.734	72.28			
51	Sb	8.641	16.53	25.3	44.2	56	108		
52	Te	9.009	18.6	27.96	37.41	58.75	70.7	137	
53	I	10.451	19.131	33					

(*con't*)

TABLE 1.2 (*continued*)

Z	Element	I_1	I_2	I_3	I_4	I_5	I_6	I_7	I_8
54	Xe	12.130	21.21	32.1					
55	Cs	3.894	25.1						
56	Ba	5.212	10.004						
57	La	5.577	11.06	19.177	49.95				
58	Ce	5.47	10.85	20.198	36.758				
59	Pr	5.422	10.55	21.624	38.98				
60	Nd	5.489	10.73	22.1	40.41				
61	Pm	5.554	10.90	22.3	41.1				
62	Sm	5.631	11.07	23.4	41.4				
63	Eu	5.666	11.241	24.9	42.6				
64	Gd	6.141	12.09	20.63	44.0				
65	Tb	5.85	11.52	21.91	39.8				
66	Dy	5.927	11.67	22.8	41.5				
67	Ho	6.02	11.80	22.84	42.5				
68	Er	6.10	11.93	22.74	42.6				
69	Tm	6.184	12.05	23.68	42.7				
70	Yb	6.254	12.18	25.03	43.7				
71	Lu	5.426	13.9	20.96	45.19				
72	Hf	6.6	14.9	23.3	33.33				
73	Ta	7.89							
74	W	7.98							
75	Re	7.88							
76	Os	8.7							
77	Ir	9.1							
78	Pt	9.0	18.563						
79	Au	9.225	20.5						
80	Hg	10.437	18.756	34.2					
81	Tl	6.108	20.428	29.83					
82	Pb	7.416	15.032	31.937	42.32	68.8			
83	Bi	7.289	16.69	25.56	45.3	56.0	88.3		
84	Po	8.42							
85	At								
86	Rn	10.748							
87	Fr								
88	Ra	5.279	10.147						
89	Ac	5.2							
90	Th	6.1							
91	Pa	5.9							
92	U	6.05							
93	Np	6.2							
94	Pu	6.06							
95	Am	5.99							
96	Cm	6.02							
97	Bk	6.23							
98	Cf	6.30							
99	Es	6.42							
100	Fm	6.50							
101	Md	6.58							
102	No	6.65							
103	Lr								

[a]From De Kock and Gray (1980).

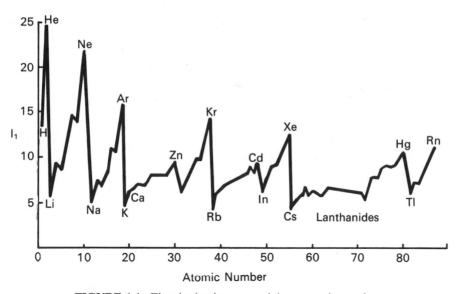

FIGURE 1.1. First ionization potential *vs.* atomic number.

1.5. THE CONCEPT OF ELECTRONEGATIVITY

We have already referred in passing to the concept of electronegativity in the introduction to this chapter. However, this concept is so central in chemical physics that we must set it out in somewhat more detail here, especially its motivation and its utility. Mulliken and Pauling [see Pauling (1960)] pointed out that in considering bonds between unlike atoms, it is helpful to introduce a quantity, called electronegativity and denoted by x, which represents the electron-attracting power of the atoms forming the bond. Let us consider this idea somewhat further.

It has already been seen how thermochemical data can be employed to determine bond energies. For homonuclear diatomic molecules containing a single bond, e.g., H_2, the bond energy is equal to the dissociation energy of the gaseous molecule into isolated atoms. Experimentally it is found that the bond energy between unlike atoms is greater than the energy expected for a truly covalent bond, i.e., one formed by the equal sharing of the bonding electrons between like atoms. The additional bond energy means a more stable bond and is sometimes referred to as the ionic resonance energy, denoted by Δ.

Consider a single covalent bond a—b between the two unlike atoms a and b. Pauling states that the bond energy of a truly covalent bond a—b should be intermediate between the bond energies of the a—a bond and the b—b bond. Hence, one can write that the bond energy $D(a—b)$ is the arithmetic mean of the bond (dissociation) energies of $D(a—a)$ and $D(b—b)$ or

$$D(a—b) = \tfrac{1}{2}[D(a—a) + D(b—b)] \tag{1.3}$$

In a slightly different form, Pauling has also taken the geometric rather than the arithmetic mean; i.e., he writes as an alternative to equation (1.3),

$$D(a—b) = [D(a—a) \times D(b—b)]^{1/2} \tag{1.4}$$

TABLE 1.3
Atomic Electron Affinities[a]

Atom	Orbital electronic configuration	Electron affinity (eV)	Orbital electronic configuration of anion
H	$1s^1$	0.754	[He]
F	$[He]2s^22p^5$	3.34	[Ne]
Cl	$[Ne]3s^23p^5$	3.61	[Ar]
Br	$[Ar]4s^23d^{10}4p^5$	3.36	[Kr]
I	$[Kr]5s^24d^{10}5p^5$	3.06	[Xe]
O	$[He]2s^22p^4$	1.47	$[He]2s^22p^5$
S	$[Ne]3s^23p^4$	2.08	$[Ne]3s^23p^5$
Se	$[Ar]4s^23d^{10}4p^4$	2.02	$[Ar]4s^23d^{10}4p^5$
Te	$[Kr]5s^24d^{10}5p^4$	1.97	$[Kr]5s^24d^{10}5p^5$
N	$[He]2s^22p^3$	0.0 ± 0.2	$[He]2s^22p^4$
P	$[Ne]3s^23p^3$	0.77	$[Ne]3s^23p^4$
As	$[Ar]4s^23d^{10}4p^3$	0.80	$[Ar]4s^23d^{10}4p^4$
C	$[He]2s^22p^2$	1.27	$[He]2s^22p^3$
Si	$[Ne]3s^23p^2$	1.24	$[Ne]3s^23p^3$
Ge	$[Ar]4s^23d^{10}4p^2$	1.20	$[Ar]4s^23d^{10}4p^3$
B	$[He]2s^22p^1$	0.24	$[He]2s^22p^2$
Al	$[Ne]3s^23p^1$	(0.52)	$[Ne]3s^23p^2$
Ga	$[Ar]4s^23d^{10}4p^1$	(0.37)	$[Ar]4s^23d^{10}4p^2$
In	$[Kr]5s^24d^{10}5p^1$	(0.35)	$[Kr]5s^24d^{10}5p^2$
Be	$[He]2s^2$	≤ 0	$[He]2s^22p^1$
Mg	$[Ne]3s^2$	≤ 0	$[Ne]3s^23p^1$
Zn	$[Ar]4s^23d^{10}$	≤ 0	$[Ar]4s^23d^{10}4p^1$
Cd	$[Kr]5s^24d^{10}$	≤ 0	$[Kr]5s^24d^{10}5p^1$
Li	$[He]2s^1$	0.62	$[He]2s^2$
Na	$[Ne]3s^1$	0.55	$[Ne]3s^2$
Cu	$[Ar]4s^23d^9$	1.28	$[Ar]4s^23d^{10}$
Ag	$[Kr]5s^14d^{10}$	1.30	$[Kr]5s^24d^{10}$
Au	$[Xe]4f^{14}5d^{10}6s^1$	2.31	$[Xe]4f^{14}5d^{10}6s^2$
He	$1s^2$	≤ 0	$1s^22s^1$
Ne	$[He]2s^22p^6$	≤ 0	$[He]2s^22p^63s^1$

[a]From DeKock and Gray (1980).

The difference between $D(a\!-\!b)$, using either the arithmetic or the geometric mean, and the experimentally determined bond energy will then lead to an approximate estimate of the so-called ionic resonance energy.

Returning to the definition of electronegativity as the ability of an atom in a molecule to attract electrons, we now give a brief and simplified description of how it can be estimated in practice, following Mulliken (1934, 1935).

As in equation (1.2), the electron affinity of an atom a for electrons can be expressed through

$$a + e^- = a^- + A_a \tag{1.2a}$$

and similarly for atom b:

$$b + e^- = b^- + A_b \tag{1.2b}$$

The first ionization potential I of an atom, discussed earlier in Section 1.4, is the energy necessary to pull an electron out of atom a to form a^+, i.e.,

$$a + I_a = a^+ + e^- \tag{1.5}$$

and similarly for atom b:

$$b + I_b = b^+ + e^- \tag{1.6}$$

What we are interested in is the molecule a—b and the interplay of a with b to form the polar state a^+b^- or a^-b^+. The electrons that are in the bond, i.e., between a and b, require us to consider the relative energies of

$$(i) \quad a + b \rightarrow a^+ + b^-, \qquad E_i = I_a + A_b$$
$$(ii) \quad a + b \rightarrow a^- + b^+, \qquad E_{ii} = I_b + A_a \tag{1.7}$$

If $E_i = E_{ii}$, we would expect the bonding electrons to be equally shared in the a—b bond; i.e., we would have a homopolar covalent bond:

$$I_a + A_b = I_b + A_a$$

or $\tag{1.8}$

$$I_a - A_a = I_b - A_b$$

However, if $I_a - A_a > I_b - A_b$, then a^-b^+ will be the net result as opposed to a^+b^-. Consequently, $I - A$ is a measure of the ability of an atom to attract electronic charge density to itself. So, following Mulliken, we shall write the electronegativity, say, x, of an atom, as

$$x = \text{constant}(I + |A|) \tag{1.9}$$

The so-called Pauling scale (see Pauling's book, *The Nature of the Chemical Bond*) and the Mulliken scale of electronegativity discussed above yield essentially equivalent results. In Chapter 4, we shall give a deeper insight into the result (1.9), and shall argue that the constant is usefully adopted as $\frac{1}{2}$, i.e., $x = \frac{1}{2}(I + |A|)$. The results thus obtained for several atoms are given in Table 1.4.

In Pauling's book, arguments are presented whereby the (rough) equality (1.3) is replaced by

$$D(a - b) = \frac{1}{2}[D(a - a) + D(b - b)] + \text{constant}(x_a - x_b)^2 \tag{1.10}$$

The considerable merit of this equation is that it relates the ab molecule directly to the homonuclear cases aa and bb, plus properties (x_a and x_b) derivable from experimental information on atoms, via equation (1.9). Some of these points will be taken up again later in this volume [see Atkins (1991) for other electronegativity scales].

TABLE 1.4

Electronegativity Scale of Elements[a]

												H 2.1						
Li 1.0	Be 1.5												B 2.0	C 2.5	N 3.0	O 3.5	F 4.0
Na 0.9	Mg 1.2												Al 1.5	Si 1.8	P 2.1	S 2.5	Cl 3.0
K 0.8	Ca 1.0	Sc 1.3	Ti 1.5	V 1.6	Cr 1.6	Mn 1.5	Fe 1.8	Co 1.8	Ni 1.8	Cu 1.9	Zn 1.6	Ga 1.6	Ge 1.8	As 2.0	Se 2.4	Br 2.8
Rb 0.8	Sr 1.0	Y 1.2	Zr 1.4	Nb 1.6	Mo 1.8	Tc 1.9	Ru 2.2	Rh 2.2	Pd 2.2	Ag 1.9	Cd 1.7	In 1.7	Sn 1.8	Sb 1.9	Te 2.1	I 2.5
Cs 0.7	Ba 0.9	La 1.1	Hf 1.3	Ta 1.5	W 1.7	Re 1.9	Os 2.2	Ir 2.2	Pt 2.2	Au 2.4	Hg 1.9	Tl 1.8	Pb 1.8	Bi 1.9	Po 2.0	At 2.2
Fr 0.7	Ra 0.9	Ac 1.1														

The lanthanide elements: Ce—Lu
1.1–1.2

The actinide elements: Th Pa U   Np—No
1.3 1.5 1.7 1.3

[a]After Pauling (1960).

1.6. DIPOLE MOMENTS AND ELECTRONEGATIVITY

We must discuss briefly here a matter that is taken up in Chapter 2, namely, the connection between dipole moments of heteronuclear diatomic molecules and electronegativity.

An obvious thing that one can do is to extract an "empirical charge" q, say, by dividing the measured dipole moments μ by the observed equilibrium bond length r_e.

In Appendix A4.9, such an empirically derived charge is compared with a model calculation in which a calculated charge is derived using Mulliken's approximation (1.9) for the electronegativity. The point we want to emphasize, at this stage, is that the charge calculated in this way correlates strongly with that derived empirically in the manner discussed above. This, of course, directly demonstrates an intimate connection between observed dipole moments and electronegativity and this is the qualitative point to be stressed here.

1.7. BOND TYPE AND LENGTH: ASSOCIATED FORCE CONSTANTS

It will be useful at this point to refer briefly to another matter to be discussed at some length in Chapter 5, that is, the connection between bond type and associated force constant. Qualitatively, what is meant here is related to the fact that molecules exhibit vibrational degrees of freedom in which the nuclei, in a diatomic molecule, to be specific, vibrate about their equilibrium internuclear separation. It seems clear that if the chemical bond between the nuclei is very strong, it will be like a very strong "spring" connecting the nuclei. In such a case, when the nuclei are displaced, one would expect rapid restoration* of the configuration toward equilibrium, i.e., a high vibrational frequency. This qualitative point is made quantitative in Chapter 5, where there is a discussion of the way vibrational spectra can be observed. Specifically, the reader who wants to observe the correlation between the restoring force constant (usually denoted by k) and the bond order immediately can refer to Table 5.2 now, where results are collected for this force constant k for eight diatomic molecules. From the earlier discussion of bond type and bond order it will be clear that the strength of the "spring" between the atoms in these diatomic molecules increases with increasing bond order.

We conclude this section, by way of summary, with Figure 1.2, taken from Pimentel and Spratley (1969), in which trends in bond properties and predicted bond orders in the first-row homonuclear diatomic molecules are displayed: the figures, beginning at the top, show successively, bond order, bond energy, bond length, and force constant *vs.* the number of valence electrons.

At this point we shall attempt to give the reader some flavor of the experimental techniques that are available for probing the basic properties of molecules. Some of these techniques will then be considered in somewhat more detail in later chapters, when they relate directly to predictions made by wave mechanics.

---

*While the atomic masses enter the vibrational frequency and reduce it as the masses are increased, the force constant is a property of the bond itself.

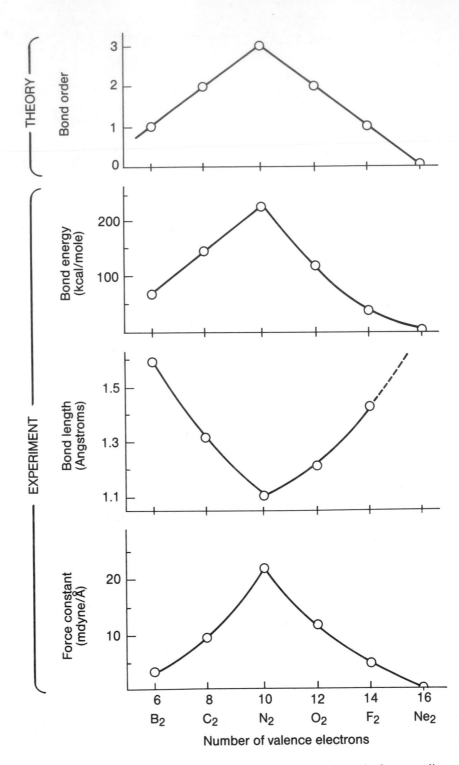

FIGURE 1.2. Trends in bond properties and predicted bond orders in first-row diatomic molecules [redrawn from Pimentel and Spratley (1969), p. 115].

1.8. SOME IMPORTANT EXPERIMENTAL METHODS

1.8.1. Background

In discussing some of the important experimental methods used in the study of the chemical physics of free molecules, it is useful to start by considering the energy relationships involved when dealing with ionic substances by utilizing the so-called Born–Haber cycle. This will be followed by an account of techniques used to probe electronic structure, such as photoelectron spectroscopy and core level splitting in paramagnetic molecules. We shall also briefly point out the way in which various molecular properties can be determined by molecular spectroscopy, a topic to be considered in depth in Chapter 5. We then consider various aspects of the use of lasers, with some introduction to basic concepts. This experimental survey will help to set the scene for the topics to be covered in the chapters that follow.

1.8.2. Mainly Thermodynamics: The Born–Haber Cycle

The purpose of this section is to consider the relation between ionic bonding and the energy of a crystalline lattice of, say, an alkali halide.

First, we set up a model for such a material, say, NaCl, following the early work of Born (1919). The potential energy curve of the Na^+Cl^- pair is the sum of a long-range Coulomb attraction, denoted by V_C in Figure 1.3, and a repulsive

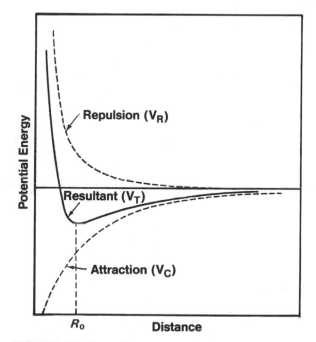

FIGURE 1.3. Potential energy *vs.* internuclear distance.

short-range contribution V_R which comes into play when the charge clouds of the ions begin to overlap (see also Chapter 3 for further development of this term). Thus, for ions carrying unit charges as in NaCl the total potential energy V_T can be expressed as

$$V_T = V_C + V_R = -e^2/R + B/R^n \qquad (1.11)$$

where e is the unit of electric charge and R is the internuclear distance. The constant B can be related to the equilibrium spacing R_0, as determined by the minimum in the potential energy curve, i.e.,

$$\left(\frac{\partial V_T}{\partial R}\right)_{R_0} = 0 = \frac{e^2}{R_0^2} - \frac{nB}{R_0^{n+1}} \qquad (1.12)$$

Solving for B and inserting the result into equation (1.11) readily yields at the equilibrium spacing R_0.

$$V_T = -(e^2/R_0)(1 - 1/n) \qquad (1.13)$$

(a) Lattice Energy

The next step is to assemble the ionic crystal, say, NaCl. One needs to calculate, by numerical summation, the total electrostatic energy of the appropriate crystal structure, which is expressed customarily through a number A, the so-called Madelung constant. Then one writes for the lattice energy the result (with charges on the ions of $+e$ and $-e$)

$$V_{\text{crys}} = -(e^2/R_0)AN(1 - 1/n) \qquad (1.14)$$

If one is dealing with 1 mole, then N is Avogadro's number.

(b) Use of Compressibility Data to Determine n

The compressibility κ is given by the second derivative of V_T with respect to volume. This is obtained [see, e.g., Rosenbaum (1970) or Day and Selbin (1969)] as

$$\kappa = 18R^4/Ae^2(n-1) \qquad (1.15)$$

(c) Some Further Refinements

Further refinements have been made of equation (1.14) leading to a more general, semiempirical formula. One can write

$$V = -(e^2/R_0)AN + NB \exp(-kR) - NC/R^6 + NE_0 \qquad (1.16)$$

the last term, for example, specifically taking account of the zero-point energy E_0 of the ions; k and C are two additional parameters [see Day and Selbin (1969)].

Let us now consider an experimental test of the electrostatic model presented above via the so-called Born–Haber cycle.

As a specific example, consider the formation of 1 mole of solid alkali metal halides from the solid metal M and $\frac{1}{2}$ mole of gaseous elemental halogen X_2. The overall reaction for this to take place is

$$M(s) + \tfrac{1}{2}X_2(g) \rightarrow MX(s) \tag{1.17}$$

which may be broken down into several steps. We start with the sublimation of the solid metal atoms to give the positive ion. The heat of sublimation S and the energy of ionization I must be supplied:

$$M(s) \xrightarrow{+S} M(g) \xrightarrow{+I} M^+(g) + e^- \tag{1.18}$$

In turn, the halogen may be dissociated into atoms: one supplies one-half of the dissociation energy D, and an electron must be added to the atom, thereby releasing energy equal to the electron affinity A:

$$\tfrac{1}{2}X_2(g) \xrightarrow{\tfrac{1}{2}D} X(g) \xrightarrow[+e^-]{-A} X^-(g) \tag{1.19}$$

Finally the ions come together to form the solid, and there will be a release of energy (the so-called lattice energy) as follows:

$$M^+(g) + X^-(g) \xrightarrow{\text{Lattice energy}} MX(s) \tag{1.20}$$

To summarize, the complete process may be represented on one diagram:

$$
\begin{array}{ccc}
M(s) + \tfrac{1}{2}X_2(g) & \xrightarrow[\tfrac{1}{2}D]{S} & M(g) + X(g) \\
\Big\downarrow {\scriptstyle -\Delta H} & & \Big\downarrow {\scriptstyle I} \quad \Big\downarrow {\scriptstyle -A} \\
MX(s) & \xleftarrow{-V} & M^+(g) + X^-(g)
\end{array}
\tag{1.21}
$$

where ΔH is the heat of formation and V is the lattice energy. From the above diagram (1.21), one can readily deduce that

$$V = \Delta H + S + \tfrac{1}{2}D + I - A \tag{1.22}$$

and in equation (1.22) we have expressed all the experimental quantities that are required to determine the crystal lattice energy V. Table 1.5 has been constructed to demonstrate the comparison by using equations (1.14), (1.16), and the Born–Haber cycle result (1.22).

TABLE 1.5
Comparison of Theory and Experiment for Lattice Energies of
Alkali Halides (kcal/mole)[a]

	$V_{\text{calculated}}$		$V_{\text{experimental}}$
Compound	Equation (1.14)	Equation (1.16)	Born–Haber cycle (1.22)
LiF	238.9	240.1	242.8
NaF	213.8	213.4	216.6
KF	189.2	189.7	191.8
RbF	180.6	181.6	184.6
CsF	171.6	173.7	176.0
LiCl	192.1	199.2	201.7
NaCl	179.2	184.3	183.9
KCl	163.2	165.4	168.3
RbCl	157.7	160.7	162.8
CsCl	147.7	152.2	157.2
LiBr	181.9	188.3	191.0
NaBr	170.5	174.6	175.5
KBr	156.6	159.3	160.7
RbBr	151.3	153.5	157.1
CsBr	142.3	146.3	151.2
LiI	169.5	174.1	178.4
NaI	159.6	163.9	164.8
KI	147.8	150.8	151.5
RbI	143.0	145.3	147.9
CsI	134.4	139.1	143.7

[a]Data taken from Ketelaar (1958).

1.8.3. Techniques for Probing Electronic Structure

We next consider photoelectron spectroscopy (PES). This technique provides a way of determining electronic energies of molecules. We shall see when molecular orbital theory is developed in Chapters 2 and 6 that this technique can be brought to bear on the predictions of this theory. Many predictions are vindicated in at least a semiquantitative way.

The ensuing discussion will embrace ultraviolet photoelectron spectroscopy (UPS) and X-ray photoelectron spectroscopy (XPS), as well as Auger electron spectroscopy (AES), electron impact and Penning ionization spectroscopy. Though somewhat anticipating later results presented in this volume, we shall include in this survey an interpretation of core level splitting in paramagnetic molecules, in terms of molecular orbitals and Koopmans' theorem (see Appendix A6.2) with special reference to O_2.

(a) Photoelectron Spectroscopy

The subject of PES has developed almost completely since the early 1960s, and has demonstrated somewhat dramatically the utility of the concepts of atomic and molecular orbitals. In this brief account we shall set out the basic principles of the method; however, in the space available, it will only be possible to present a few, somewhat arbitrarily selected, examples to illustrate just what can be learned from this approach.

PES is about the removal of electrons from molecules following bombardment by monochromatic photons. The electrons ejected are termed photoelectrons.

The photoelectric effect, which was of central importance in the origins of quantum theory, was observed initially on surfaces of metals such as the alkalis, which are readily ionized. When such a surface is bombarded with photons, whose frequency can be varied, photoelectrons are produced after a threshold frequency v_t has been reached. Then one is at the point where the photon energy hv_t is just sufficient to overcome the work function W of the metal, so that $hv_t = W$. As the photon frequency is increased, the excess energy goes into kinetic energy of the photoelectrons, and one can write the obvious energy equation

$$hv = \tfrac{1}{2}mv^2 + W \qquad (1.23)$$

where v is the velocity of the photoelectrons. As the work functions of the surfaces of the alkali metals are a few electron volts, (near) ultraviolet radiation is appropriate for causing ionization.

PES can be thought of as directly analogous to the photoelectric effect, the theory of which was given by Einstein in 1905. It involves (a) the use of higher-energy photons and (b) the use (frequently) of samples in the gaseous phase. Equation (1.23) is applicable to photoelectrons ejected from a gaseous atom or molecule provided the work function W is replaced by the ionization potential I.

In spite of the early history of the photoelectric effect, the experimental difficulties proved to be substantial, and it was not until around 1960 that photoelectron studies became a major branch of spectroscopy.

(b) Interpretation in Terms of Molecular Orbitals

In Chapters 4 and 6, we shall see that the generation of molecular orbitals (MOs) $\psi_i(\mathbf{r})$, belonging to the molecule as a whole, via a one-body potential energy $V(\mathbf{r})$, lies deep in the many-electron problem. The MOs satisfy the Schrödinger equation

$$\nabla^2\psi_i + (2m/\hbar^2)[\varepsilon_i - V(\mathbf{r})]\psi_i = 0 \qquad (1.24)$$

The essential point to be discussed fully in Chapters 4 and 6 is then the construction of $V(\mathbf{r})$. This, it turns out, must include not only contributions from the nuclei in the molecule and the potential created by the ground-state electron density, but it must also incorporate significant many-electron contributions (termed exchange and correlation).

Assuming that we have generated MOs and their corresponding energy levels ε_i in some such fundamental manner from equation (1.24), it will be very useful to adopt the picture this gives us in interpreting the PES results. Figure 1.4 shows an orbital energy diagram for filled nondegenerate valence and core orbitals of an atom or molecule. One can consider each orbital energy as measured relative to a level, the zero of energy, say, corresponding to the removal of an electron from that orbital.

From a technical standpoint, it is obvious from Figure 1.4 that an incident photon will need more energy (typically soft X-rays are required) to remove an electron from a core orbital than from a valence orbital (needing typically ultraviolet radiation). Thus the levels of the core electrons are labeled in Figure 1.4 as associated with XPS, and those of the valence electrons with UPS. AES is also included in Figure 1.4. As in XPS, an initial step involved in AES is that in which a core electron is ejected by a photon. Following this process, a valence electron falls down to fill the core vacancy, and this releases sufficient energy to eject a valence electron, termed an Auger electron. Since a doubly charged ion in an excited state is produced in this process, one may write

$$M + h\nu \rightarrow (M^{2+})^* + e^- \text{ (photoelectron)} + e^- \text{ (Auger)} \qquad (1.25)$$

It is also relevant to mention here electron impact and Penning ionization spectroscopy. In the former, an electron transfers translational energy to M, thereby exciting it. The energy of such excitation is measured by the loss of energy of the electron

$$M + e^- \rightarrow M^* + e^- \qquad (1.26)$$

In the latter, an excited atom A^* ionizes M on collision and we can write

$$A^* + M \rightarrow A + M^+ + e^- \qquad (1.27)$$

The development of gas-phase UPS is associated with the names of Al-Joboury and Turner, and with Vilesov, Kurbatov, and Terenin in 1961–1962. The early development of XPS is described in the book by Siegbahn *et al.* (1969).

There is a point of terminology worthy of note here. Electron spectroscopy for chemical analysis is abbreviated as ESCA. In fact this abbreviation is most

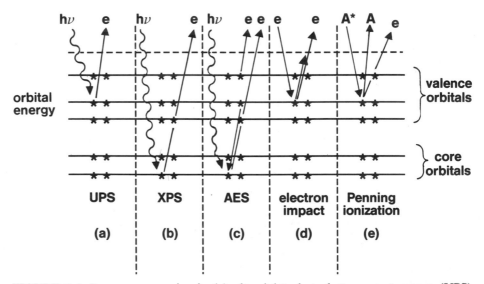

FIGURE 1.4. Processes occurring in (a) ultraviolet photoelectron spectroscopy (UPS), (b) X-ray photoelectron spectroscopy (XPS), (c) Auger electron spectroscopy (AES), (d) electron impact, (e) Penning ionization [redrawn from Hollas (1982)].

commonly used in relation to the field of XPS rather than in general electron
spectroscopy.

21
*Chemical
Concepts and
Experimental
Techniques*

(c) Photoionization Processes in UPS and XPS

UPS and XPS, as illustrated in Figure 1.4a and b, involve the removal of an electron by means of a photon of energy $h\nu$. In both these cases, a singly charged ion, say, M^+, is produced according to

$$M + h\nu \rightarrow M^+ + e^- \tag{1.28}$$

When M denotes an atom, the total change in angular momentum for the process (1.28) above must obey the electric dipole selection rule $\Delta l = \pm 1$ (see Appendix A5.5), but the electron can take away any amount of momentum. As an example, suppose the electron is removed from a d-orbital of M. It then carries away with it one or three quanta of angular momenta depending on whether $\Delta l = -1$ or $+1$, respectively. In general, a free electron can be described by a mixture of s, p, d, etc. wave functions, but in this case the electron that is ejected has just p and f character.

However, turning to molecules, although the electric dipole selection rules apply to the photoionization process, the MOs themselves are now admixtures of s, p, d, f, etc. atomic orbitals, and it follows that the electron removed must be described by a more complex admixture of s, p, d, f, etc. character.

(d) Koopmans' Theorem

As will be shown in Chapter 4, an important (though approximate) result from quantum mechanics is the so-called "theorem" of Koopmans, which relates the orbital energy of the state from which the electron is ejected to the ionization potential of that state. The reason the result is approximate is that it does not allow the orbitals of the electrons to relax when the electron is ejected. Furthermore Koopmans' result is derived by neglecting electron correlation, that is, the influence on the electronic motions in a multielectron system as a direct result of their interelectronic Coulombic repulsions (see Appendix A4.7). Finally, if core electrons are involved, then relativistic effects come into play (see Appendix A5.7 and A5.8) and these are not incorporated into the customary discussions of Koopmans' theorem. Direct use will be made of this theorem in Chapter 6, as applied to the determination of the ionization potentials of polyatomic molecules.

(e) Core Level Splitting in Paramagnetic Molecules

We introduce this topic by referring first to the ESCA technique as applied by Siegbahn *et al.* (1969) to study the spectra of N_2, NO, and O_2. The energy level diagrams of these three molecules are set out in Chapter 2. The N_2 molecule has a closed shell configuration, whereas NO, having an odd number (15) of electrons, is paramagnetic with one unpaired electron. Figure 1.5 shows the N_2 $1s$ and O_2 $1s$ lines from these three molecules.

In general terms, to be elaborated in Chapter 2, in the N_2 molecule the atomic $1s$ level is degenerate with respect to spin after emission of one of its electrons: i.e., the binding energy is the same for the spin-up (↑) and spin-down (↓) electrons. Accordingly, as is verified by reference to Figure 1.5, no energy splitting is observed for the $1s$ line in this molecule. The NO molecule has one unpaired electron and Figure 1.5 shows the spin splitting of the $1s$ levels in this paramagnetic molecule. The most interesting case, however, is O_2 with an even number of electrons as does N_2. Again, splitting is evident in the ESCA spectrum. This reflects the fact, predicted correctly by MO theory (see Chapter 2), that in the ground state of O_2 the two outermost electrons have parallel, not paired, spins. The O_2 molecule, in solid-state language, is the prototype of ferromagnetism. The power of the ESCA technique is clearly in evidence in this direct demonstration of spin splitting of core levels in paramagnetic molecules.

1.8.4. Molecular Spectroscopy and Molecular Properties

In addition to the techniques already considered in this chapter, other spectroscopic methods are available that yield important information about molecular properties. The atoms in a molecule vibrate around an equilibrium position, and, of course, the molecule as a whole rotates. These modes of motion give rise to various types of spectroscopic results. Furthermore, there are electronic transitions as well as changes in electron and nuclear spins in a magnetic field. Table 1.6 lists the types of spectroscopy, the molecular energy involved, and the information extracted from each type of study. We will consider these in some depth in Chapter 5, but we include the table here to summarize the scope of molecular spectroscopy in the determination of various molecular properties.

This also is the point at which to review important experimental techniques using lasers.

1.8.5. Coherent Radiation: Laser Studies

(a) Background

The advent of laser spectroscopy has had considerable impact on our understanding of the chemical physics of free molecules. We remind the reader that unlike conventional sources of radiation, laser emission is entirely via the process of stimulation, as opposed to spontaneous emission (see also Appendix A5.5). A common feature of all lasers is the coherence of the radiation emitted; conventional sources, in contrast, are incoherent in nature, i.e., the electromagnetic waves associated with any two photons of the same wavelength bear no phase relationship. Coherence of laser radiation produces a source of intense local heating.

Of course, for induced emission to occur, one needs population inversion (from the normal Boltzmann distribution). The process by which population inversion is brought about is known as pumping and an active medium is produced (i.e., a system in which population inversion has been accomplished). For every

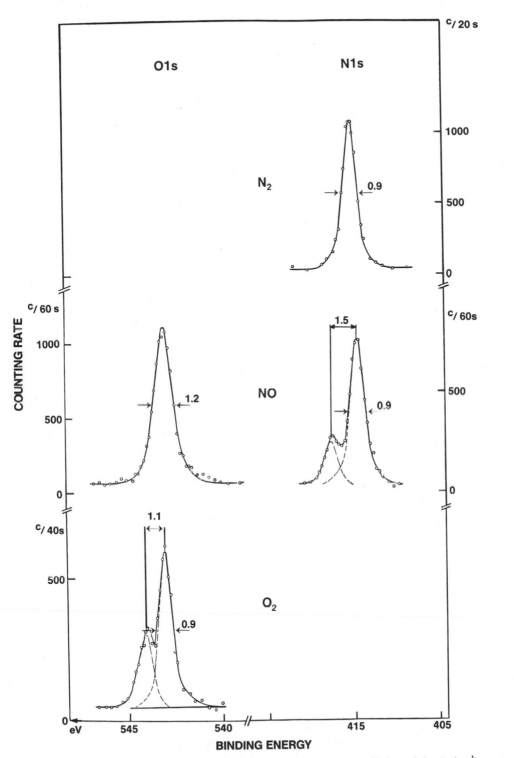

FIGURE 1.5. ESCA spectra from N_2, NO, and O_2 showing spin splitting of the 1*s* levels in the two paramagnetic molecules [redrawn from Siegbahn *et al.* (1969), p. 57].

TABLE 1.6
Molecular Properties and Molecular Spectroscopy[a]

Type of spectroscopy	Type of molecular energy	Information extracted
Microwave	Rotation of heavy molecules	Interatomic distances, dipole moments, bond angles
Far infrared	Rotation of light molecules, vibrations of heavy molecules	Interatomic distances, bond force constants
Infrared	Vibrations of light molecules; vibration–rotation	Interatomic distances, bond force constants, molecular charge distributions
Raman	Rotation, vibrations	Interatomic distances, bond force constants, charge distributions (for energy changes not observable with infrared)
Visible, ultraviolet	Electronic transitions	All properties above, plus bond dissociation energies
Electron spin resonance (ESR)	Energy required to reverse the direction of electron spin in magnetic field	Position of spectrum, shape of spectral lines, hyperfine structure
Nuclear magnetic resonance (NMR) (radio frequency)	Transition between different energy states of nuclei	Chemical shifts, chemical identification

[a]Adapted from L. Hollenberg (1970).

photon entering the active medium, two are emitted, which results in the desired energy. We shall consider the details of these important aspects of laser studies in what follows.

The radiation is coherent since stimulated emission synchronizes the radiation of different excited molecules. When one uses a chemical laser, a chemical reaction is utilized to provide the energy, and to maintain an adequate concentration of excited species for the photon cascade to take place. Many different chemical reactions suitable for this purpose exist.

Below we shall summarize a number of the important concepts needed to appreciate the use of lasers. However, let us first remind the reader that laser is shorthand for "light amplification by stimulated emission of radiation." If the energy involved is in the microwave region then we have a maser.

(i) Absorption and Emission of Radiation. Let us start by reviewing briefly the absorption and emission of radiation in general. To do so, consider a two-state system, with energy levels designated by E_1 and E_2. The frequency v associated with a transition between these two levels is evidently given by

$$hv = E_2 - E_1 = \Delta E \tag{1.29}$$

One can now identify several possible processes.

(ii) Induced Absorption. A molecule M absorbs a photon and is excited from E_1 to E_2, such that

$$M + hv \rightarrow M^* \tag{1.30}$$

where M^* denotes an excited state of the molecule M.

(iii) Spontaneous Emission. The excited molecule M* spontaneously emits a photon:

$$M^* \rightarrow M + hv \tag{1.31}$$

(iv) Induced or Stimulated Emission. This process differs from (iii) since the excited molecule M* is induced or stimulated such that we have

$$M^* + hv = M + 2hv \tag{1.32}$$

It appears from equation (1.32) that we are putting in one photon and that two are produced. However, it is essential to remember that energy has to be supplied initially to excite molecule M to the state M*. There is, of course, no overall gain in energy.

(v) Rate of Change of Population N_2 of State 2: Einstein Coefficients. Following Einstein, let us write the rate of change with time t of the population N_2 of state 2 due to the induced absorption as

$$dN_2/dt = N_1 B_{12} \rho(\bar{v}): \ v = c\bar{v} \tag{1.33}$$

Here $\bar{v}$ is the wave number $(1/\lambda)$, where λ is the wavelength of the radiation. In equation (1.33), B_{12} is an Einstein coefficient while $\rho(\bar{v})$ denotes the radiation density (see also Appendix A5.5). This is given according to Planck's radiation law by

$$\rho(\bar{v}) = \frac{8\pi hv^3}{\exp(hc\bar{v}/k_B T) - 1} \tag{1.34}$$

The induced emission, by analogy with the case of absorption in equation (1.33), causes a rate of change of population given by

$$dN_2/dt = -N_2 B_{21} \rho(\bar{v}) \tag{1.35}$$

where $B_{21} = B_{12}$ (see also Appendix A5.5).

In contrast to equations (1.33) and (1.35), whose right-hand sides are proportional to the radiation density, for spontaneous emission one has

$$dN_2/dt = -N_2 A_{21} \tag{1.36}$$

where A_{21} is another Einstein coefficient, to be discussed further below.

Now it is to be stressed that when we have equilibrium in populations all three processes (1.33), (1.35), and (1.36) are taking place at the same time and we can then write

$$dN_2/dt = (N_1 - N_2)B_{21} \rho(\bar{v}) - N_2 A_{21} = 0 \tag{1.37}$$

It is important at this point to recall the Boltzmann distribution, and then at equilibrium we have

$$N_2/N_1 = (g_2/g_1) \exp(-\Delta E/k_B T) \tag{1.38}$$

where g_1 and g_2 represent the degrees of degeneracy of the two states E_1 and E_2, respectively. Using equation (1.38) in conjunction with equations (1.34) and (1.37), we identify A_{21} as

$$A_{21} = 8\pi h \bar{v}^3 B_{21} \qquad (1.39)$$

a relation also referred to in Appendix A5.5.

Equation (1.39) is of considerable importance. In particular, it indicates clearly that spontaneous emission increases rapidly relative to induced emission as $\bar{v}$ increases, and lasers operate through the process of induced emission. Let us now turn to consider briefly the uses of laser spectroscopy in chemical physics.

(vi) Population Inversion. We recall next equation (1.32) for an induced or stimulated process. Laser radiation, as already noted, is entirely a product of stimulated emission, in contrast to older sources of radiation, which are generated by spontaneous emission.

Induced emission from an upper level, E_{upper}, of a two-level system is possible when there exists what is termed population inversion between the two energy levels, i.e., E_{upper} to E_{lower}. Then

$$N_{\text{upper}} > N_{\text{lower}} \qquad (1.40)$$

where N_i is the population of the state i. This requirement for population inversion means that one must disturb the normal Boltzmann population, which is usually such that

$$N_{\text{upper}} < N_{\text{lower}} \qquad (1.41)$$

The population as in equation (1.41) is disturbed by the input of energy. The process that brings about population inversion is termed "pumping." From equation (1.32), it thus seems that for every photon that enters, two photons are emitted.

Amplification in lasers is brought about by the medium under study being between two mirrors, both of them reflecting. One of these mirrors reflects less than the other to allow some of the stimulated radiation to leak out and form a laser beam. Returning to population inversion, we recall that it appears that we are putting in one photon and getting out two. Thus there must be an input of energy to excite molecule M to the state M*, so that there is no overall gain in energy, as must be the case.

Lasers have the important characteristics that they are monochromatic and coherent. However, most lasers operate at very low efficiency ($\sim$0.1%). There are, by now, quite a variety of lasers, e.g., the medium may be ruby, helium–neon, dye, nitrogen, or any one of several others. Each of these has slightly different properties, which allows the selection of desired characteristics. The laser properties of coherence, monochromaticity, and high intensity mean that atoms and molecules can be selectively excited to a greater extent than in classical spectroscopy. In addition, lasers allow atoms and molecules to become excited in a stepwise absorption of photons. Hence molecules can be selectively excited and/or excited in a stepwise fashion.

(b) Raman Spectroscopy and Lasers

As laser radiation has much greater intensity than that from conventional sources, much weaker Raman processes can be studied than was possible hitherto. In addition, the resolution available is much higher because of narrower line widths. Consequently, not only conventional Raman studies (which we discuss below) are possible using lasers, but much wider studies of Raman scattering become feasible. Let us turn immediately to a consideration of Raman spectroscopy in general and the use of lasers for its implementation.

When a substance is subjected to electromagnetic radiation, the action of the oscillating electric field $\mathscr{E}$ causes the electrons in the material to oscillate. This produces an oscillating induced dipole moment μ_{ind} and the molecule is said to be polarized. The induced dipole is related to the electric field through the polarizability α of the substance by the expression

$$\mu_{\text{ind}} = \alpha \mathscr{E} \tag{1.42}$$

The proportionality constant (i.e., α) in this relationship between $\boldsymbol{\mu}$ and $\mathscr{E}$ defines the polarizability (see also Chapter 3 for more discussion of α). The induced oscillating dipole $\boldsymbol{\mu}$, in turn, emits electromagnetic radiation of the same frequency ν_0, say, as the incident radiation. This phenomenon is known as Rayleigh scattering, in which there is no net gain or loss of energy by the molecule.

In addition to this Rayleigh scattering, it is observed that there is scattered radiation which can have a frequency above and below the incident frequency ν_0. This is known as Raman scattering. The scattered light is further classified as follows: (a) Scattered radiation with frequency $(\nu_0 - \nu_m)$ is termed Stokes radiation, and the radiation emitted obviously corresponds to a lower photon energy than the incident radiation. (b) Scattered light with frequency $(\nu_0 + \nu_m)$ is called anti-Stokes radiation and the scattered light has now gained energy in the process. It is to be noted that ν_m is a frequency characteristic of the substance under investigation.

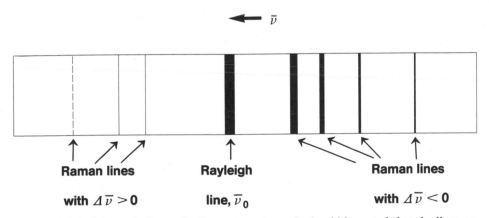

FIGURE 1.6. Schematic form of a Raman spectrum. It should be noted that the lines on the high-frequency side of the Rayleigh line are weak. $\bar{\nu}_0$ is the Rayleigh line wave number.

The Rayleigh frequency (equal to the incident radiation frequency v_0) corresponds to a Rayleigh line in the spectrum. Lines due to the Raman effect with frequencies $(v_0 - v_m)$ and $(v_0 + v_m)$ can be found on the appropriate side of the Rayleigh line. These lines give rise to what is known as the Raman spectrum of the scattering substance (see Figure 1.6). Furthermore, the frequency difference v_m between the Rayleigh line and a Raman line is independent of the frequency of the incident radiation. It is characteristic of the scatterer and related to its vibrational or rotational transitions (see Chapter 5 for a full discussion of these molecular degrees of freedom). This difference v_m is referred to as the Raman shift.

It should be stressed at this point that rotational Raman scattering occurs only if the polarizability of the molecule changes during molecular rotation. This is another way of saying that rotational scattering occurs only if the polarizability is anisotropic (i.e., dependent on the orientation of the molecule with respect to the direction of the incident beam). This means that diatomic molecules exhibit rotational Raman spectra whereas symmetric molecules such as methane do not.

Vibrational Raman scattering occurs only if the polarizability changes during a vibration. Hence the vibrations of homonuclear diatomic molecules (see Chapter 5), which are not infrared-active, are Raman-active.

High-intensity lasers are important in obtaining Raman spectra. One of the major disadvantages of early Raman spectroscopy was that the weakness of its spectra led to poor resolution compared to microwave and infrared techniques, but laser sources have greatly enhanced the resolution possible with Raman spectroscopy.

It will be clear from this discussion that the uses of Raman studies are quite similar to those of infrared and microwave spectroscopy. Raman lines allow one to determine molecular parameters as in microwave spectroscopy. Vibrational Raman lines provide information on molecular symmetry and interatomic forces. Referring to the Table 1.6 in Section 1.8.4, we can see what molecular information can be obtained by these various techniques. In Chapter 5, we shall provide further discussion of infrared and microwave spectroscopies.

With the use of laser radiation, further refinements in Raman spectroscopy are possible, such as are found in hyper-Raman, coherent anti-Stokes Raman, and stimulated Raman effects. The details of these advances can be found in the books by Steinfeld (1985) and Hollas (1987).

(c) Laser-Induced Fluorescence and Phosphorescence Spectroscopy

It is important first to recall just what is the basic process of fluorescence (we shall also treat the closely related phenomenon of phosphorescence). In fact, the light emitted by an excited molecule may be of the fluorescence or the phosphorescence type. Let us therefore consider these in turn.

(i) Fluorescence. Incident energy, depicted in (a) in Figure 1.7, excites a molecule in its ground state singlet S_0 to the excited singlet S_1. We have vibrational excitation during the transition (a) (as predicted by the Franck–Condon principle treated in Appendix A5.4). Collisions with other molecules in the medium under study induce vibrational transitions which imply a stepwise lowering from one vibrational level to another, as shown by (b) in the figure. After the molecule

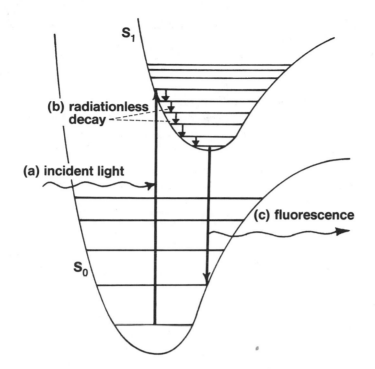

FIGURE 1.7. The processes leading to fluorescence [redrawn from Atkins (1991) p. 128].

reaches the lowest vibrational level, one of two things can occur:

1. Electronic energy may be carried away by the medium as thermal energy.
2. A fluorescent decay of the excited state may take place when a photon is emitted, and the molecule falls into a lower electronic state, which is shown as (c) in Figure 1.7. This emission is known as fluorescence.*

(ii) Phosphorescence. A phosphorescent substance will emit energy when it is illuminated. This emission of energy may persist for some time after the illumination has ceased. It is in this latter respect that phosphorescence differs from fluorescence. Let us consider the mechanism involved in phosphorescence.

1. From Figure 1.8, incident energy (a) excites a ground-state molecule from S_0 to S_1.
2. As in fluorescence we have a vibrational deactivation in S_1; however, if the vibrational deactivation is not too fast another process may be dominant.
3. Assume that a triplet state T_1 of the excited molecule exists as shown in Figure 1.8. The molecule will then have the opportunity to switch from S_1 to T_1 as it descends in its stepwise deactivation, as illustrated in the figure (i.e., by intersystem crossing).
4. Coupling between orbital and spin that is present (see Appendix A5.6) is strong enough to break down the selection rule that singlet–triplet transitions are forbidden. It is such singlet–triplet transitions $T_1 \rightarrow S_0$ that pro-

---

*Note added in proof: For a fluorescence laser study of Ca_2, see G. R. Freeman and N. H. March, *J. Molecular Structure* (*Theochem*) (1993).

ceed sufficiently slowly to cause emission of light to persist even after illumination has ceased.

5. Phosphorescence, therefore, involves a change of multiplicity (unpairing of spins, as a singlet has paired electrons and a triplet unpaired spins) at an intermediate step in the mechanism as depicted. This constitutes a major difference from fluorescence.

6. Hence, in summary, phosphorescence is possible if an appropriate triplet state is present in the vicinity of an excited singlet state.

Lasers, as emphasized already, provide high-intensity monochromatic light sources. If the laser emission frequency matches a vibrational absorption line of the molecule, it is then possible to excite selectively a significant fraction of the molecules into the upper level and then to measure the collisional relaxation of the particular level. The population changes can be followed by infrared fluorescence.

For example, consider the laser-induced fluorescence and the $V - V'$ transfer (see Chapters 5 and 7 for details) for the reaction:

$$HCl(v = 1) + DCl(v = 0) \rightarrow HCl(v = 0) + DCl(v = 1) \qquad \text{(1.43)}$$

The $V - V'$ transfer in this reaction (1.43) is shown in Figure 1.9. Following laser excitation of HCl $(v = 1)$ the populations can be monitored by infrared fluorescence. In addition, the time-resolved fluorescence of DCl displays (see

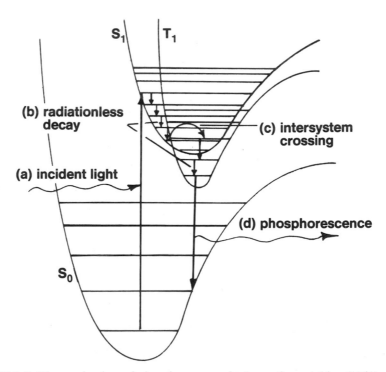

FIGURE 1.8. The mechanism of phosphorescence [redrawn from Atkins (1991), p. 276].

Figure 1.10) a very fast-rising exponential due to rapid $V - V'$ transfer (in equation 1.43) and a slower-decaying exponential caused by $V - T$ transfer.* Therefore, Figure 1.10 clearly displays the fluorescence intensity from HCl ($v = 1$) following a laser excitation pulse (lower curve) as compared to the fluorescence intensity from DCl ($v = 1$) (upper curve) that is formed in the $V - V'$ transfer.

(d) Picosecond Spectroscopy

While conventional spectroscopy (see also Chapter 5) is very useful in studying the rates of chemical reactions (see Chapter 7) its limiting factor is the time necessary for measuring the intensity of a spectral line. We usually have a typical limiting factor of the order of 1 s for conventional spectroscopy, which can be improved to microseconds with a fast-recording infrared spectrometer, for example. However, using laser pulses, the time has been reduced to picoseconds

*$V \rightarrow R$ transfer also occurs in this process (V, R, T represent vibrational, rotational, and translation processes, respectively).

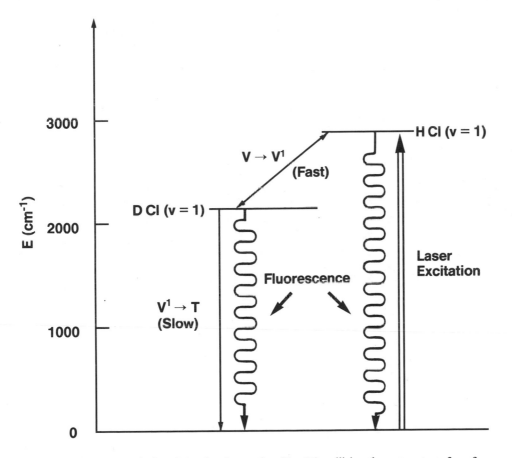

FIGURE 1.9. Laser induced level scheme for $V \rightarrow V'$ collisional energy transfers for reaction (1.43). Following laser excitation of HCl ($v = 1$), the population can be monitored by the infrared fluorescence [adapted from Chen and Moore, 1971].

(ps), i.e., 10^{-12} s. As an example, free electrons generated by the photoionization of Fe $(CN)_6^{3-}$ or I^- ions are hydrated to H_2O^- in about 3–5 ps. Picosecond pulses are being used to study several rapid processes which could not be examined by other methods because of the time-factor limitation. For example, short pulses (subnanosecond/picosecond) can be used to measure relaxation and other energy transfer processes on the time scale of the pulse width. For example, picosecond pulses can be used to measure the vibrational relaxation time in an excited electronic state of rather complicated molecules such as azulene (see Figure 1.11).

Indeed the whole field of molecular energy transfer and relaxation processes is important, and lasers can be used to determine the relaxation time for various types of transitions, e.g., electronic to electronic $(E \rightarrow E')$, vibrational to vibrational $(V \rightarrow V')$, translational to translational $(T \rightarrow T')$, electronic to vibrational $(E \rightarrow V)$, and so on.

Depending on which type of relaxation process one wishes to study, the appropriate laser pulse can be employed, and this is one important use of certain laser pulses such as picosecond techniques. As an example, reference can be made to Figure 1.11, where the energy level diagram and relaxation processes in azulene are shown. Here S_0, S_1, and S_2 refer to singlet states, while T_1 denotes the triplet state. The two relaxation times are such that τ_v is the vibrational time while τ_{1s} is the intersystem singlet–triplet relaxation time.

The vibrational states in S_1 of azulene are excited by a picosecond pulse. A second picosecond pulse (delayed from the first) excites S_1 to S_2. Then azulene

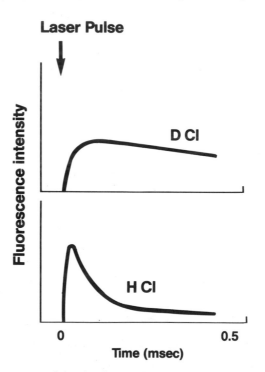

FIGURE 1.10. Fluorescence intensity from HCl ($v = 1$) following a laser excitation pulse (lower) compared to that (upper) from DCl formed in $V \rightarrow V'$ transfer as in Figure 1.9 [adapted from Chen and Moore (1971)].

fluoresces from the S_2 state. The magnitude of fluorescence, depends, of course, on the population of the S_1 state with the frequency v_0 from the laser. If one measures the intensity of the fluorescence as a function of the pulse delay between $2v_0$ and v_0, one obtains the relaxation time of the state excited by $2v_0$.

(e) Two-Photon and Multiphoton Absorption

The photon density is of course very high in lasers, and absorption of two or more photons is possible because of the huge abundance. Two-photon absorption (Raman is in this category) can be monitored by fluorescence or by counting the ions produced by further photons. Consequently, how photons are absorbed and the resulting mechanism can be explored. Multiphoton ionization is advantageous in cases where the fluorescence quantum yield is too small for the method of two-photon fluorescence.

Figure 1.12 shows multiphoton absorption. Referring first to Figure 1.12a, we note that in a two-photon absorption process the first photon (with wave number $\bar{v}_a$) takes a molecule from energy state 1 to a (virtual) state V while the second photon (wave number $\bar{v}_b$) takes the molecule from the virtual state to energy state 2. This is the process followed by a Raman scattering event.

It must be noted that the two photons can be of equal (Figure 1.12b) or unequal energies (Figure 1.12c).

It is, of course, also possible to have a process involving more than two-

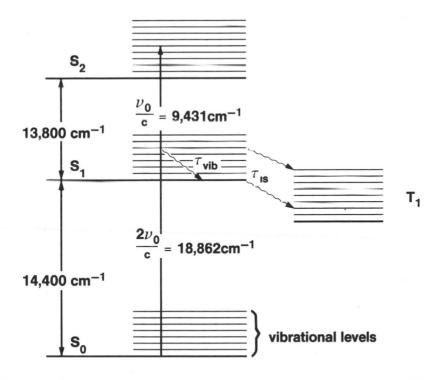

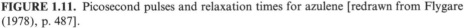

FIGURE 1.11. Picosecond pulses and relaxation times for azulene [redrawn from Flygare (1978), p. 487].

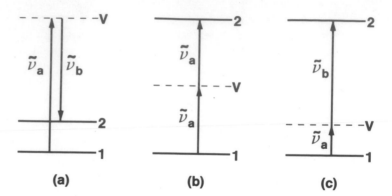

FIGURE 1.12. Multiphoton processes: (a) Raman scattering, (b) absorption of two identical photons, and (c) absorption of two different photons [redrawn from Hollas (1987), p. 337].

photon absorption following the lines already laid down for a two-photon process. Multiphoton processes are of considerable interest in chemical physics; the interested reader can refer to Steinfeld (1985) and to Hollas (1987) for details.

The above brief introduction to laser spectroscopy should give the flavor of its importance in chemical physics studies. We shall return to the use of lasers in the study of chemical reactions in Chapter 7.

PROBLEMS

1.1. We recommend that the reader consult equations (1.21) and then work out the following problem [see also Atkins (1978)] by using the Born–Haber cycle. **(a)** Consider the example of the enthalpy of hydration of H^+ and Cl^-. The overall process can be broken down into the following steps:

$$H^+(g) + Cl^-(g) \rightarrow H(g) + Cl(g) \qquad \Delta H_m^\theta(i)$$

$$H(g) + Cl(g) \rightarrow HCl(g) \qquad \Delta H_m^\theta(ii) = -DH_m^\theta(298\ K)$$

$$\underline{HCl(g) \rightarrow H^+(aq) + Cl^-(aq) \qquad \Delta H_m^\theta(iii)}$$
$$H^+(g) + Cl^-(g) \rightarrow H^+(aq) + Cl^-(aq)$$

Calculate the enthalpy of hydration at 298 K in kJ mole, given the hints, and data below. **(b)** The ionization potential of atomic hydrogen is 1310 kJ, while the electron affinity of atomic chlorine is 347 kJ. Determine from these the value of $-\Delta H_m^\theta(i)$. **(c)** Furthermore, from spectroscopic measurements one knows that $DH_m(298\ K) = -432$ kJ mole. Using Hess's law, one can also find that ΔH^θ (iii) $= -75.14$ kJ mole. [Answer: ΔH_m^θ (298 K) $= -1468$ kJ mole].

1.2. Expand the total energy $E(Z, N)$ of atomic ions in a Taylor series about the neutral atom point where the number of electrons N becomes equal to the atomic number Z. Keeping only the first three terms of the Taylor expansion, relate the coefficients of this truncated Taylor expansion to (a) the first ionization potential I and (b) the electron affinity A. Using the result obtained thereby, calculate directly the derivative $(\partial E/\partial N)_Z$ at the point $N = Z$. (We shall see in Chapter 4 that this derivative is, in fact, the chemical potential of the electron cloud of the neutral atom.) Note that the value is $(I + A)/2$, which is Mulliken's approximation for the electronegativity. (We shall press further, in Chapter 4, the use of the chemical potential as a quantitative measure that reflects, rather directly, electronegative properties of atoms and ions.)

1.3. By making some additional, plausible assumptions, show that the considerations of Problem 1.2 can be used to derive the approximate relation between nth and first ionization potentials:

$$I_n \doteq nI_1 \tag{P1.1}$$

[Consult the book by Phillips and Williams (1965) for a discussion of the usefulness of this. For a more fundamental, first-principles treatment of this relation, the more advanced reader can refer to the work of Pyper and Grant (1978) and Lawes *et al.* (1978).]

1.4. If I is the ionization potential for the hydrogen atom in units of e^2/a_0, and r is the distance from the nucleus in units of the Bohr radius a_0, use the $1s$ wave function $[\psi_{1s} = N\exp(-Zr/a_0)]$ to show that the election density $\rho(r)$ has the form

$$\rho(r) = \text{constant} \cdot \exp[-2(2I)^{1/2}r] \tag{P1.2}$$

(N.B.: Although the result (P1.2) is, of course, specific to H, it can be shown that at sufficiently large r the electron density $\rho(r)$ in a multielectron atom with measured ionization potential I has an asymptotic form like (P1.2) times a power of r.)

1.5. List some important differences between fluorescence and phosphorescence.

1.6. As we recall, the radiation density is given by

$$\rho(v) = \frac{8\pi h v^3}{c^3} \frac{1}{\exp(hv/kT) - 1}$$

Determine typical values of $\rho(v)$ in the microwave and near-ultraviolet regions (it will be, of course, necessary to know typical values of v in these regions).

1.7. Explain the mechanism of laser operation by use of mathematical expressions.

1.8. (a) What important properties of molecules can be obtained from molecular spectroscopy?
(b) What techniques can be used to probe electronic structure. Explain the significance of each.

REFERENCES

P. W. Atkins, *Physical Chemistry*, W. H. Freeman, San Francisco (1978).

P. W. Atkins, *Quanta*, 2nd Ed., Oxford University Press, New York (1991)

R. S. Berry, S. A. Rice, and J. Ross, *Physical Chemistry*, John Wiley and Sons, New York (1980).

M. Born, *Verhandl. deut. Physik. Ges.* **21**, 13 (1919).

H. L. Chen and C. B. Moore, *J. Chem. Phys.* **54**, 4072 (1971).

M. Day and J. L. Selbin, *Theoretical Inorganic Chemistry*, Reinhold, New York (1969).

R. L. De Kock and H. B. Gray, *Chemical Structure and Bonding*, Benjamin/Cummings, Menlo Park, CA (1980).

W. H. Flygare, *Molecular Structure and Dynamics*, Prentice-Hall, Englewood Cliffs, NJ (1978).

J. M. Hollas, *High Resolution Spectroscopy*, Butterworth, London (1982).

J. M. Hollas, *Modern Spectroscopy*, John Wiley and Sons, New York (1987).

J. L. Hollenberg, *J. Chem. Educ.* **47**, 2 (1970).

J. A. A. Ketelaar, *Chemical Constitution*, 2nd Ed., Elsevier, Amsterdam (1958).

G. P. Lawes, N. H. March, and M. S. Yusaf, *Physics Letters* **A67**, 342 (1978).

I. N. Levine, *Molecular Spectroscopy*, John Wiley and Sons, New York (1975).

G. N. Lewis, *J. Am. Chem. Soc.* **38**, 762 (1916).

R. S. Mulliken, *J. Chem. Phys.* **2**, 782 (1934).

R. S. Mulliken, *J. Chem. Phys.* **3**, 573 (1935); *ibid* **46**, 497 (1949).

L. Pauling, *The Nature of the Chemical Bond*, 3rd Ed., Cornell University Press, Ithaca (1960).

C. S. G. Phillips and R. J. P. Williams, *Inorganic Chemistry Vols I and II*, Oxford University Press, (1965).

G. C. Pimentel and R. D. Spratley, *Chemical Bonding Clarified Through Quantum Mechanics*, Holden-Day, San Francisco (1969).

N. C. Pyper and I. P. Grant, *Proc. Roy. Soc.* **A359**, 525 (1978).

E. J. Rosenbaum, *Physical Chemistry*, Appleton-Century Crofts, New York (1970).

R. T. Sanderson, *Science* **114**, 670 (1951).

R. T. Sanderson, *Chemical Periodicity*, Reinhold, New York (1960).

R. T. Sanderson, *Inorganic Chemistry*, Reinhold, New York (1967).

K. Siegbahn, C. N. Nordling, G. Johansson, J. Hedman, P. F. Hedén, K. Hamrin, U. Gelius, T. Bergmark, L. O. Werme, R. Manne, and Y. Baer, *ESCA Applied to Free Molecules*, North-Holland, Amsterdam (1969).

J. L. Steinfeld, *Molecules and Radiation: An Introduction to Modern Spectroscopy*, 2nd Ed., MIT Press, Cambridge, MA (1985).

Further Reading

R. L. De Kock, *J. Chem. Educ.* **65**, 934 (1987).

E. A. U. Ebsworth, D. W. H. Rankin, and S. Cradock, *Structural Methods in Inorganic Chemistry*, Blackwell Scientific, Oxford (1987).

J. H. D. Eland, *Photoelectron Spectra*, Open University Press, Milton Keynes, UK (1985).

G. A. Gallup, *J. Chem. Educ.* **65**, 671 (1988).

R. D. Levine and R. B. Bernstein, *Molecular Reaction Dynamics and Chemical Reactivity*, Oxford University Press, Oxford (1987).

S. H. Lin (ed.), *Radiationless Transitions*, Academic Press, New York (1980).

R. P. Wayne, *Principles and Applications of Photochemistry*, Oxford University Press, Oxford (1988).

Chapter Two

THE NATURE OF BONDING IN DIATOMS

2.1. VALENCE BOND AND MOLECULAR ORBITAL METHODS

Having established the nature and strength of bonds in largely classical chemical language in the previous chapter, we must now consider the way in which the wave mechanical theory of electronic structure can lead to an understanding at a first-principles level of the nature of the bonding in (a) homonuclear and (b) heteronuclear diatoms. This will be done first with particular reference to the single bond in the H_2 molecule. Here, of course, one of the major successes of the early work using quantum mechanics was the theory of Heitler and London (1927) on the nature of the covalent bond in H_2. In many respects, the ideas underlying their pioneering work remain those which, in principle, are most deeply embedded in the chemist's concepts. Their viewpoint was that molecules are built, first and foremost, out of atoms, which retain some of their own distinctive features, even when bound into a molecule. This picture, however, in spite of the efforts of researchers like Moffitt and others [see, e.g., the review by Balint-Kurti and Karplus (1974)] has proved the most difficult one to develop quantitatively from the Schrödinger equation. Therefore, a lot of emphasis has gone into developing the major alternative theory in which the electrons are thought to occupy molecular orbitals which belong to the molecule as a whole. Though, conceptually, this method is fundamentally different from the Heitler–London or valence bond method, in fact the atomic orbitals in most treatments are used as the basic building blocks from which the molecular orbitals are constructed.

2.2. NATURE OF WAVE FUNCTIONS FOR H_2^+ AND H_2

It is sadly true that analytical progress in treating either atoms or molecules remains so difficult that, at the time of writing, the nonrelativistic Schrödinger equation has only been solved exactly for the hydrogen atom and for the hydrogen molecule ion H_2^+. These are evidently both one-electron problems, whereas the major difficulties in both atomic and molecular theory reside in the treatment of more than one electron, where, of course, the Coulombic repulsion energy e^2/r_{ij} between two electrons i and j at separation r_{ij} is the major stumbling block to analytical progress.* Since the He atom itself cannot be treated exactly (but see

---

*Or $e^2/4\pi\varepsilon_0 r_{ij}$: for simplicity of notation we shall put $4\pi\varepsilon_0$ equal to unity throughout.

Appendix AVI), and as already noted in Chapter 1, this is the united atom limit of the H_2 molecule, it should occasion no surprise that the molecular problem of H_2 has to be treated by approximate methods of the type mentioned in Section 2.1.

However, before going into the problems of primary interest for chemistry, namely the neutral molecules H_2, O_2, N_2, etc., we shall outline the way the H_2^+ molecular ion can be solved. We first sketch the way in which the Schrödinger equation separates in a suitable coordinate system, which is the key to the exact solution, pioneered by Burrau (1930) and Teller (1930) and further developed later by Bates *et al.* (1953). However, because quantum chemistry, at the present stage of development, is crucially about judicious approximation, and the concepts to which such approximations lead, we shall then briefly consider the approximate way in which the H_2^+ ion can be solved by molecular orbital (MO) theory, in the approximation in which the atomic wave functions are used as building blocks of the molecular orbital, through the so-called linear combination of the atomic orbitals (LCAO) method. Then in Section 2.3.1 we generalize this LCAO–MO treatment to deal with the neutral H_2 molecule, which, of course, already involves the Coulombic interelectronic repulsion referred to above.

We shall now introduce the simplest problem of immediate chemical interest by discussing briefly the exact form of the wave function of H_2^+. Just as for the H atom, where the wave functions are obtained by separating the variables in the three-dimensional Schrödinger equation, in that case in spherical polar coordinates (r, θ, ϕ) shown in Figure A1, so in the two-center problem under discussion now, it is possible to separate the variables. Thus, writing the Schrödinger equation in units in which $e = m = \hbar = 1$, as

$$\nabla^2 \Psi + 2[E + 1/r_1 + 1/r_2]\Psi = 0 \tag{2.1}$$

where r_1 measures the distance of the point under discussion from nucleus 1 and r_2 that from nucleus 2, the appropriate coordinates, with R the internuclear separation also in atomic units, are

$$\lambda = (r_1 + r_2)/R, \qquad \mu = (r_1 - r_2)/R \tag{2.2}$$

together with the angle ϕ about the internuclear axis. Writing the total wave function Ψ in the separable form

$$\Psi(\phi, \mu, \lambda) = \Phi(\phi)X(\mu)Y(\lambda) \tag{2.3}$$

and putting $p^2 = -\frac{1}{4}R^2E$, one obtains (cf. Appendix A1.1)

$$\Phi(\phi) = \begin{cases} \cos(m\phi), \\ \sin(m\phi), \end{cases} \qquad m = 0, 1, 2 \text{ etc.} \tag{2.4}$$

and that $X(\mu)$ and $Y(\lambda)$ satisfy

$$\frac{d}{d\mu}\left[(1 - \mu^2)\frac{dX}{d\mu}\right] + \left(-A + p^2\mu^2 - \frac{m^2}{1 - \mu^2}\right)X = 0 \tag{2.5}$$

and

$$\frac{d}{d\lambda}\left[(\lambda^2 - 1)\frac{dY}{d\lambda}\right] + \left(A + 2R\lambda - p^2\lambda^2 - \frac{m^2}{\lambda^2 - 1}\right)Y = 0 \tag{2.6}$$

Equation (2.5) is such that, if m and p are specified, solutions that have physical significance exist only for particular values of the separation constant A. The equations above were given by Baber and Hasse (1935), as well as other researchers; the detailed solutions of equations (2.5) and (2.6) are extensively discussed by Bates *et al.* (1953), where references to earlier work are also included. As the solutions cannot be written in simple closed form, we shall be content to use the numerical results from solving equations (2.5) and (2.6) for comparison purposes. However, because of the separation of variables effected above, one has reduced Schrödinger's partial differential equation (2.1) in three variables to the ordinary differential equations (2.5) and (2.6), plus the elementary solutions (2.4) in the azimuthal angle ϕ. These ordinary differential equations are exactly soluble. Unfortunately, this problem of H_2^+ is the only electronic problem of real chemical interest that has proved amenable to exact numerical solution so far. However, as we shall see, approximate solutions are of great interest in themselves, when chemically motivated, especially when their consequences can be shown through simple analytical expressions. We shall return to the exact wave functions obtained for H_2^+ by solution of equations (2.5) and (2.6) when we have set up a simple approximate theory of this molecular ion problem.

2.3. THE LCAO APPROXIMATION TO GROUND STATES OF H_2^+ AND H_2

With $\mathscr{H}$ the full Hamiltonian* of the H_2^+ molecular ion, excluding the nuclear–nuclear potential energy, we construct the ground-state wave function as a linear combination of hydrogen $1s$ orbitals centered on nuclei a and b:

$$\psi_g = N_g[1s_a + 1s_b] \tag{2.7}$$

N_g being as usual the normalization factor of this wave function. The wave function is denoted in equation (2.7) by the subscript g, meaning gerade, i.e., even, because of the plus sign; the difference function $\psi_u = N_u[1s_a - 1s_b]$ is ungerade or odd.

The energy corresponding to these two wave functions is calculated according to the variation principle (see Appendix A2.1) as the expectation value of $\mathscr{H}$ with respect to the LCAO wave function. One then obtains, after a short calculation, that

$$E_g = \int \psi_g \mathscr{H} \psi_g \, d\tau = (H_{aa} + H_{ab})/(1 + S) \tag{2.8}$$

---

*Classically $\mathscr{H} = E$, with energy E, but $\mathscr{H}$ is written in terms of coordinates and (canonically conjugate) momenta. In quantum mechanics, $\mathscr{H}$ is an operator and $\mathscr{H}\Psi = E\Psi$ is the Schrödinger equation [see equations (1.24) and (2.1)].

whereas the average with respect to the ungerade wave function ψ_u yields

$$E_u = \int \psi_u \mathcal{H} \psi_u \, d\tau = (H_{aa} - H_{ab})/(1 - S) \qquad (2.9)$$

Here the notation is as follows:

$$H_{aa} = \int 1s_a \mathcal{H} 1s_a \, d\tau \quad (=H_{bb})$$

$$\qquad (2.10)$$

$$H_{ab} = \int 1s_a \mathcal{H} 1s_b \, d\tau \quad (=H_{ba})$$

while the overlap integral $S = S_{ab}$ is given by

$$S = \int 1s_a 1s_b \, d\tau \qquad (2.11)$$

In writing these results, use has been made of the explicit forms of the normalization factors N_g and N_u, namely,

$$N_g = (2 + 2S)^{-1/2}$$

$$\qquad (2.12)$$

$$N_u = (2 - 2S)^{-1/2}$$

Explicit expressions for H_{aa}, H_{ab}, and S are collected in Appendix A2.2, where use is made of the hydrogenic form of the $1s$ orbital. These expressions are, of course, functions of the internuclear distance R, and may be used to plot the total energy E as a function of R for these two states of H_2^+. The comparison with the exact numerical results from the method of Section 2.2 shows that, as dictated by use of the variational method, the energy is somewhat higher than the exact energy in each case. Nevertheless, the variational treatment clearly accounts for the binding of the H_2^+ molecular ion and gives a sensible value for its equilibrium bond length. We can therefore have some confidence in the energy curve predicted by the LCAO–MO method when we come to apply it to molecules where the exact solutions are not known. This is already the case for H_2, and therefore obviously for all the neutral homonuclear diatoms.

However, as is inevitable from the use of the variational principle, the wave function is of poorer quality than the energy, and to see this we again return to the exact treatment of the wave function in Section 2.2. In Table 2.1 samples of exact numerical values are recorded for states corresponding to the LCAO bonding orbital*

$$1s\sigma_g = [\exp(-r_1) + \exp(-r_2)][2\pi(1 + S)]^{-1/2} \qquad (2.13)$$

for two values of internuclear separation R. The wave function is also shown for

---

*For details of nomenclature, the reader is referred to Section 2.6 and Appendix AIV.

TABLE 2.1

Comparison of Exact and LCAO Wave Functions for Bonding ($1s\sigma_g$)
and Antibonding ($2p\sigma_u$) States of H_2^+ [a]

Distance from center along internuclear axis	Wave function		Wave function	
	Exact	LCAO	Exact	LCAO
	$1s\sigma_g$ at $R = 2$		$1s\sigma_g$ at $R = 4$	
0	0.315	0.233	0.127	0.099
1	0.458	0.360	0.175	0.153
2	0.120	0.132	0.378	0.373
3	0.030	0.049	0.123	0.137
4	0.007	0.018	0.039	0.050
	$2p\sigma_u$ at $R = 2$		$2p\sigma_u$ at $R = 4$	
0	0.000	0.000	0.000	0.000
1	0.453	0.536	0.146	0.141
2	0.199	0.197	0.436	0.435
3	0.078	0.073	0.158	0.160
4	0.029	0.027	0.055	0.057

[a] All distances, including the internuclear separation R, are measured in atomic units [after Bates *et al.* (1953)].

comparison. The antibonding orbital is also used in Table 2.1 (lower half), the form being explicitly

$$2p\sigma_u = [\exp(-r_1) - \exp(-r_2)][2\pi(1 - S)]^{-1/2} \tag{2.14}$$

to compare with the exact values from Section 2.2. As expected, the errors in the LCAO wave functions in comparison with the exact values are more substantial than those in the corresponding energy values. However, the LCAO–MO method remains attractive for many purposes; it is frequently possible to gain valuable insight into the problem, which is often not the case with heavy numerical computations, even though the latter may prove well nigh completely quantitative. The merit of the exact solution of the H_2^+ problem is that it allows us to confront the approximate LCAO theory directly with the precise values of both wave function and energy. We must bear in mind the sizable quantitative errors thereby exposed, especially in wave functions, that might be expected in the use of the LCAO–MO method later, in a variety of contexts (see also particularly Chapter 6).

2.3.1. LCAO Treatment of the H_2 Molecule

In Chapter 1 we outlined the way in which bonding can be viewed, in particular, in the homonuclear diatoms H_2, O_2, and N_2 in classical chemical language. Here we make a start on the wave-mechanical theory by studying the simplest such homonuclear diatom H_2, again employing the LCAO–MO method, illustrated above in the approximate treatment of H_2^+. Then, later in the chapter, we shall apply the same method to other homonuclear diatomic molecules in the first row of the Periodic Table. Specifically, of prime concern in this chapter will be ground-state molecular energies, electron configurations, bond lengths, dissociation energies, spectroscopic symbols, and the so-called correlation diagram for the homonuclear diatoms. The same properties will then be considered for heteronuclear diatomic molecules, along with a study of dipole moments and a further development of the concept of electronegativity already stressed in Chapter 1.

2.3.2. The Molecular Energy of the H_2 Molecule in Its Ground State

Consideration of the homonuclear H_2 molecule using the LCAO–MO approximation follows essentially the lines adopted for the H_2^+ molecule described just above. In the H_2 molecule we have the simplest two-electron bond (the covalent bond) and an understanding of the nature of this bond is important as the concepts involved can be extended to a study of more complex molecules containing covalent bonds and their properties.

In the case of H_2 we shall obtain expressions (within the Born–Oppenheimer approximation, see Appendix A2.3) that result from the motion of two electrons in a field of two fixed nuclei (i.e., two protons) separated by a distance R, which is a parameter in the expression for the molecular energy E. The equilibrium distance in the molecule is the value of R that yields the minimum energy, i.e.,

$$dE/dR = 0 \qquad (2.15)$$

In order to find an expression for E we employ the nonrelativistic Hamiltonian with fixed nuclei for the H_2 molecule. This reads, in atomic units (a.u.),

$$\mathscr{H} = (-\tfrac{1}{2}\nabla_1^2 - 1/r_{a1} - 1/r_{b1}) + (-\tfrac{1}{2}\nabla_2^2 - 1/r_{a2} - 1/r_{b2}) + 1/r_{12} \qquad (2.16)$$

where 1 and 2 refer to the electrons and a and b to the two nuclei. It should be noted that we have omitted the nuclear repulsion term $1/R$ in equation (2.16). This is because it is merely a constant for fixed nuclei which can simply be added later. The notation employed in equation (2.16) is shown explicitly in Figure 2.1. Using this Hamiltonian (2.16), we must solve the Schrödinger wave equation

$$\mathscr{H}\Psi(1, 2) = E_{el}\Psi(1, 2) \qquad (2.17)$$

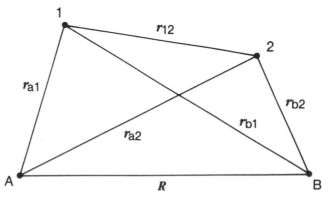

FIGURE 2.1. Coordinates for the hydrogen molecule H_2.

where the electronic energy E_{el} is related to the total energy E of the molecule by

$$E = E_{el} + R^{-1} \tag{2.18}$$

We next note that the Hamiltonian (2.16) is the sum of two H_2^+-like Hamiltonians plus the electron–electron interaction term r_{12}^{-1}. If we define

$$H^h(1) = -\tfrac{1}{2}\nabla_1^2 - 1/r_{a1} - 1/r_{b1} \tag{2.19}$$

with an analogous definition for $H^h(2)$, then equation (2.17) can be written

$$[H^h(1) + H^h(2) + r_{12}^{-1}]\Psi(1, 2) = E_{el}\Psi(1, 2) \tag{2.20}$$

If, to gain orientation, the electron repulsion is at first omitted, solutions of equation (2.20) can be expressed as a product of orbitals that describe the motion of one electron in the field of the two nuclei, where the orbitals satisfy the Schrödinger equation

$$H^h(i)\phi_k(i) = E_k\phi_k(i) \tag{2.21}$$

where i refers to either of the two electrons. Equation (2.21) is, of course, the same equation that was solved in the case of H_2^+. Hence, we can use the solution obtained there to aid in solving the H_2 molecule problem, but in the case of H_2 the repulsion term r_{12}^{-1} must also be incorporated. The solution of the Schrödinger equation for the He atom (cf. Appendix AVI), where two electrons move in the field of a nucleus of charge $2e$, indicates the way to include the r_{12}^{-1} term. The Hamiltonian for the He atom is

$$\mathcal{H}_{He} = -\tfrac{1}{2}\nabla_1^2 - \tfrac{1}{2}\nabla_2^2 - 2r_1^{-1} - 2r_2^{-1} + r_{12}^{-1} \tag{2.22}$$

where r_1, r_2, and r_{12} are the distances as represented in Figure 2.2.

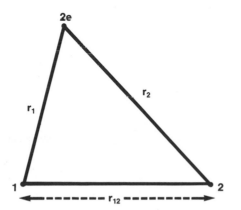

FIGURE 2.2. Coordinates for the helium atom used in equation (2.22). r_{12} is the interelectronic distance.

2.3.3. The United Atom: He

The ground state of the He atom has an electronic configuration $1s^2$ (see Appendix A1.2 for the configuration of any atom) and therefore the complete (space × spin functions) normalized wave function Ψ_0 is*

$$\Psi_0 = 2^{-1/2} 1s(1)1s(2)[\alpha(1)\beta(2) - \beta(1)\alpha(2)] \tag{2.23}$$

where $1s(1)$ and $1s(2)$ are the atomic orbitals of electrons 1 and 2, and the ground-state energy E is determined by the expression

$$E = \int \Psi_0 \mathscr{H}_{He} \Psi_0 \, d\tau \tag{2.24}$$

We recall that the spin functions α and β are normalized and orthogonal, and utilizing this fact together with equations (2.23) and (2.24), we find that

$$E = \int 1s(1)1s(2)\mathscr{H}_{He}1s(1)1s(2) \, d\tau \tag{2.25}$$

If equation (2.22) is expressed as

$$\mathscr{H}_{He} = H(1) + H(2) + 1/r_{12} \tag{2.26}$$

then the ground-state energy for the He atom is

$$E = \int 1s(1)H(1)1s(1) \, d\tau_1 \int 1s(2)1s(2) \, d\tau_2$$

$$+ \int 1s(2)H(2)1s(2) \, d\tau_2 \int 1s(1)1s(1) \, d\tau_1$$

$$+ \int 1s(1)1s(2)(1/r_{12})1s(1)1s(2) \, d\tau \tag{2.27}$$

The first two terms on the right-hand-side of equation (2.27) are evidently equal and hence

$$E = 2E_{1s} + J \tag{2.28}$$

where

$$J = \int 1s(1)1s(2)(1/r_{12})1s(1)1s(2) \, d\tau$$

J measuring the Coulombic repulsion between the electrons and being termed the Coulomb integral, and the normalization conditions

$$\int 1s(1)1s(1) \, d\tau_1 = \int 1s(2)1s(2) \, d\tau_2 = 1$$

---

*For a more detailed discussion of the spin wave function, see Section 2.4. Of course, the product form for the spatial part represents an approximation (see Appendix A4.6).

having been employed in reaching equation (2.28). We see immediately that the introduction of the r_{12}^{-1} term into the Hamiltonian leads to an energy E involving the Coulomb integral. This fact is important in the determination of E for H_2, where more integrals are involved. We turn next to consider this point for H_2.

2.3.4. Coulomb and Multicenter Integrals in H_2

If we employ the technique just used for the He atom coupled with equation (2.21), then the following expression is obtained for the ground-state energy E_0 of the H_2 molecule:

$$E_0 = 2E_{1\sigma_g} + J_{1\sigma_g,1\sigma_g} \tag{2.29}$$

Here we have

$$E_{1\sigma_g} = \int \psi_{1\sigma_g}(i) H^h(i) \psi_{1\sigma_g}(i) \, d\tau_i \tag{2.30}$$

and

$$J_{1\sigma_g,1\sigma_g} = \int \psi_{1\sigma_g}^2(1) r_{12}^{-1} \psi_{1\sigma_g}^2(2) \, d\tau_1 \, d\tau_2 \tag{2.31}$$

As we have already encountered $E_{1\sigma_g}$ in the case of H_2^+, we turn now to a more detailed consideration of $J_{1\sigma_g,1\sigma_g}$ in order to demonstrate the nature of the integrals involved. These embody the new features added by the electron–electron repulsion term r_{12}^{-1} in H_2. We recall from Section 2.3 that, in LCAO form:

$$\psi_{1\sigma_g} = [2(1 + S)]^{-1/2} [1s_a(1) + 1s_b(1)] \tag{2.32}$$

With equation (2.32) in equation (2.31), it follows after some manipulation that

$$J_{1\sigma_g,1\sigma_g} = [2(1 + S)^2]^{-1} \left[\int 1s_a(1)^2 r_{12}^{-1} 1s_a(2)^2 \, d\tau_1 \, d\tau_2 \right.$$

$$+ \int 1s_a(1)^2 r_{12}^{-1} 1s_b(2)^2 \, d\tau_1 \, d\tau_2$$

$$+ 4 \int 1s_a^2(1) r_{12}^{-1} 1s_a(2) 1s_b(2) \, d\tau_1 \, d\tau_2$$

$$+ \left. 2 \int 1s_a(1) 1s_b(1) r_{12}^{-1} 1s_a(2) 1s_b(2) \, d\tau_1 \, d\tau_2 \right] \tag{2.33}$$

Conventionally, the integrals appearing in equation (2.33) are referred to as follows:

1. One-center integral: the first integral in the square bracket and written as (aa | aa)
2. Two-center integrals: the second integral in equation (2.33), written in short-hand notation as (aa | bb)
3. Two-center mixed integrals: the third integral, written (aa | ab)
4. Two-center exchange integrals: (ab | ab)

$$\left. \right\} \quad \text{(2.34)}$$

Analytical expressions for the integrals are available in the literature: We will not consider their form at this time as our purpose here was to demonstrate the effect of treating electron–electron interactions in a simple two-electron system and to alert the reader to the necessity of more sophisticated ways of dealing with many-electron molecules because of the number of integrals which will then be involved (see Appendix A2.4). We shall return to this point in our discussion of the self-consistent-field method later on in the book (see especially Section 4.5 and Appendixes A4.6 and A4.7).

The molecular energy for H_2 in its ground state using equation (2.29) [and understanding more of its details by the introduction of equation (2.34)] including the $1/R$ term is

$$E = 2E^{\text{h}}_{1\sigma_g} + J_{1\sigma_g,1\sigma_g} + R^{-1} \qquad \text{(2.35)}$$

Equation (2.35), using available analytic functions, allows one to plot the ground-state energy $E(R)$ *vs.* R. The resulting curve using equation (2.35) yields a bond length for the H_2 molecule quite near to that observed experimentally. The dissociation energy can also be calculated, and it is only about 60% of the observed value. These discrepancies from experimental results (especially the dissociation energy) must be accounted for and will be dealt with later, but this simple LCAO–MO treatment points the way to an understanding of the nature of chemical bonding. Figure 2.3 shows (schematically) the shape of the $E(R)$ *vs.* R curve using equation (2.35) and demonstrates that a minimum (i.e., $dE/dR = 0$) is obtained.

2.4. VALENCE BOND THEORY OF THE H_2 MOLECULE

In the simple LCAO–MO treatment of the H_2 molecule, dealt with in some detail above, the electrons were assigned to molecular orbitals embracing both nuclei. In the further approximation adopted in that presentation, these MOs were formed by taking linear combinations of $1s$ atomic orbitals.

The valence bond (VB) or Heitler–London (1927) method is based on the quite different philosophy of "atoms in molecules," which considers two H atoms a and b, each in its ground state (with atomic orbitals denoted again by $1s_a$ and $1s_b$), approaching each other from a supposedly large internuclear separation R. A possible wave function describing the molecule would be

$$\Psi(1, 2) = 1s_a(1)1s_b(2) \qquad \text{(2.36)}$$

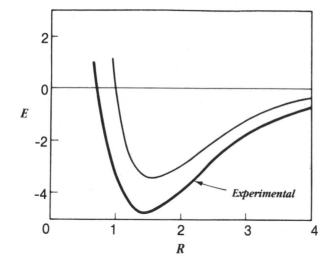

FIGURE 2.3. Bonding energies calculated using equation (2.35) and experimental values for ground state H_2.

Equation (2.36) implies that electron 1 is associated with nucleus a and electron 2 with nucleus b, but this wave function does not take into account the fact that the electrons are indistinguishable from each other. Hence, a wave function with equal validity to that in equation (2.36) would be

$$\Psi(2, 1) = 1s_a(2)1s_b(1) \tag{2.37}$$

Heitler and London constructed a linear combination of (2.36) and (2.37) such that the normalized wave function is

$$\Psi_g = \frac{1s_a(1)1s_b(2) + 1s_a(2)1s_b(1)}{(2 + 2S^2)^{1/2}} \tag{2.38}$$

where S as usual denotes the overlap integral.

The spin functions α and β will not be included for the moment, this aspect of the problem being considered a little later. The total nonrelativistic Hamiltonian is

$$\mathscr{H} = -\tfrac{1}{2}\nabla_1^2 - \tfrac{1}{2}\nabla_2^2 - r_{a1}^{-1} - r_{b2}^{-1} - r_{a2}^{-1} - r_{b1}^{-1} + r_{12}^{-1} + R^{-1} \tag{2.39}$$

where the notation is the same as that employed in the LCAO–MO treatment. The ground-state energy is therefore

$$E_0 = \int \Psi_g \mathscr{H} \Psi_g \, d\tau_1 \, d\tau_2 \tag{2.40}$$

where Ψ_g is given by equation (2.38). After some manipulation, use of equations (2.38) and (2.39) in equation (2.40) yields

$$E_0 = 2E_{1s} + (Q + A)/(1 + S^2) \tag{2.41}$$

where E_{1s} denotes the hydrogen atom ground-state energy with

$$Q = -\int 1s_a(1)^2 r_{b1}^{-1} d\tau_1 - \int 1s_b(2)^2 r_{a2}^{-1} d\tau_2$$

$$+ \int 1s_a(1)1s_b(2)r_{12}^{-1}1s_a(1)1s_b(2)\, d\tau_1\, d\tau_2 + R^{-1} \tag{2.42}$$

$$A = -S\int 1s_a(1)r_{b1}^{-1}1s_b(1)\, d\tau_1 - S\int 1s_b(2)r_{a2}^{-1}1s_a(2)\, d\tau_2$$

$$+ \int 1s_a(1)1s_b(2)r_{12}^{-1}1s_a(2)1s_b(2)\, d\tau_1\, d\tau_2 + S^2 R^{-1}$$

Using the potential energy curve obtained from equation (2.41), one finds that the dissociation energy is about 70% of the observed experimental value. This result represents a slight improvement over the LCAO–MO theory. We shall compare the results of both the MO and VB treatments in more detail below and also discuss ways they can be improved. First, however, let us turn to the inclusion of the spin functions α and β into the wave function (2.38) in the VB theory and its importance in the formation of the chemical bond. First, however, we note that we can also calculate E for the H_2 molecule with a trial (antisymmetric spatial) wave function

$$\psi = (2 - 2S)^{-1/2}[1s_a(1)1s_b(2) - 1s_a(2)1s_b(1)] \tag{2.43}$$

and by precisely the same method used to arrive at equation (2.41) one obtains

$$E = 2E_{1s} + (Q - A)/(1 - S^2) \tag{2.44}$$

When E is calculated from equation (2.44), it is found to be higher than the value obtained using equation (2.41); hence the wave function in equation (2.43) yields a higher energy or repulsive state for H_2.

In the construction of the spin wave functions, we first must take account of the spin functions α and β* associated with the space functions in equations (2.38) and (2.43), and we proceed as follows: For the two electrons in the H_2 molecule

---

*Spin wave function α refers to a state with spin quantum number equal to $+\frac{1}{2}$, whereas β corresponds to $-\frac{1}{2}$.

we have four possible combinations of α and β:

$$\left.\begin{array}{c} \alpha(1)\alpha(2) \\ \beta(1)\beta(2) \\ \alpha(1)\beta(2) + \alpha(2)\beta(1) \\ \alpha(1)\beta(2) - \alpha(2)\beta(1) \end{array}\right\} \quad \text{(2.45)}$$

The most general statement of the Pauli exclusion principle (cf. Section 1.1) demands that the resulting complete wave function (i.e., the space function times the spin function) be antisymmetrical in the interchange of electron coordinates, both space and spin. Using equations (2.38) and (2.45), we find that the combinations which satisfy this requirement are:

$$\Psi_{S=0} = N[1s_a(1)1s_b(2) + 1s_a(2)1s_b(1)] \times [\alpha(1)\beta(2) - \alpha(2)\beta(1)] \quad \text{(2.46)}$$

and

$$\Psi_{S=1} = N[1s_a(1)1s_b(2) - 1s_a(2)1s_b(1)] \times \begin{cases} \alpha(1)\alpha(2) \\ \beta(1)\beta(2) \\ \alpha(1)\beta(2) + \alpha(2)\beta(1) \end{cases} \quad \text{(2.47)}$$

where as usual N is the normalization constant.

The function in equation (2.46) represents the state in which the spins of the two electrons are opposed (i.e., the total spin $S = 0$). Since the multiplicity is $(2S + 1)$, this is a singlet state. Equation (2.46), therefore, represents the wave function that describes the covalent bond and yields a stable molecule. The three spin functions appearing in equation (2.47) all correspond to a total spin of 1 and represent the state of higher energy in this case. These are the wave functions with multiplicity 3, and represent the triplet state.

As the next step, we shall compare the VB and the MO predictions for H_2 and shall subsequently refer to a refined approach to obtain results closer to those observed experimentally.

2.5. COMPARISON OF VALENCE BOND AND MOLECULAR ORBITAL THEORIES FOR H_2

Let us begin this comparison of the simple VB and MO theories we have used in calculating the ground-state energy of H_2 by examining the relationship between their respective wave functions. We need not include the spin parts of the wave functions as it has already been seen that they are the same in both cases. Furthermore, we shall also omit the normalization constants as they do not affect the essentials of the comparison. The two space wave functions are explicitly

$$\Psi_{MO} = [1s_a(1) + 1s_b(1)][1s_a(2) + 1s_b(2)] \quad \text{(2.48)}$$

or equivalently

$$\Psi_{MO} = \underbrace{1s_a(1)1s_a(2) + 1s_b(1)1s_b(2)}_{(i)} + \underbrace{1s_a(1)1s_b(2) + 1s_b(1)1s_a(2)}_{(ii)} \qquad (2.49)$$

and

$$\Psi_{VB} = 1s_a(1)1s_b(2) + 1s_b(1)1s_a(2) \qquad (2.50)$$

Referring to equations (2.49) and (2.50), we can see immediately that the two wave functions differ by the contribution labeled (i) in equation (2.49). Consequently the MO wave function contains the wave function Ψ_{VB} [i.e., terms labeled (ii) in equation (2.49)].

In Ψ_{MO}, the first two terms, i.e., (i), are called ionic since they correspond to both electrons on the same nucleus; the last two terms in equation (2.49) are the covalent terms. Ψ_{MO} is an equal admixture of ionic and covalent terms since the coefficients of (i) and (ii) in equation (2.49) are the same. Ψ_{VB} contains only covalent terms. It appears that the MO theory gives too much weight to the ionic terms and the VB theory does not include them at all. We might feel intuitively

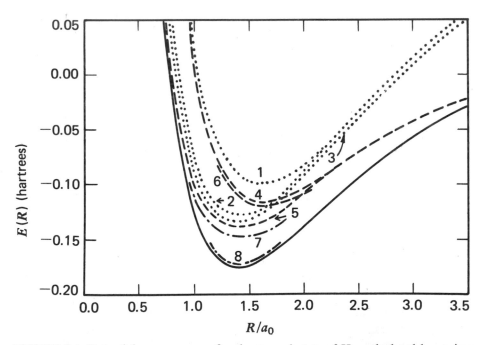

FIGURE 2.4. Potential energy curves for the ground state of H_2, calculated by various approximations; energy zero corresponds to $H(1s) + H(1s)$. (1) Simple LCAO–MO calculation. (2) Same as (1) but with scaling. (3) Self-consistent field calculations (see Chapter 4). (4) Heitler–London treatment. (5) Same as (4) with scaling. (6) Mixed MO and VB method. (7) Same as (6) with scaling. (8) Eleven-term variation function including terms in r_{12}. The solid line is an experimental curve [after Berry *et al.* (1980), p. 233].

that a wave function intermediate between VB and MO might be desirable if we can give proper weight to the ionic and covalent terms (compare also the Coulson–Fischer trial wave function in Appendix A2.5). Therefore, we define

$$\Psi_{cov} \equiv 1s_a(1)1s_b(2) + 1s_b(1)1s_a(2)$$
$$\Psi_{ionic} \equiv 1s_a(1)1s_a(2) + 1s_b(1)1s_b(2) \tag{2.51}$$

If we mix the functions in equation (2.51) we can write as an improved trial function

$$\Psi = C_1\Psi_{cov} + C_2\Psi_{ionic} \tag{2.52}$$

where C_1 and C_2 are constants for a particular internuclear separation R. Using the variation principle (cf. Appendix A2.1), we would obtain a better wave function by varying the ratio of C_1 and C_2 to obtain a minimum energy $E_0(R)$.

When Ψ in equation (2.52) is used, one does indeed improve the $E_0(R)$-vs.-R curve and the corresponding values of the dissociation energy and the bond length (see Figure 2.4). Though it is clear that by means of equation (2.52) there is improvement over the simple VB and MO treatments, it can be seen from Figure 2.4 that significant discrepancies from the experimental curve still remain.

We can extend this treatment further by using the wave function Ψ_{MO} for the $1\sigma_u^2$ state (i.e., one puts both electrons in $1\sigma_u$). Thus we adopt the following wave function:

$$\Psi_{MO}(1\sigma_u)^2 = [1s_a(1) - 1s_b(1)][1s_a(2) - 1s_b(2)] \tag{2.53}$$

In equation (2.53) the normalization constant and the spin part of the wave function are again omitted. Expanding equation (2.53) yields

$$\Psi_{MO}(1\sigma_u)^2 = 1s_a(1)1s_a(2) + 1s_b(1)1s_b(2) - 1s_a(1)1s_b(2) - 1s_b(1)1s_a(2) \tag{2.54}$$

Using equation (2.51), we can evidently write

$$\Psi_{MO}(1\sigma_u)^2 = \Psi_{ionic} - \Psi_{cov} \tag{2.55}$$

The following linear combination is then a possible improved trial wave function:

$$\Psi_{MO} = A\Psi_{MO}(1\sigma_g)^2 + B\Psi_{MO}(1\sigma_u)^2 \tag{2.56}$$

where A and B are constants depending on R. By varying A and B for each value of R this wave function can be used to obtain a minimum energy.

Equation (2.56) includes what is termed configuration interaction. By using more configurations, the wave function can be still further improved* (see Figure 2.4), but we will not pursue this here.

---

*But see Appendix AVI for the helium atom.

2.6. CHEMICAL BONDING, MOLECULAR ORBITAL ENERGY LEVELS, AND CORRELATION DIAGRAMS FOR HOMONUCLEAR DIATOMS

Having considered the ground-state molecular energy of H_2 resulting from the VB and LCAO–MO methods at some length, we now turn to consider other properties of homonuclear diatoms. In particular, we shall emphasize those properties which can, specifically, be extracted by using the MO wave functions employed in Section 2.3. We shall then extend the treatment to more complex homonuclear diatomic molecules.

We start by recalling that ψ^2 is the probability density (or $\psi^*\psi$ if ψ is complex) associated with a normalized wave function ψ. Writing the normalized LCAO–MOs for the H_2 molecule as

$$\psi_{\pm} = N_{\pm}(1s_a \pm 1s_b) \tag{2.57}$$

$N_{\pm}$ being the approximate normalizing constant for bonding and antibonding orbitals ψ_+ and ψ_-, respectively, we have that

$$\psi_{\pm}^2 = N_{\pm}^2[(1s_a)^2 + (1s_b)^2 \pm 2(1s_a 1s_b)] \tag{2.58}$$

which yields the electron probability density. A schematic diagram representing the molecular orbitals resulting from combining s orbitals is shown in Figure 2.5.

2.6.1. Symmetry Classification

We note that when we use the linear combination $(1s_a + 1s_b)$ we obtain build-up of electron density between the nuclei resulting in the bonding orbital σ_g; whereas in the case of $(1s_a - 1s_b)$ the result is an antibonding orbital σ_u. The symbol σ refers to the symmetry of the MO; a σ orbital is unchanged by rotation around the molecular axis. The subscripts g and u indicate the symmetry type of the orbital with respect to inversion through the midpoint of the two nuclei. A g subscript is used for an orbital that does not change when there is an inversion (a symmetric orbital), and a u subscript indicates that the orbital wave function changes sign on inversion (an antisymmetric orbital). In the case of H_2, both

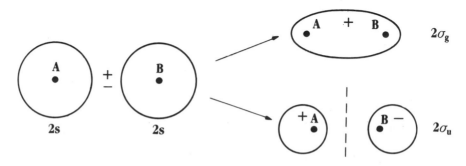

FIGURE 2.5. Diagram for the combining of two $2s$ atomic orbitals to yield bonding $(2\sigma_g)$ and antibonding $(2\sigma_u)$ molecular orbitals [redrawn from McWeeny (1979), p. 94].

electrons (of opposite spin) occupy the σ_g MO (i.e., the MO of lowest energy resulting in an electron density build-up between the nuclei, and hence chemical bonding).

Following the same procedure, one would expect that when considering more complex homonuclear diatomics, the linear combinations of various types of atomic orbitals would yield a variety of electron density cloud distributions. However, one must know which combinations of atomic orbitals will lead to the maximum overlapping necessary for bond formation. For example, p-type orbitals combining with s-orbitals must overlap in a head-on manner as illustrated in Figure 2.6.

On the left side of Figure 2.6 we note that there is no net overlap, while on the right overlap does result in the favorable overlap condition for bonding. In addition, the wave functions of the atomic orbitals that combine must be of the same sign (which puts a restriction on the symmetries of the atomic orbitals that can combine) and must have comparable energies. These points are summarized for the combinations of atomic orbitals that result in forming MOs with x as the bonding axis in Table 2.2. With these restrictions, Figure 2.7 demonstrates the electron density build-up for some allowable combinations. The σ symbol denotes that there is cylindrical symmetry about the molecular axis. The π symbol denotes the different symmetry displayed in Figure 2.7. In particular, one notes that the

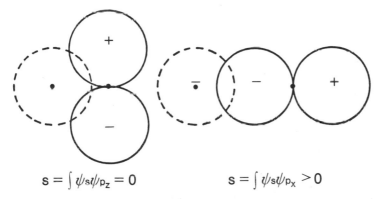

$$s = \int \psi_s \psi_{p_z} = 0 \qquad\qquad s = \int \psi_s \psi_{p_x} > 0$$

FIGURE 2.6. Overlap of atomic s and p orbitals. Overlap integral S is zero for left configuration.

TABLE 2.2
Atomic Orbital Combinations For AOs ϕ_a and ϕ_b on the z-Axis[a]

ϕ_a	ϕ_b	
	Allowed	Forbidden
s, p_z	s, p_z, d_{z^2}	$p_x, p_y, d_{xy}, d_{xz}, d_{yz}, d_{x^2-y^2}$
p_x, d_{xz}	p_x, d_{xz}	$s, p_y, p_z, d_{xy}, d_{yz}, d_{z^2}, d_{x^2-y^2}$
d_z^2	s, p_z, d_{z^2}	$p_x, p_y, d_{xy}, d_{yz}, d_{xz}, d_{x^2-y^2}$
$d_{x^2-y^2}$	$d_{x^2-y^2}$	$s, p_x, p_y, p_z, d_{xy}, d_{xz}, d_{yz}, d_{z^2}$

[a]Cumper (1966).

π-bonding orbital is composed of $+$ and $-$ lobes which are separated by a nodal plane; hence its sign change on rotation by half a turn (i.e., 180°).

The number preceding the symmetry symbols ($\sigma, \pi, \ldots$) designates MOs of a particular symmetry in order of energy. It has been found "experimentally" that the customary energy ordering for MOs is

$$1\sigma_g < 1\sigma_u < 2\sigma_g < 2\sigma_u < 1\pi_u < 3\sigma_g < 1\pi_g < 3\sigma_u < \cdots \qquad (2.59)$$

A comparison of the way the MO energies are related to those of the AOs that comprise them is shown schematically in Figure 2.8. We note from this figure that all π MOs are doubly degenerate (i.e., $1\pi_u$ has the pair $1\pi_{xu}$ and $1\pi_{yu}$ resulting from the p_x and p_y AOs).

2.6.2. Correlation Diagrams

Molecular orbital energy levels are often considered by means of a correlation diagram which demonstrates how the orbital energies change as a function of the internuclear distance R between the two nuclei. This means we must study the limits of separability of the two nuclei (i.e., at $R = \infty$ to $R = 0$).

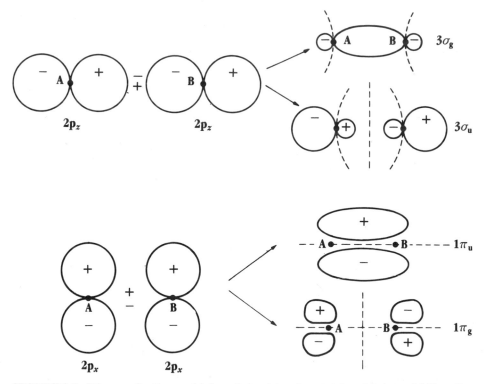

FIGURE 2.7. Diagram for the combining of: (top) two $2p_z$ atomic orbitals to yield bonding ($3\sigma_g$) and antibonding ($3\sigma_u$) molecular orbitals; (bottom) two $2p_z$ atomic orbitals to yield bonding ($1\pi_u$) and antibonding ($1\pi_g$) molecular orbitals [redrawn from McWeeny (1979), pp. 94–95].

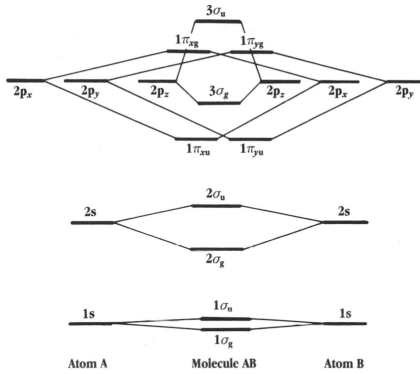

FIGURE 2.8. The molecular orbital energy levels for a homonuclear diatomic molecule AB in relation to the energy of separated atoms A and B.

For example, if the two nuclei of H_2^+ are allowed to coalesce, we form what is termed a united atom (see Section 2.3.3), which is an isoelectronic ion (He^+) with a nuclear charge of $+2e$. This means that each MO of H_2^+ becomes an AO of the united atom. As the nuclei of H_2^+ approach each other we go continuously from a molecular state to an atomic state, i.e., the ground state correlates with a H(1s) in the separated atom limit and He^+(1s) in the united atom limit. Also, as another example consider the limiting behavior of the $1\sigma_g$ orbital in the case of H_2. Both electrons in H_2 are paired in the σ_g orbital, yielding an electronic configuration of $1\sigma_g^2$. For large values of R, H_2 dissociates into two atoms; hence the limiting form of the $1\sigma_g$ MO for $R = \infty$ must turn into the 1s atomic orbitals (one centered on each nucleus). The $1\sigma_g$ orbital for the molecule is therefore said to correlate with the two 1s AOs for the separated atom case. These correlations can be systematically studied for more complicated diatomics because the AOs and MOs taking part can be characterized by the projection of angular momentum along the bond axis (say, the z-axis). It can be shown from Section 2.2 that for a linear molecule of one electron (H_2^+) the angular momentum L_z satisfies the expression

$$L_z\psi = \lambda\psi, \qquad \lambda = 0, \pm 1, \pm 2, \ldots \text{ (in units of } \hbar) \qquad (2.60)$$

where ψ is the electronic wave function of the molecule and λ is a quantum number (λ is a good quantum number due to axial symmetry). We recall that the

λ in a molecular electronic structure is similar to that of the quantum number m in the electronic structure of atoms. Hence in the case of a many-electron linear molecule

$$L_z \psi = \Lambda, \qquad \Lambda = 0, \pm 1, \pm 2, \ldots \tag{2.61}$$

We therefore classify states by angular momentum. Following the $s, p, d, f \ldots$ notation for atoms, we usually write

$$|\lambda| = \begin{Bmatrix} 0, & 1, & 2, & 3 \\ \sigma, & \pi, & \delta, & \phi \end{Bmatrix} \tag{2.62}$$

and when the degenerate pairs of MOs are unequally occupied

$$|\Lambda| = \begin{Bmatrix} 0, & 1, & 2, & 3 \\ \Sigma, & \Pi, & \Delta, & \Phi \end{Bmatrix} \tag{2.63}$$

The spin operators S_z and S^2 also commute with the molecular electronic Hamiltonian (omitting spin–orbit interaction); therefore, we can write the spectroscopic symbol of a linear molecule as

$$^{(2S+1)}\Lambda_{g \, or \, u}^{\pm} \tag{2.64}$$

The subscript in (2.64) refers to the inversion symmetry of the electron wave function, as usual, and the superscript applies only to Σ-states (i.e., $\Lambda = 0$) and refers to the reflection symmetry of the electronic wave function with respect to a plane containing the bond axis (+ indicates no change of sign; − means that a sign change occurs; see Table 2.3).

   With this background, we consider first a correlation diagram for homonuclear diatomics followed by an examination of the electronic configuration, the dissociation energy, the bond length, and the spectroscopic terms of the ground state of several homonuclear diatomic molecules.

   The correlation diagram for a homonuclear diatomic molecule is shown schematically in Figure 2.9. As an aid to its interpretation, we refer to Figure 2.10, which demonstrates a correlation resulting from the combination of the two $2p_x$ AOs of the separated atoms in producing $1\pi_u$ and $1\pi_g$ MOs, and these go over into p_x and d_{xz} AOs, respectively, on the united atom.

   The lines connecting the orbitals on both sides of Figure 2.9 indicate correlations like those demonstrated in Figure 2.10. The correlation diagram shows the way the levels change with R. The vertical line (which represents molecular formation) illustrates very clearly the energy levels shown in Figure 2.8. It should be noted that the correlation diagram follows the noncrossing rule, which asserts that two states of the same symmetry cannot cross as R changes. We are now in a position to follow the aufbau principle in arriving at the electron configuration and spectroscopic symbol of homonuclear diatomic molecules. Table 2.3 is constructed by using the principles just discussed and also gives dissociation energies and bond lengths of some homonuclear diatomics.

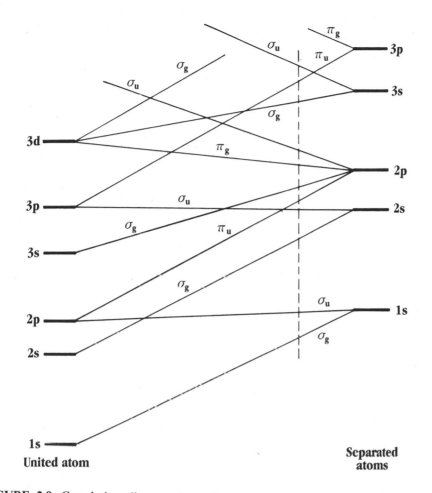

FIGURE 2.9. Correlation diagram for a homonuclear diatomic molecule comparing united-atom levels to those of the separated atoms [redrawn from McWeeny (1979), p. 98].

We now turn to a discussion of the properties displayed in Table 2.3 and begin by examining the electron configuration of some of the molecules and their properties listed in the table. It should be noted that the bond order is defined as the number of bonding pairs of electrons less the number of antibonding pairs.

2.6.3. Electron Configurations

We now enumerate the electron configurations for "molecules" from He_2 to Ne_2:

1. He_2 $(1\sigma_g)^2(1\sigma_u)^2$: it is easy to see why He_2 is unstable in the ground state. When a bonding and antibonding orbital (both from the same AOs) are occupied by two electrons, the result leads to no bond formation (i.e.,

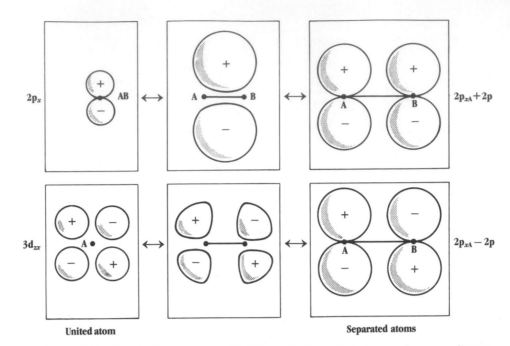

FIGURE 2.10. Correlation between orbital forms in the united atom and separated atom for homonuclear diatomic molecule: (top) the π_u combination of two separated atom p orbitals (right) correlates with a united atom p-orbital (left); (bottom) the π_g combination (right) goes to a united atom d-orbital (left) [redrawn from McWeeny (1979), p. 99].

TABLE 2.3
Electron Configurations and Spectroscopic Symbols of Some Homonuclear Diatomic Molecule Ground States[a]

Molecule	Electron configuration	Bond order	Dissociation energy (eV)	Bond length (Å)	Spectroscopic symbol
H_2	$(1\sigma_g)^2$	1	4.75	0.74	$^1\Sigma_g^+$
He_2	$(1\sigma_g)^2(1\sigma_u)^2$	0	—	—	$^1\Sigma_g^+$
Li_2	$KK(2\sigma_g)^2$	1	1.14	2.67	$^1\Sigma_g^+$
Be_2	$KK(2\sigma_g)^2(2\sigma_u)^2$	0	—	—	$^1\Sigma_g^+$
B_2	$KK(2\sigma_g)^2(2\sigma_u)^2(1\pi_u)^2$	1	3.0	1.59	$^3\Sigma_g^-$
C_2	$KK(2\sigma_g)^2(2\sigma_u)^2(1\pi_u)^4$	2	6.24	1.24	$^1\Sigma_g^+$
N_2	$KK(2\sigma_g)^2(2\sigma_u)^2(1\pi_u)^4(3\sigma_g)^2$	3	9.76	1.09	$^1\Sigma_g^+$
O_2	$KK(2\sigma_g)^2(2\sigma_u)^2(3\sigma_g)^2(1\pi_u)^4(1\pi_g)^2$	2	5.12	1.21	$^3\Sigma_g^-$
F_2	$KK(2\sigma_g)^2(2\sigma_u)^2(3\sigma_g)^2(1\pi_u)^4(1\pi_g)^4$	1	1.60	1.41	$^1\Sigma_g^+$
Ne_2	$KK(2\sigma_g)^2(2\sigma_u)^2(3\sigma_g)^2(1\pi_u)^4(1\pi_g)^4(3\sigma_u)^2$	0	—	—	$^1\Sigma_g^+$

[a]KK denotes configuration $(1\sigma_g)^2(1\sigma_u)^2$. Dissociation energies are taken from B. de B. Daurent (1970).

repulsion results as the corresponding energy of the molecule is greater than that of the separate atoms).

2. Li_2 $KK(2\sigma_g)^2$: the two K shells are complete and do not enter into the bonding situation expressed by $(2\sigma_g)^2$ and a stable species exists.

3. Be_2 $KK(2\sigma_g)^2(2\sigma_u)^2$: an analogous situation to that in (1) above and Be_2 is unstable.

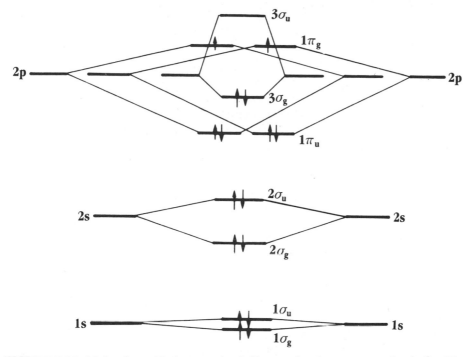

FIGURE 2.11. Molecular orbital energy level diagram for the oxygen molecule O_2. The molecule is paramagnetic as can be observed from the two parallel spin electrons in $1\pi_g$ [redrawn from McWeeny (1979), p. 103].

4. B_2 $KK(2\sigma_g)^2(2\sigma_u)^2(1\pi_u)^2$: one electron in each of the two degenerate π-orbitals as demanded by Hund's rules* results in an energy lowering and a stable molecule.

5. C_2 $KK(2\sigma_g)^2(2\sigma_u)^2(1\pi_u)^4$: with six bonding and two antibonding electrons, a stable molecule results.

6. N_2 $KK(2\sigma_g)^2(2\sigma_u)^2(1\pi_u)^4(3\sigma_g)^2$: a single bond $(3\sigma_g)^2$ and two bonds of π-type, i.e., $(1\pi_{xu})^2(1\pi_{yu})^2$, resulting in strong bonding properties.

7. O_2 $KK(2\sigma_g)^2(2\sigma_u)^2(3\sigma_g)^2(1\pi_u)^4(1\pi_g)^2$: eight bonding electrons and four antibonding (two of which have the same spin) as ground state is paramagnetic (hence a triplet as indicated). The bonding energy in O_2 is approximately that of a double bond (i.e., four electrons). It is worth examining the MO diagram (see Figure 2.11) of this electron configuration as its MO treatment predicted a paramagnetic molecule, a major success of the MO theory: It will be noted that all levels have two electrons with opposed spin until one reaches the $1\pi_g$ level, which is degenerate. The two unpaired electrons in these orbitals lead to an explanation of the triplet state $(2S + 1) = 3$ and O_2's paramagnetic nature.

8. F_2 $KK(2\sigma_g)^2(2\sigma_u)^2(3\sigma_g)^2(1\pi_u)^4(1\pi_g)^4$: $(2\sigma_g)^2$ and $(2\sigma_u)^2$ represent the bonding and antibonding situation considered in (1) above; and in addition one has the same bonding–antibonding result in the case of $(1\pi_u)^4$

---

*For degenerate orbitals (i) electrons avoid the same orbital and (ii) 2 electrons, singly occupying a pair of degenerate orbitals like $2p_x$ and $2p_y$ have their spins parallel.

and $(1\pi_g)^4$. Consequently, only $(3\sigma_g)^2$ is left for bonding and is equivalent to a single bond.

9. Ne_2 $KK(2\sigma_g)^2(2\sigma_u)^2(3\sigma_g)^2(1\pi_u)^4(1\pi_g)^4(3\sigma_u)^2$: unstable because all bonding–antibonding pairs lead to repulsion.

Not only do the electron configurations uphold the spectroscopic symbol shown in Table 2.3 but also the nature of the bonding result in each case helps to account for the magnitudes of the dissociation energies and the bond lengths. The result of the properties displayed in Table 2.3 can also be accounted for by a consideration of contour maps of electron densities of the orbitals and the total electron density as shown in Figure 2.12. We also note, referring to Table 2.3, and the electron configurations, that the σ^2, $\sigma^2\pi^2$, and $\sigma^2\pi^4$ configurations correspond closely to single, double, and triple bonds, respectively.

2.7. HETERONUCLEAR DIATOMIC MOLECULES

In homonuclear diatomic molecules the two atoms have the same (i.e., equivalent) attraction on the electron distribution. However, in heteronuclear diatomics the two different atoms (e.g., a and b) will have a polarizing effect on the electron distribution. The electron distribution will be shifted toward the atom with the greater electronegativity (i.e., the one whose electron cloud has the stronger attractive power). Electronegativity will be considered in a little more detail later on in this section, but for the present let us turn to the changes we must make in the LCAO–MO formulation in order that it be applicable to heteronuclear diatomic molecules.

The wave functions will in fact have to be weighted to account for the differences in electronegativity of the two atoms forming the molecule. The wave function can therefore be written in the same manner as for the homonuclear diatomic case provided one includes a weighting coefficient W to account for the fact that the coefficients in the LCAO expression will not be equal in heteronuclear diatomics:

$$\psi_\pm = N(\phi_a \pm W\phi_b) \tag{2.65}$$

where N is the normalizing factor and ϕ_a and ϕ_b are the AOs corresponding to atom a and b, respectively.

The electron probability density in the bonding MO, therefore, is

$$\rho = \psi_+^2 = N^2(\phi_a + W\phi_b)^2 \tag{2.66}$$

or

$$\rho = N^2(\phi_a^2 + W^2\phi_b^2 + 2\phi_a W\phi_b) \tag{2.67}$$

and the corresponding N can be determined as follows:

$$\int \psi^2 \, d\tau = \int N^2(\phi_a^2 + W^2\phi_b^2 + 2\phi_a W\phi_b) \, d\tau = 1 \tag{2.68}$$

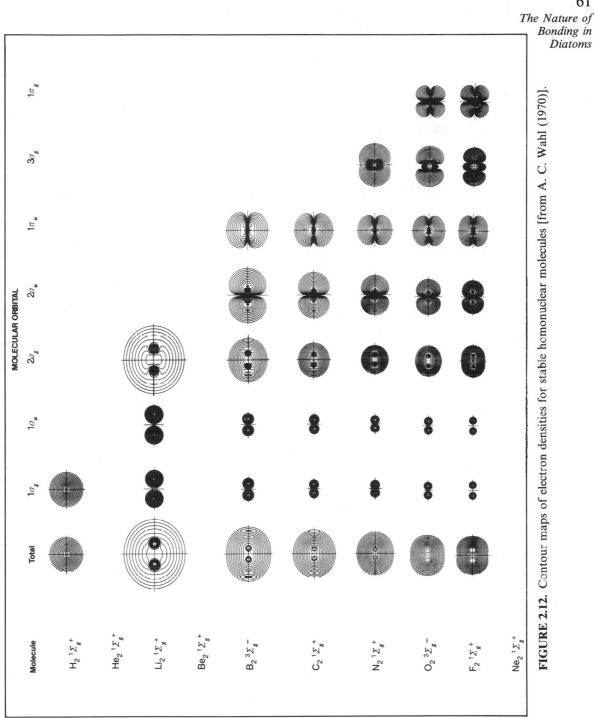

FIGURE 2.12. Contour maps of electron densities for stable homonuclear molecules [from A. C. Wahl (1970)].

and hence

$$N^{-2} = \int \phi_a^2 \, d\tau + W^2 \int \phi_b^2 \, d\tau + 2W \int \phi_a \phi_b \, d\tau \qquad (2.69)$$

or

$$N^{-2} = 1 + W^2 + 2WS \qquad (2.70)$$

W can be determined from the appropriate secular equations (see especially Chapter 6) and S is the familiar overlap integral.

We see immediately that the normalized ψ-function is different from that in the homonuclear diatomic case and a corresponding change will take place in the molecular energy as well as in the distribution of the electron density between and around the two nuclei. Let us turn now to the correlation diagram for a heteronuclear diatomic molecule and then compare and contrast it with the correlation diagram in the homonuclear diatomic case (see Figure 2.9).

As a somewhat oversimplified but nevertheless useful starting point, one can assume that each MO correlates only with an AO of one of the two atoms. The correlation diagram for a heteronuclear diatomic molecule (Figure 2.13) does not

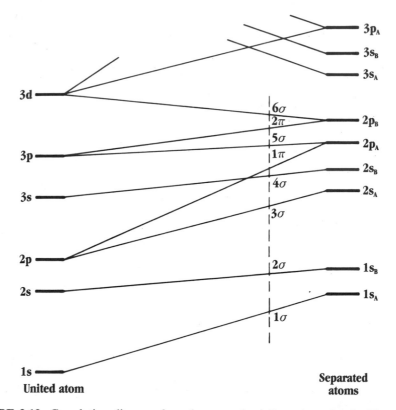

FIGURE 2.13. Correlation diagram for a heteronuclear diatomic molecule. The vertical broken line corresponds to the sequence of the MO energies in the NO molecule [redrawn from McWeeny (1979), p. 104].

show clearly that the nature of atoms a and b determines the separated-atom energy levels. For example, for a particular n and l value, the atom with the higher nuclear charge will have the lower atomic energy levels. Nevertheless, the diagram does give useful guidance in the same way as that for homonuclear diatomics in that one gains some idea of the MO energy sequence for a particular a and b combination. Furthermore, when ψ is evaluated by taking into account the magnitudes of the coefficients [i.e., the value of W in equation (2.65)] in the LCAO–MO electron density, contours for total probabilities can be determined. Figure 2.14 is an illustration of this point obtained by use of improved MO calculations. We stated earlier that the electron pulling power (i.e., the electronegativity) plays an important part in determining the electron distribution in the molecule. We note in Figure 2.14 that H must have an electronegativity between the metallic and nonmetallic elements (substantiated in Table 2.4). Let us then look more closely into the concept of electronegativity, as it apparently plays such an important part in determining the properties of heteronuclear diatomics.

Referring to Table 2.4, we are interested first in the nature of bonding in the heteronuclear diatomics shown. As we have seen, a bonding orbital has a higher electron density along the bond axis than that of the separated atoms. In all the hydrides of Table 2.4, the 1σ-orbital is like the $1s$-orbital in the heavy atom and, as for all core orbitals, does not contribute to the bonding. Similarly the π-orbital is not a bonding one as it is composed of the heavy-atom $2p$-orbitals and therefore does not bond to the H atom. On the other hand, the 2σ-orbital is a bonding one in LiH, but not in HF, while the 3σ-orbital is a principal contributor to the bonding.

We note that the dissociation energy has a tendency to increase in most cases as we go from top to bottom in Table 2.4. The dipole moment changes due to the difference in polarity as the heavier atom of the molecule becomes electronegative, and the bond length decreases steadily as the electronegativity of the atom bonded to hydrogen increases.

We now turn to a consideration of dipole moments, bond lengths, percentage ionic character and the importance of electronegativity in these properties of heteronuclear diatomics (cf. Section 1.6).

We have already noted that, due to electronegativity, there is an unequal sharing of the bonding electrons in heteronuclear diatomics, which means that the bond possesses both ionic and covalent character. We can make a good estimate of the amount of ionic character for a diatomic molecule that has a single bond as follows:

In the LCAO–MO treatment, the wave function for a bonding electron

$$\psi = C_1\phi_a + C_2\phi_b \tag{2.71}$$

where, from equation (2.65), $C_2/C_1 = W$. The ionic character (IC) can be defined as

$$IC = |C_1^2 - C_2^2| \tag{2.72}$$

and we see immediately from equation (2.72) that $IC = 0$ if the electron density of the bond is shared equally, and $IC = 1$ (i.e., 100%) if either C_1 or $C_2 = 0$. In

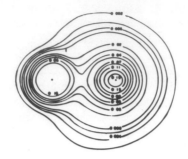

LiH $^1\Sigma^+$

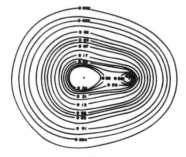

BeH $^2\Sigma^+$

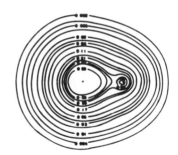

BH $^1\Sigma^+$

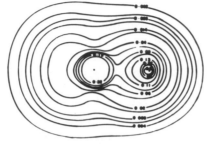

CH $^2\Pi_2$

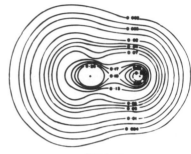

NH $^3\Sigma^-$

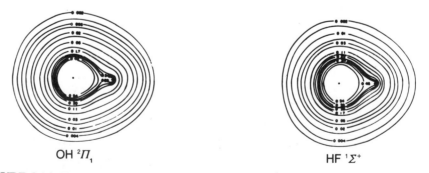

OH $^2\Pi_1$

HF $^1\Sigma^+$

FIGURE 2.14. Equal electron-density contours for total probability densities of the first row diatomic hydrides [from Bader *et al.* (1967)].

TABLE 2.4
Some Properties of Diatomic Hydrides in Ground State

Molecule	Electron configuration	Dipole moment (D)	Bond energy (eV)	Bond length (Å)	Spectroscopic symbol
LiH	$1\sigma^2 2\sigma^2$	-6.00	2.52	1.60	$^1\Sigma^+$
BeH	$1\sigma^2 2\sigma^2 3\sigma$	-0.28	2.40	1.30	$^2\Sigma^+$
BH	$1\sigma^2 2\sigma^2 3\sigma^2$	1.73	3.54	1.24	$^1\Sigma^+$
CH	$1\sigma^2 2\sigma^2 3\sigma^2 1\pi$	1.57	3.65	1.12	$^2\Pi$
NH	$1\sigma^2 2\sigma^2 3\sigma^2 1\pi^2$	1.63	3.40	1.05	$^3\Sigma^-$
OH	$1\sigma^2 2\sigma^2 3\sigma^2 1\pi^3$	1.78	4.62	0.97	$^2\Pi$
HF	$1\sigma^2 2\sigma^2 3\sigma^2 1\pi^4$	1.94	6.11	0.92	$^1\Sigma^+$

the latter case we have an ionic molecule a^+b^- or a^-b^+. Further, if $C_1 = C_2$ then the dipole moment $\mu = 0$. Since $\mu = er$ for a 100% ionic compound, where e is the electronic charge and r is the bond length of the molecule a—b, the dipole moment of a molecule with a particular IC is modified to read

$$\mu = er|C_1^2 - C_2^2| = er(\text{IC}) \tag{2.73}$$

In Table 2.5 we list results for several molecules and those of their corresponding properties that are pertinent to this discussion. The quantity Δx in the table is the difference in the electronegativities of the atoms from which the molecule is built. The IC values given are those calculated from equation (2.72). We note that on chemical grounds, KBr, KCl, CsI, CsCl, and KI are known to be almost completely ionic, and by using equation (2.72) we find IC values for these molecules of 65% or more. The difference may be accounted for by the fact that

TABLE 2.5
Dipole Moments, Ionic Character and Electronegativity Differences Δx
in Heteronuclear Diatomic Molecules[a]

Molecule	Bond length (Å)	Dipole moment (D)	er_e (D)	Percent ionic character	Δx
HF	0.92	1.91	4.41	43.0	1.7
HCl	1.27	1.03	6.12	16.8	0.95
HBr	1.41	0.78	6.76	11.5	0.75
HI	1.61	0.38	7.72	4.9	0.45
FCl	1.63	0.88	7.82	11.3	0.75
FBr	1.76	1.29	8.43	15.3	0.85
ClBr	2.44	0.57	10.27	5.6	0.2
ClI	2.32	0.65	11.15	5.8	0.5
TlCl	2.54	4.44	12.20	36.4	1.75
LiI	2.39	7.07	11.44	61.6	1.65
LiF	1.51	6.60	7.25	91.0	3.00
CsI	3.32	10.20	15.92	64.1	1.95
CsCl	2.91	10.42	13.96	74.7	2.45
KBr	2.82	10.41	13.55	76.8	2.15
KCl	2.67	10.48	12.81	81.8	2.35

[a]Constructed from data given by Orville-Thomas (1966).

the electron density around each ion is not spherical, thereby producing a polarizing effect. The table also clearly reveals the major role of electronegativity in determining the electronic distribution along the bond axis.

PROBLEMS

2.1. Use the virial theorem (see Appendix A2.6) in the form appropriate to a general internuclear separation R of a diatomic molecule, namely

$$2T + U = - R \, dE/dR \tag{P2.1}$$

where T is kinetic energy, U potential energy and $E = T + U$ the total molecular energy, to derive separate expressions for $T(R)$ and $U(R)$ given that $E(R)$ has the so-called Lennard-Jones 12-6 form (see Chapter 3):

$$E(R) = E(\infty) + A/R^{12} - B/R^6 \tag{P2.2}$$

Show, in particular, that in this model, as the atoms are brought up from infinite separation (i.e., R tends to infinity), there is an initial decrease in kinetic energy. [In fact, this is often important for molecular bonding [see, e.g., Ruedenberg (1964)]. (N.B.: The reader not familiar with the virial theorem may usefully consult the first part of Appendix A2.6, where the theorem is derived by simple classical mechanical arguments.] To go to the quantum-mechanical case under discussion, simply replace classical time averages by quantum mechanical averages. Finally, in the language of Appendix A2.6, the term on the right-hand side of (P2.1) is the virial of the forces required to hold the nuclei in position at distance R when R differs from its equilibrium value, at which separation, of course dE/dR equals zero. Then the virial theorem at equilibrium reduces to $2T + U = 0$ or $T = - E$, as for atoms.)

2.2. Following Nalewajski (1979), assume that the change in kinetic energy of a diatomic molecule from the sum of the kinetic energies of the separated atoms, say, $\Delta T(R)$, is given by

$$\Delta T(R) = a \exp[- 2c(R - b)] - 2a \exp[- c(R - b)] \tag{P2.3}$$

where a, b, and c are parameters. **(a)** Show that for this (Morse) form of $\Delta T(R)$, the kinetic energy change becomes zero when $R = R_0$, where

$$R_0 = b - (1/c) \ln 2 \tag{P2.4}$$

(b) Writing the virial theorem (P2.1) in the form

$$\Delta T(R) = - d(R\Delta E)/dR \tag{P2.5}$$

where ΔE is the total energy change from the sum of the energies of the two constituent atoms at infinite separation, show that, quite generally, the position of the node in $\Delta T(R)$ can be predicted from the total energy curve $\Delta E(R)$ by a simple construction, namely at R_0 the tangent to the $\Delta E(R)$ curve has the ΔE intercept $2\Delta E(R_0)$.

2.3. Consider a HX molecule, say, a hydrogen halide, in a case where one can ignore departures from covalency. With identical building blocks ϕ_H and ϕ_X use **(a)** the LCAO–MO description, with wave function

$$\psi_{MO}(1, 2) = N_{MO}[\phi_H(1) + \phi_X(1)][\phi_H(2) + \phi_X(2)] \tag{P2.6}$$

and **(b)** the valence bond model with

$$\psi_{VB}(1, 2) = N_{VB}[\phi_H(1)\phi_X(2) + \phi_X(1)\phi_H(2)] \tag{P2.7}$$

to prove that (see also Problem 2.2):

$$\Delta T_{VB}(R) = \frac{S + S^2}{1 + S^2} \Delta T_{MO}(R) \tag{P2.8}$$

where S is the overlap integral.

REFERENCES

W. G. Baber and H. R. Hasse, *Proc. Camb. Phil. Soc.* **31**, 564 (1935).

R. F. W. Bader, J. Keaveney, and P. E. Cade, *J. Chem. Phys.* **47**, 3381 (1967).

G. G. Balint-Kurti and M. Karplus, in: *Orbital Theories of Molecules and Solids* (N. H. March, ed.) Clarendon Press, Oxford (1974).

D. R. Bates, K. Ledsham, and A. L. Stewart, *Phil. Trans. Roy. Soc.* **A246**, 28 (1953).

R. S. Berry, S. A. Rice, and J. Ross, *Physical Chemistry*, John Wiley and Sons, New York (1980).

M. Born and J. R. Oppenheimer, *Ann. Physik* **84**, 457 (1927).

O. Burrau, *Det. Kgl. Danske Vid Selgkab* **7**, 458 (1930).

C. W. N. Cumper, *Wave Mechanics for Chemists*, Academic Press, New York (1966).

B. de B. Daurent, Bond Dissociation Energies in Single Molecules, NSRDS-NBS31, U.S. Department of Commerce, U.S. Government Printing Office, Washington, D.C. (1970).

W. Heitler and F. London, *Z. Phys.* **44**, 455 (1927).

R. McWeeny, *Coulson's Valence*, 3rd ed., Clarendon Press, Oxford (1979).

R. F. Nalewajski, *J. Phys. Chem.* **82**, 1439 (1979).

W. J. Orville-Thomas, *The Structure of Small Molecules*, Elsevier, Amsterdam (1966).

K. Ruedenberg, *J. Phys. Chem.* **68**, 1676 (1964).

E. Teller, *Zeit. Phys.* **61**, 458 (1930).

A. C. Wahl, *Scientific American* **322**, No. 4, 54 (April, 1970).

Further Reading

P. W. Atkins, *Molecular Quantum Mechanics*, 2nd ed., Oxford University Press, Oxford (1983).

H. F. Hameka, *Quantum Theory of the Chemical Bond*, Hafner Press, MacMillan, New York (1975).

J. N. Murrell, S. F. A. Kettle, and J. M. Tedder, *The Chemical Bond*, 2nd Ed., John Wiley and Sons, New York (1985).

J. P. Lowe, *Quantum Chemistry*, Academic Press, New York (1978).

L. Pauling, *The Nature of the Chemical Bond*, 3rd Ed., Cornell University Press, Ithaca (1960).

J. C. Slater, *Quantum Theory of Molecules and Solids*, Vol. I, *Electronic Structure of Molecules*, McGraw-Hill, New York (1963).

B. Webster, *Chemical Bonding Theory*, Blackwell Scientific Publications, Oxford (1990).

A. F. Wells, *Structural Inorganic Chemistry*, Clarendon Press, Oxford (1984).

MOLECULAR INTERACTIONS

3.1. INTRODUCTION

So far we have been concerned with forces within an isolated molecule. We now turn to the forces between molecules; that is to intermolecular forces. Compelling qualitative evidence that such forces must exist comes from the very existence of condensed phases (i.e., liquids and solids) and the fact that such phases resist further compression. It is clear from all this that both attractive and repulsive forces must be operating. It is important in this context to note that:

1. Attractive forces prove to be dominant at long range. Here, the meaning of long range is such that the overlap of electronic distributions in the molecules that are interacting can be neglected.
2. Repulsive contributions to the intermolecular forces are dominant when molecules are forced to within small separations of each other, i.e., into regions where there is significant overlap of the respective electron charge clouds in the molecules.

The study of intermolecular forces—as of the time of writing—remains a topic of central interest in chemical physics. In the treatment below we will first set out the basic aspects of the subject and then follow with some comments on the directions that investigations are taking in this field.

Attractive interactions between molecules are usually grouped under the general heading of van der Waals forces. Let us turn first to the basic origin of these forces.

3.2. LONG-RANGE FORCES

We note at the outset the useful separation of forces into two categories: Coulombic and exchange. Coulombic forces are due to the direct interaction between charged systems while exchange forces are quantum-mechanical in origin. We consider both contributions in the present chapter.

Starting with Coulombic forces, we find useful to classify the following long-range interactions: (1) Ion–ion, (2) Ion–dipole, (3) Ion–induced-dipole, (4) Dipole–dipole, and (5) Dipole–induced-dipole.

Table 3.1 lists some typical magnitudes for these interactions, following Berry *et al.* (1980). These values at separations of both 5 and 10 Å have been expessed in electron volts for typical values of charge, dipole moment, and polarizability.

TABLE 3.1

Typical Magnitudes of Interactions of Various Types of
Long-Range Forces[a]

	Interaction energy (eV)	
Interaction	Separation 5 Å	Separation 10 Å
Ion–ion	2.9	1.4
Ion–dipole	0.2	0.04
Ion–induced-dipole	0.03	0.002
Dipole–dipole	0.02	0.003
Dipole–induced-dipole	5×10^{-4}	8×10^{-6}

[a]Constructed from Berry *et al.* (1980) with the following typical values: $Q = \pm e$ (single charge), $\mu = 1.5$ debye and polarizability α taken from $\alpha/4\pi\varepsilon_0 = 3$ Å^3 (typical of small molecules). The energy unit used is the electron volt, which is equal to 22 kcal/mole.

The equations for the five long-range interactions listed above are collected in Table 3.2, while their derivations, which involve classical electrostatic considerations are outlined in Berry *et al.* (1980). What is instructive in Table 3.1 is the relative magnitude of each type of interaction at a given separation and also the decrease of the interaction energy as the separation is doubled.

In addition to the Coulombic-type interactions listed in Tables 3.1 and 3.2, we must also consider the long-range interaction between molecules known as the London dispersion force and we will now take up its basic origin.

3.2.1. The London Dispersion Force

The interactions displayed in Tables 3.1 and 3.2 result from classical electrostatic treatment of intermolecular interactions. In contrast, the London dispersion

TABLE 3.2

Interaction Potential Formulas[a]

Interaction type	Notation	Interaction energy
Ion–ion	$Q_1 \quad Q_2$ $\bullet \quad \bullet$ $\leftarrow r \rightarrow$	$V(r) - V(\infty) = \int_r^\infty -F(r)\, dr = Q_1 Q_2/4\pi\varepsilon_0 r$ F is force: Coulomb's law
Ion–dipole	$\|\!\leftarrow\!\!r\!\longrightarrow\!\| \; +Q$ $Q \bullet$ ion $-Q$ dipole	$V(r, \theta) = \mu Q \cos\theta/4\pi\varepsilon_0 r^2$ μ is the dipole moment
Ion–induced-dipole	Same as the figure above	$V(r) = \alpha Q^2/32\pi^2\varepsilon_0^2 r^4$
Dipole–dipole	Two permanent dipoles a and b, each oriented at angle θ and ϕ relative to the a–b axis	$V(r, \theta_a, \theta_b, \phi_a, \phi_b)$ $= (\mu_a\mu_b/4\pi\varepsilon_0 r^3)[-2\cos\theta_a\cos\theta_b$ $+ \sin\theta_a\sin\theta_b\cos(\phi_b - \phi_a)]$
Dipole–induced-dipole	Same as for dipole–dipole	$V(r, \theta_a) = -\alpha_b\mu_a^2(3\cos^2\theta_a + 1)/2(4\pi\varepsilon_0)^2 r^6$ $\alpha =$ polarizability

[a]Detailed derivations of the formulas are given in Berry *et al.* (1980).

force arises from quantum mechanical considerations. For comparison with the magnitudes summarized in Table 3.1, it is worth noting here that typical values for the magnitude of the dispersion forces are:

$$2 \times 10^{-3} \text{ eV} \quad \text{at 5 Å}$$

$$3 \times 10^{-5} \text{ eV} \quad \text{at 10 Å}$$

To understand the origin of dispersion forces, it is important to emphasize that electrons in a molecule are in a state of motion. While this motion may lead to an average dipole moment that is identically zero, it is also true that, at any instant, a dipole moment can be present. This in turn induces a dipole moment in another molecule in its proximity and this results in classical induction forces which turn out to lead to a net attraction. What we have here then is an induced-dipole–induced-dipole interaction. It is useful to begin the quantum-mechanical treatment of this dispersive interaction by referring to the so-called Drude model. The treatment below follows that of Rigby *et al.* (1986).

3.2.2. The Drude Model

Let us first consider the following one-dimensional model (see Figure 3.1) where

1. Each molecule is represented by two charges $+Q$ and $-Q$.
2. The positive charges are held fixed, while the negative charges perform simple harmonic oscillations.

Denoting the displacements on molecules a and b by z_a and z_b respectively, we have for the instantaneous dipole moments at time t:

$$\mu_a = Qz_a(t) \tag{3.1}$$

and

$$\mu_b = Qz_b(t) \tag{3.2}$$

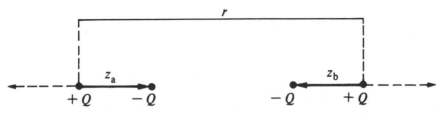

FIGURE 3.1. Depicts the one-dimensional Drude model of the dispersion energy [redrawn from Rigby *et al.* (1986)].

Evidently this is also a case where the average dipole moment is zero, as noted above. If k denotes the force constant and M the mass of the oscillating charge on each molecule, then the classical frequency v is given by

$$v = (1/2\pi)(k/M)^{1/2} \tag{3.3}$$

If we write the Schrödinger equation for molecule a as

$$(1/M)\, \partial^2\psi/\partial z_a^2 + 8\pi^2/h^2(E_a - \tfrac{1}{2}kz_a^2)\psi = 0 \tag{3.4}$$

we arrive at the usual level spectrum

$$E_a = (n_a + \tfrac{1}{2})hv \tag{3.5}$$

with an analogous result for molecule b.

Of course, in the above argument we have assumed that molecules a and b are sufficiently far apart that the charges oscillate independently, which is why we can write equation (3.4). If we include the dipole–dipole interaction at separation r shown in Table 3.2, then it is necessary to add to the potential energy in equation (3.4) a term which is inversely proportional to r^3 for the "parallel dipole interaction," yielding the Schrödinger equation for the interacting pair of molecules

$$\frac{1}{M}\left(\frac{\partial^2\Psi}{\partial z_a^2} + \frac{\partial^2\Psi}{\partial z_b^2}\right) + \frac{8\pi^2}{h^2}\left(E - \tfrac{1}{2}kz_a^2 - \tfrac{1}{2}kz_b^2 - \frac{2z_a z_b Q^2}{(4\pi\varepsilon_0)r^3}\right)\Psi = 0 \tag{3.6}$$

Introducing sum and difference coordinates defined by

$$Z_1 = (z_a + z_b)/2^{1/2} \qquad \text{and} \qquad Z_2 = (z_a - z_b)/2^{1/2} \tag{3.7}$$

we can rewrite the Schrödinger equation (3.6) in harmonic oscillator form with potential energy $\tfrac{1}{2}k_1 Z_1^2 + \tfrac{1}{2}k_2 Z_2^2$, where the modified force constants k_1 and k_2 are given by

$$k_1 = k - 2Q^2/(4\pi\varepsilon_0)r^3, \qquad k_2 = k + 2Q^2/(4\pi\varepsilon_0)r^3 \tag{3.8}$$

As we often do for notational compactness, we put $4\pi\varepsilon_0 = 1$, and recalling equation (3.3) for the frequency v of the noninteracting oscillators we can rewrite equation (3.8) as

$$v_1 = v(1 - 2Q^2/kr^3)^{1/2}, \qquad v_2 = v(1 + 2Q^2/kr^3)^{1/2} \tag{3.9}$$

For large r we can now expand $(1 \pm 2Q^2/kr^3)^{1/2}$ in equation (3.9) to obtain

$$v_1 = v(1 - Q^2/kr^3 - \tfrac{1}{8}(2Q^2/kr^3)^2 + \cdots)$$
$$v_2 = v(1 + Q^2/kr^3 - \tfrac{1}{8}(2Q^2/kr^3)^2 + \cdots) \tag{3.10}$$

The zero-point energy can be expressed in terms of the frequencies v_1 and v_2 as

$$E_0 = \tfrac{1}{2}h(v_1 + v_2) \tag{3.11}$$

to be compared with the result $\tfrac{1}{2}hv$ per oscillator without interaction. Using equation (3.10) one then obtains for the correction due to interaction

$$E_0 - hv = -\tfrac{1}{2}hv(Q^4/k^2r^6) \tag{3.12}$$

Hence one reaches the result for the dispersion energy, denoted $\phi_{\text{disp}}(r)$:

$$\phi_{\text{disp}}(r) = -\tfrac{1}{2}hv(Q^4/k^2r^6) \tag{3.13}$$

It is to be noted that the negative sign in equation (3.13) represents a lowering of energy owing to the dispersion interaction; i.e., the interaction is attractive. Of course, in real molecular systems, the r^{-6} behavior in equation (3.13) is only the leading term at large r; for smaller r, terms proportional to r^{-8}, r^{-10}, etc. begin to play a significant role. Furthermore, since one is really dealing with an interaction mediated by electromagnetism, one should strictly take account of the fact that electromagnetic waves propagate with a finite velocity c, that of light. This leads, as shown by Casimir and Polder (1948), to a modification of the r^{-6} term at very large r (in fact to r^{-7}). Fortunately, this "retardation" effect alters the interaction from the r^{-6} form only at separations which are so large that they are scarcely ever of interest in chemical physics.

3.2.3. Introduction of Molecular Polarizability into the Dispersion Interaction

The above treatment was based on an elementary model, and to cast the result into a more useful form it is important to introduce the molecular polarizability α into the dispersion interaction.

To do so, consider the total force, which for an harmonic oscillator is $-kz$, when an electric field $\mathscr{E}$ is applied in the z-direction. This force for charge Q can be written as

$$F = -kz - Q\mathscr{E} \tag{3.14}$$

The equilibrium position of the oscillator is evidently given by the value of z at which the force F is zero. Hence,

$$z_{\text{equil}} = -Q\mathscr{E}/k \tag{3.15}$$

The induced dipole moment is thus

$$\mu_{\text{ind}} = -Qz_{\text{equil}} = Q^2\mathscr{E}/k \tag{3.16}$$

The molecular polarizability α can now be introduced through

$$\mu_{\text{ind}} = \alpha \mathscr{E} \tag{3.17}$$

and hence in the above model we can replace the quantity Q^2/k by α. This allows the dispersion force to be written in terms of the polarizability α. Specifically, from equation (3.13), the potential energy $\phi_{\text{disp}}(r)$ takes the form

$$\phi_{\text{disp}}(r) = -\tfrac{1}{2}h\nu\alpha^2/r^6 \tag{3.18}$$

Finally, we must modify equation (3.18) due to the three-dimensional character of the real problem of dispersion forces as opposed to the one-dimensional model on which the above calculation was based. This changes the quantity $\frac{1}{2}$ in equation (3.18) to $\frac{3}{4}$. As the final step, we wish to identify the characteristic energy $h\nu$ in the model with a real molecular energy. The appropriate choice turns out to be the ionization potential I, to yield finally the desired result

$$\phi_{\text{disp}}(r) = -3\alpha^2 I/4r^6 \tag{3.19}$$

While the above argument has been presented for like molecules, the generalization to an unlike A–B pair can be written, in an obvious notation, as

$$\phi_{\text{disp}}^{\text{AB}}(r) = -\tfrac{3}{2}(\alpha_A\alpha_B/r^6)[I_A I_B/(I_A + I_B)] \tag{3.20}$$

If we take the limit of equation (3.20) when the two molecules A and B become identical, i.e., $\alpha_A \to \alpha_B \to \alpha$ etc., it becomes clear that we regain the original result equation (3.19).

The important conclusions from the above admittedly oversimplified model are:

1. The decrease with separation r of the long-range dispersion force as r^{-6}.
2. The importance of polarizabilities and ionization potentials in estimating the coefficient of the r^{-6} interaction energy arising from the dispersion interactions discussed above.

3.2.4. Van der Waals Equation of State Related to the London Dispersion Force

The perfect gas equation of state $pV = nRT$ was modified to allow for intermolecular forces by van der Waals, who wrote

$$(p + an^2/V^2)(V - nb) = nRT \tag{3.21}$$

where the constant a is a measure of the strength of the attraction while b is related to the "excluded volume" arising from strong repulsive interactions between molecules at short distances.

Because of our present concern with long-range attractive forces, it is of interest to enquire whether there is a connection between the van der Waals constant a in equation (3.21) and molecular polarizability α, the latter quantity having entered the equation for the long-range dispersion force derived above in an important way. Thus, in Figure 3.2 we have plotted essentially the log a vs. log α. For numerous atoms and molecules, it can be seen that this log–log plot is remarkably linear, giving evidence of a marked correlation between the two quantities, which supports the gist of the arguments presented above. In the light of this evidence for the importance of polarizability in determining long-range attractive forces, we have collected data for several atoms and molecules in Table 3.3.

Having treated long-range forces in some detail, we must now turn to a consideration of the nature of the short-range interactions between molecules.

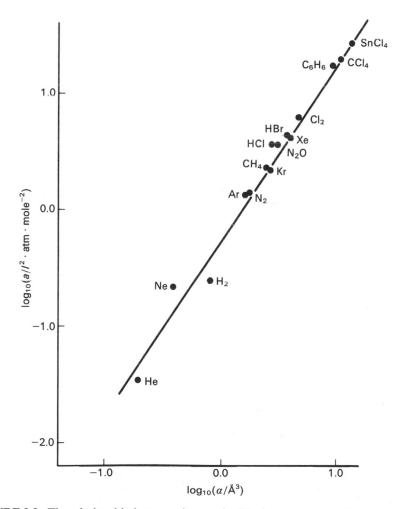

FIGURE 3.2. The relationship between the van der Waals constant a and the polarizability α for some selected molecules [redrawn from Lesk (1982), p. 287].

TABLE 3.3
Polarizabilities of Some Neutral Atoms and Molecules in Their Electronic
Ground States[a]

System	Polarizability (Å^3)	System	Polarizability (Å^3)	System	Polarizability (Å^3)
He	0.204956	Xe	4.04	O_2	1.60
Ne	0.3946	H_2	0.819	HCl	2.60
Ar	1.64	D_2	0.809	NH_3	2.22
Kr	2.48	CO	1.98	C_6H_6	10.4

[a]Constructed from Berry *et al.* (1980).

3.3. SHORT-RANGE FORCES

We recall first that by "short range" we refer to distances where electron distributions of the separate molecules that interact with one another have appreciable overlap, which causes the molecules to repel each other at such distances. This repulsive interaction presents more substantial difficulties for theory than the long-range forces dealt with above. While detailed quantum chemical calculations, referred to in later chapters, have been used with some success, they are still very laborious and only modestly successful at a fully quantitative level. It is therefore worthwhile to consider various models of empirical or semiempirical character which can provide useful representations of these repulsive interactions. Let us therefore start with the simplest model, that of hard spheres, representing the "excluded volume" referred to above in connection with the van der Waals equation of state.

3.3.1. The Hard-Sphere Model

In this model, we have hard spheres of diameter σ. The two molecules under consideration will experience elastic collisions when their centers approach to a distance equal to σ; otherwise they do not interact. Therefore we can write for the potential energy ϕ representing this hard-sphere model

$$\phi(r) = \infty \qquad (r \leq \sigma)$$

$$\phi(r) = 0 \qquad (r > \sigma)$$

(3.22)

A plot of this model potential energy is shown in Figure 3.3.

As to the equation of state, there is only one parameter in this model, the so-called packing fraction η, which is the ratio of the volume occupied by hard spheres to the total volume. While computer simulation has given the equation of state for this model in precise numerical terms, it is of interest that the results are well represented by the so-called Carnahan–Starling (1969) form:

$$\frac{pV}{Nk_BT} = (1 + \eta + \eta^2 - \eta^3)/(1 - \eta)^3$$

(3.23)

Needless to say, it can be demonstrated that real fluids exhibit departures from the predictions of this model, as it totally neglects the presence of attractive forces.

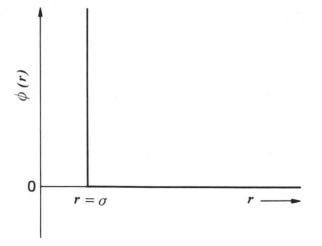

FIGURE 3.3. Schematic form of hard-sphere potential.

3.3.2. Point Centers of Inverse Power Law Repulsion

In this model, the potential energy $\phi(r)$ is given by

$$\phi(r) = ar^{-\delta} \tag{3.24}$$

where a is a constant and δ is an index of repulsion. The values chosen for δ are greater than 3 and very often between 9 and 15. It is to be noted that:

1. Equation (3.24) contains the hard-sphere model as the limit as δ tends to infinity.
2. Thermodynamics is simple with the form (3.24), as discussed especially by Hoover and Ross (1971). In particular, there is scaling with a combination of thermodynamic variables.

The form of equation (3.24) is sketched in Figure 3.4. We note here that the model incorporates the fact that for the case of finite δ the more energetic

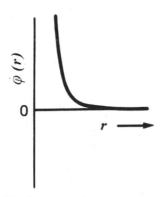

FIGURE 3.4. The inverse power law repulsion potential.

molecules can approach each other more closely before they are repelled. It is this feature which makes the model quite useful for those properties that are dependent on the temperature. While it is a relatively easy model to use, it must be recognized that, as with the hard-sphere representation, which it embraces, this model neglects the attractive part of the intermolecular interaction.

3.3.3. The Square-Well Model

This model of short-range forces takes into account both attractive and repulsive aspects. We note that it has a hard-core part, as in Section 3.3.1, now though surrounded by an attractive well of a constant depth, as shown in Figure 3.5. It can be seen that $\phi(r)$ is represented by

$$\phi(r) = \infty \qquad (r \leq \sigma_1)$$
$$\phi(r) = -\varepsilon \qquad (\sigma_1 < r < \sigma_2) \qquad \text{(3.25)}$$
$$\phi(r) = 0 \qquad (r \geq \sigma_2)$$

The adjustable parameters in equation (3.25) are evidently σ_1, σ_2, and ε. Again, this model is mathematically simple, and in the sense that it embodies an attractive component, it is somewhat more realistic than the other models considered so far.

3.3.4. The Lennard-Jones Potential

A valuable semiempirical potential, the so-called Lennard-Jones (6-12) form, may be written

$$\phi(r) = 4\varepsilon[(\sigma/r)^{12} - (\sigma/r)^6] \qquad \text{(3.26)}$$

and is sketched in Figure 3.6. It can be seen that the depth of the well is ε, while the distance at which $\phi = 0$ is simply σ. The minimum occurs at a distance given by $2^{1/6}\sigma$. The values of ε and σ for a number of substances are listed in Table

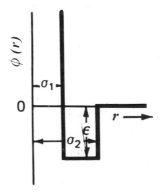

FIGURE 3.5. The square-well potential.

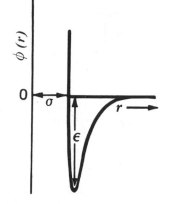

FIGURE 3.6. Form of the Lennard-Jones potential.

TABLE 3.4
Lennard-Jones Potential Parameters ε and σ in equation (3.26)[a]

Substance	ε/k_B (K)	σ (nm)	Substance	ε/k_B (K)	σ (nm)	Substance	ε/k_B (K)	σ (nm)
He	10.41	0.2602	CO_2	245.3	0.3762	$n-C_4H_{10}$	285.6	0.5526
Ne	42.0	0.2755	CH_4	161.3	0.3721	$i-C_4H_{10}$	260.9	0.5629
Ar	141.6	0.3350	CF_4	156.5	0.4478	C_2H_4	244.3	0.4070
Kr	199.8	0.3581	SF_6	207.7	0.5252	H_2O	266.8	0.3703
Xe	281.0	0.3790	C_2H_6	241.9	0.4371	CCl_3F	267.4	0.5757
N_2	104.2	0.3632	C_3H_8	268.5	0.4992	$CHClF_2$	288.3	0.4649
O_2	126.3	0.3382						

[a]After Rigby *et al.* (1986).

3.4. This useful potential clearly reduces at sufficiently large r to the r^{-6} behavior characteristic of the attractive dispersion interaction derived in Section 3.2.1, and we shall return to it later in this chapter.

3.3.5. The Exponential −6 Potential

Though the Lennard-Jones potential remains useful, it is more fundamental to represent the short-range interaction by an exponential term than by an inverse power law. The former, as already noted, reflects the fact that densities and wave functions decay exponentially with distance. One of the most widely used potentials, which is a modification of a linear combination of an exponential term proportional to $\exp(-\beta r)$ and a dispersion term proportional to r^{-6} has the form

$$\phi(r) = \frac{\varepsilon}{1-6/\alpha} \left\{ \frac{6}{\alpha} \exp\left[\alpha\left(1 - \frac{r}{r_m}\right)\right] - \left(\frac{r_m}{r}\right)^6 \right\} \tag{3.27}$$

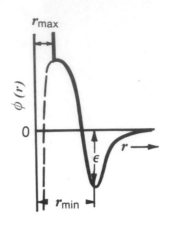

FIGURE 3.7. Shape of the exponential -6 potential.

r_m being the value of r at the minimum. This model interaction is characterized by three parameters, ε, r_m, and α, and yields the function sketched in Figure 3.7; α is an adjustable parameter with values usually between 12 and 15.

As an example of the use of this form, Ross (1986) has compared such a potential with shock wave results for Ar and his results are shown in Figure 3.8. Quantum chemical calculations, to be referred to below, have also been made. They are, not surprisingly, less satisfactory than a semiempirical form based on equation (3.27) [see also Ross (1989)].

3.4. SHORT-RANGE ENERGIES AND QUANTUM-CHEMICAL STUDIES

In nonbonding systems, it is not a simple matter to treat force fields theoretically. However, a description of repulsive forces that are important in the case of the ground state of the H—H system, as represented by the Heitler–London (valence bond) theory, can provide some insight into the origin of the repulsive interactions.

It will be recalled from Chapter 2 that the Heitler–London wave function for symmetric (plus) and antisymmetric (minus) spatial states can be written in terms of atomic wave functions ψ_A and ψ_B as

$$\Psi_\pm = \psi_A(1)\psi_B(2) \pm \psi_A(2)\psi_B(1) \tag{3.28}$$

It was shown in Chapter 2 that the interaction energy ΔE is of the form

$$\Delta E_\pm = (Q \pm J)/(1 \pm S^2) \tag{3.29}$$

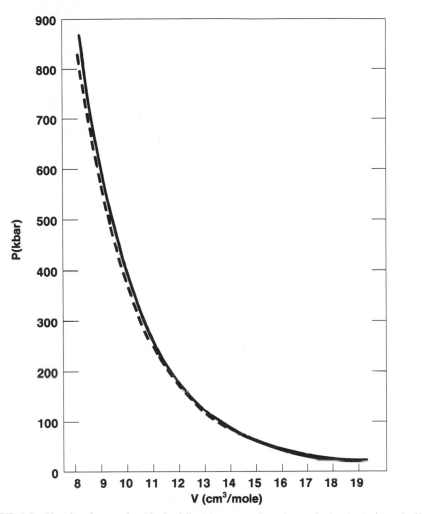

FIGURE 3.8. Shock tube results (dashed line) compared to theoretical calculations (solid line) using an exponential −6 potential for dense argon to calculate pressure and energy [redrawn from Ross *et al.* (1986)].

where, Q, J, and S are the Coulomb, exchange, and overlap integrals, respectively:

$$Q = \int \psi_A(1)\psi_B(2)V_e \, d\tau_1 \, d\tau_2$$

$$J = \int \psi_A(1)\psi_B(1)\psi_A(2)\psi_B(2)V_e \, d\tau_1 \, d\tau_2 \qquad (3.30)$$

$$S = \int \psi_A(1)\psi_B(1) \, d\tau_1$$

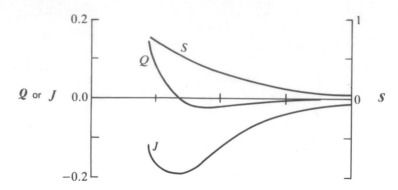

FIGURE 3.9. The integrals Q, J, and S for two hydrogen atoms [adapted from Rigby *et al.* (1986), p. 14].

and V_e is given by

$$V_e = (e^2/4\pi\varepsilon_0)(1/r_{a2} + 1/r_{b1} - 1/r_{12} - 1/r) \tag{3.31}$$

We further recall that the positive signs in the wave function yield positive signs in the energy equations (3.29), and as J is large and negative, this gives a negative interaction energy.

What is to be stressed in the present context is that negative signs in the wave function (3.28) lead to a nonbonding or repulsive state. This repulsive interaction is analogous to that which occurs in the case of the interaction between two closed-shell atoms. It is therefore helpful to plot the way in which the various energies encountered in equations (3.29) and (3.20) vary with separation and this is done in Figure 3.9. These quantities lead to the interaction energies ΔE as shown in Figure 3.10. The major contribution to the energy turns out to arise from the exchange integral J.

For numerous small molecules in nonbonding situations, *ab initio* calculations have been used (see Chapter 6). Simple molecular orbital theory also yields the same type of results. For details we refer the reader to Rigby *et al.* (1986). The simple treatment for H—H sketched above is to specifically draw attention to the value of quantum-chemical methods in studying short-range forces, even though it may be some time before semiempirical treatments can be dispensed with.

3.5. THE HYDROGEN BOND

The so-called hydrogen bond results when a hydrogen atom covalently bound to a highly electronegative element such as O, N, and F forms a weak bond to another O (etc.); for example,

$$\text{O—H---O} \qquad \text{or} \qquad \text{N—H---N}$$

$$\text{N—H---O}$$

The hydrogen bond, while weaker than the covalent bond, is nevertheless stronger than other intermolecular forces. For example, the enthalpy associated with a hydrogen bond is about 15 to 20 kJ/mole while that of a covalent bond is around 400 kJ/mole. The hydrogen bond is especially important in determining the structure of proteins and nucleic acids, as well as influencing the properties of water and other related systems.

The nature of the hydrogen bond has been studied by spectroscopic methods which use infrared and nuclear magnetic resonance (NMR) in the liquid and solid states (see Chapter 5), microwave and radiofrequency spectroscopic methods being appropriate in the vapor phase.

A theoretical description of the hydrogen bond has led to current thinking which indicates that the major part of this interaction is electrostatic in origin, with some covalent contribution. It should be noted that in the A—H bond (a polar bond due to large electronegativity of A), the proton of the hydrogen atom is more or less "exposed." The small size of the hydrogen atom (no inner shells of course) allows another appropriate atom to approach closely to the A—H system. There can be no doubt that the hydrogen bond represents an important and unique intermolecular force and therefore we shall expand a little as well as giving references to more detailed treatments.

Infrared and Raman spectra also demonstrate clear evidence for the existence of hydrogen bonding. For example, spectral bands corresponding to the vibrations associated with the A—H bond show changes in energy, intensity, and shape when the A—H bond is involved in hydrogen bonding. We would anticipate that stretching vibrations along the hydrogen-bond axis would decrease in frequency as the bending vibrations, which are at right angles to the hydrogen-bond axis,

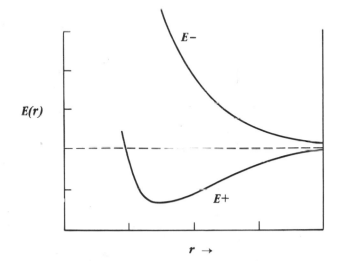

FIGURE 3.10. The potential energy curves for two hydrogen atoms resulting from the Q, J, and S integrals of Figure 3.9: E_+ (attraction), E_- (repulsion) [adapted from Rigby *et al.* (1986), p. 14].

increase in frequency. Figure 3.11 is a typical example of this effect, which plainly demonstrates the characteristic changes in peak intensity and band shape. This type of experimental method provides information about the strength of the hydrogen bond.

Analogous information about hydrogen bonding can be obtained by NMR (see also Chapter 5). The chemical shifts of the protons, discussed in that chapter, are extremely sensitive to hydrogen bonding. For instance, Figure 3.12 shows the change in the chemical shift of the OH peak of *t*-butanol as a function of the concentration of the alcohol in CCl_4.

While hydrogen bonds have been studied primarily in solution or in crystals, some cases have been investigated in the gas phase using microwave and radio-frequency spectroscopic methods. Table 3.5 records some dimerization energies for gas-phase hydrogen-bonded dimers.

Further details may be found, e.g., in Cumper (1966), Eisenberg and Kauzmann (1969), and Murrell *et al.* (1985).

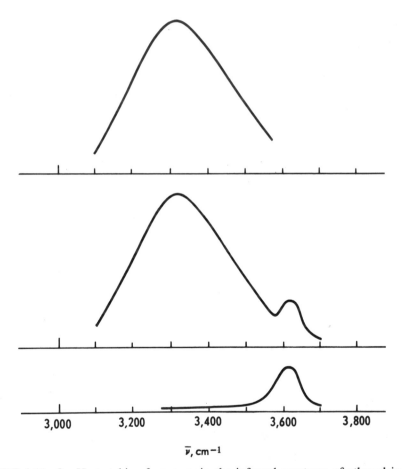

$\bar{\nu}$, cm^{-1}

FIGURE 3.11. O—H stretching frequency in the infrared spectrum of ethanol in CCl_4: (top) pure ethanol, (middle) 0.42 mole/liter, (bottom) 0.006 mole/liter [redrawn from Royer (1968), p. 205].

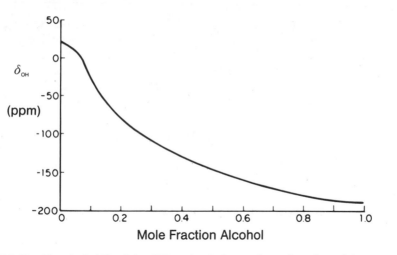

FIGURE 3.12. Chemical shift of the OH peak of *t*-butanol as a function of the concentration of the alcohol in CCl_4. Shifts are relative to the CH_3 resonance peak [redrawn from Davis (1965), p. 546].

TABLE 3.5
Diamerization Energies for Some
Gas-Phase Hydrogen-Bonded
Dimers[a]

Dimer	Dimerization energy (kJ/mole)
$(HF)_2$	29 ± 4
$(H_2O)_2$	22 ± 6
$(NH_3)_2$	19 ± 2
$(HCl)_2$	9 ± 1
$(H_2S)_2$	7 ± 1

[a]After J. N. Murrell *et al.* (1985).

3.6. THE RELATION TO EXPERIMENTAL STUDIES

The purpose of this section is to summarize the relation of the above to some of the experimental methods that are used to study intermolecular forces. As each of the methods to be referred to represents a field of its own, the interested reader is referred to advanced texts for detailed analyses (see, e.g., Rigby *et al.* 1986).

3.6.1. The Virial Equation: Imperfect Gases

The virial expansion of the equation of state of an imperfect gas can be written

$$pV/RT = 1 + B(T)/V + C(T)/V^2 + \cdots \qquad (3.32)$$

where V is the molar volume while $B(T)$ and $C(T)$ are the so-called second and third virial coefficients, respectively. The second virial coefficient depends on the interaction between molecules. Consequently, as we are in the dilute gas limit, $B(T)$ contains information on pair potentials. The third virial coefficient is dependent on the interactions of molecules in triplet clusters. Hence deviations from the perfect gas equation of state are evidently carrying significant information on intermolecular forces.

What is important now is to express $B(T)$ in equation (3.32) in terms of the pair potential. This can be accomplished by using the partition function (see also Chapter 7). For example, we consider an assembly of N identical particles, each of mass M in a volume V. If this volume is occupied by a perfect gas, there will be, by definition, no interaction between the particles (or strictly an infinitesimally weak interaction to allow equilibrium at temperature T, say), and the partition function Z for the assembly is given by

$$Z_N = (2\pi M k_B T/h^2)^{3N/2}(V^N/N!) \tag{3.33}$$

where this expression is simply $Z_1^N/N!$, with Z_1 the translational partition function for a single particle:

$$Z_1 = V(2\pi M k_B T/h^2)^{3/2} \tag{3.34}$$

However, when one "switches on" the pair interactions between the particles in the gas, then equation (3.33) is modified to read

$$Z_N = [(2\pi M k_B T/h^2)^{3N/2}]Q(N) \tag{3.35}$$

where $Q(N)$ is referred to as the configurational partition function. It is shown, e.g., in Kauzmann (1967), that $Q(N)$ can be evaluated in a low-density gas, and inserting the result into equation (3.35) enables Z_N to be written in the form of a density N/V expansion to leading order as

$$N!Z_N = [(2\pi M k_B T)/h^2]^{3N/2}V^N(1 + bN^2/V + \cdots) \tag{3.36}$$

where the quantity b is given in terms of the pair potential $\phi(r)$ via the so-called Mayer function [see Mayer and Mayer (1940)]:

$$f(r_{ij}) = [\exp(-\phi(r_{ij})/k_B T) - 1] \tag{3.37}$$

as

$$b = 2\pi \int_0^\infty f(r)r^2 \, dr \tag{3.38}$$

We also recall from introductory statistical mechanics* that the pressure p of the gas is given in terms of Z_N by

$$p = k_B T (\partial \ln Z_N / \partial V)_T \qquad (3.39)$$

Consequently, using equations (3.36) and (3.39), one can write after a short calculation, in which the approximation $(1 + x)^{-1} = 1 - x + x^2 - \cdots$ is invoked:

$$pV/Nk_B T = 1 - b(N/V) + \text{terms of higher order in the density } N/V \qquad (3.40)$$

Introducing Avogadro's number explicitly through $N = nN_0$, where n is the number of moles, one can readily relate $B(T)$ to b in equations (3.38) and (3.40) to find the so-called second virial coefficient $B(T)$ in terms of the intermolecular potential as

$$B(T) = -N_0 b = 2\pi N_0 \int_0^\infty \{1 - \exp[-\phi(r)/k_B T]\} r^2 \, dr \qquad (3.41)$$

Equation (3.41) in conjunction with equation (3.32) is important in relating deviations from the perfect-gas equation of state to the intermolecular pair interaction $\phi(r)$. This formula will now be evaluated for a number of the model potentials enumerated in Section 3.3.

(a) Model Results for the Second Virial Coefficient $B(T)$

For the hard-sphere potential, we recall from Section 3.3 that $\phi(r)$ is infinite for r less than the hard-sphere diameter σ and zero otherwise. Substituting this into equation (3.41), one readily finds

$$B(T) = 2\pi N_0 \int_0^\sigma r^2 \, dr = \tfrac{1}{3} 2\pi N_0 \sigma^3 \qquad (3.42)$$

This "excluded volume" result shows that for hard spheres the virial coefficient $B(T)$ is a positive constant, independent of temperature. This latter property is, of course, a consequence of the infinite "barrier" potential inside the hard-sphere diameter σ. Note that equation (3.42) is contained within equation (3.23).

It is now of interest to compare the above result (3.42) with that derived from the Lennard-Jones potential, which we now consider.

(b) Lennard-Jones Potential: The Virial Coefficient

It is useful [see, e.g., Hirschfelder *et al.* (1954)] to define the following dimensionless variables:

$$r^* = r/\sigma \qquad (3.43)$$

$$T^* = k_B T/\varepsilon \qquad (3.44)$$

---

*Briefly, the Helmholtz free energy $F = -Nk_B T \ln Z_N$ while $p = -(\partial F/\partial V)_T$.

and

$$B^* = \frac{B(T)}{(2\pi/3)N_0\sigma^3} \tag{3.45}$$

Substituting the Lennard-Jones potential $\phi(r) = 4\varepsilon[(\sigma/r)^{12} - (\sigma/r)^6]$ into equation (3.41) for $B(T)$, the result for $B^*(T^*)$ takes the form

$$B^*(T^*) = -(4/T^*)\int_0^\infty r^{*2}(-12/r^{*12} + 6/r^{*6})\exp[-(4/T^*)(1/r^{*12} - 1/r^{*6})]\,dr^* \tag{3.46}$$

where equation (3.46) results from equation (3.41) after integration by parts. $B^*(T^*)$ can be tabulated once for all, and the numerical results can be found in Hirschfelder *et al.* (1954).

The conclusions, then, from this short calculation are as follows:

1. B^* is a universal function of T^*.
2. The diameter σ and the measure of the well depth ε can be chosen for an optimal fit for each gas considered.
3. Agreement between theory and experiment is then remarkably good, as illustrated in Figure 3.13 for different gases.

Other pair potentials can, of course, be employed in equation (3.41), representing refinements of the above results, but we shall not pursue the details further here.

3.6.2. Transport Properties of Gases

We have clearly demonstrated above that the second virial coefficient $B(T)$ can be used effectively as a measure of intermolecular interaction via a pair potential representation. We now consider transport, rather than equilibrium thermodynamics, as a further means of characterizing such interactions. If a study is made of the transport of momentum, energy, or mass through a gas that is subjected to gradients of velocity, temperature, or concentration, the results will contain useful information on intermolecular forces. For instance, viscosity, thermal conductivity, and diffusion are related to the pair potential $\phi(r)$ of a dilute gas. We take these three properties in turn below and demonstrate how one can make this association with pair potentials quite explicit for these properties of matter.

(a) Viscosity

From simple kinetic theory, it can be shown that the viscosity η can be expressed in the form

$$\eta = \tfrac{1}{3}nM\bar{u}\lambda \tag{3.47}$$

where n is the number density of molecules with mass M, $\bar{u}$ the average speed of the molecules, and λ the average distance traveled by a molecule between

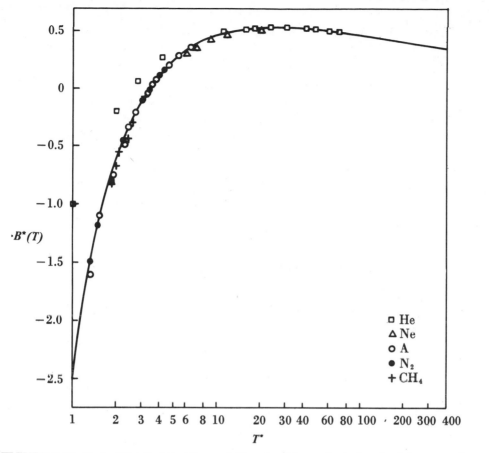

FIGURE 3.13. Test of the Lennard-Jones 6-12 potential as a basis for obtaining second virial coefficients in simple gases [redrawn from W. Kauzmann (1966), p. 87].

collisions, termed the mean free path. This can be written in the approximate form

$$\lambda = 1/2^{1/2}\pi n\sigma^2 \tag{3.48}$$

where equation (3.48) assumes the Maxwell–Boltzman distribution of velocities. A slightly improved calculation would replace the factor $1/3$ in equation (3.47) by $5\pi/32$.

(b) Thermal Conductivity

At the level of treatment of equation (3.47), one can write for thermal conductivity κ the result

$$\kappa = \tfrac{1}{3}nM\bar{u}c_v\lambda \tag{3.49}$$

where c_v is the specific heat of the gas at constant volume. A somewhat refined treatment analogous to that described for viscosity would replace the factor $1/3$ in equation (3.49) by $25\pi/64$. From equations (3.47) and (3.49), $\kappa = \eta c_v$.

TABLE 3.6

Viscosity, Thermal Conductivity, and Diffusion Coefficients (at 0°C and 1 atm) and
Molecular Diameters Calculated from Expressions from Kinetic Theory[a]

Gas	η (micropoise)	κ (cal/cm · s · deg) $\times 10^6$	D (cm²/s)	σ (Å) from η	from D	from κ
Ne	297	110.0	0.452	2.58	2.42	2.57
Ar	210	39.0	0.156	3.64	3.47	3.65
N₂	166	55.0	0.185	3.75	3.48	4.30
O₂	192	58.4	0.187	3.61	3.35	4.06
CH₄	103	73.4	0.206	4.14	3.79	4.79
CO₂	137	34.9	0.0974	4.63	4.28	5.76

[a]Data from J. O. Hirschfelder *et al.* (1954).

(c) Diffusion

In the case of gaseous diffusion, simple kinetic theory yields the result

$$D = \tfrac{1}{3}\bar{u}\lambda \tag{3.50}$$

with the refined coefficient as $3\pi/16$.

Using this hard-sphere model of the gas, which, as we have emphasized, ignores attractive forces between molecules, we find the predictions in reasonable agreement with experiment. For example, one can extract hard-sphere diameters from measurements of D, κ, and η, respectively, to give values spreading by only 10 to 20% (see Table 3.6).

3.6.3. Pair Potentials and Transport Coefficients

In order to transcend the accuracy of the hard-sphere model, it is, of course, necessary to take account of the attractive forces between molecules through the use of more realistic pair potentials $\phi(r)$. In particular, the existence of attractive forces, plus the lack of a sharply defined distance at which the repulsive forces set in, are responsible for a major part of the temperature dependence of the transport coefficients. This clearly implies that information about intermolecular interactions is contained in the temperature dependence of the transport coefficients of gases.

(a) Sutherland's Model

Sutherland proposed adding an attractive potential of the form

$$\phi(r) = -\varepsilon(\sigma/r)^n \qquad (r > \sigma) \tag{3.51}$$

to the hard-sphere repulsion treated above, where ε and n are positive constants. With this choice of pair potential, he was able to show that at high temperatures the viscosity can be expressed in the form

$$\eta = \eta_0/(1 + s/T) \tag{3.52}$$

where η_0 represents the viscosity for the hard-sphere model alone. The constant s can be written in the form

$$s = i(n)\varepsilon/nk_B \tag{3.53}$$

where $i(n)$ is a weakly n-dependent quantity, with a value of approximately 0.2 when n lies in the range 2 to 8 (e.g., $n = 6$ to reflect long-range dispersion interactions).

The above treatment has proved useful for representing the viscosity of gases as a function of temperature over specific temperature ranges. Table 3.7 lists some empirical values of s for a number of gases. It is to be noted that s is largest for gases with the highest boiling points (i.e., those with the strongest attractive pair potential), which in turn corresponds to the largest values of ε in equations (3.51) and (3.53).

(b) Inverse Power Repulsive Potentials

Returning briefly to the potential energy $\phi(r) = ar^{-\delta}$ in equation (3.24), we note that Chapman (1939) studied the temperature behavior of the viscosity and thermal conductivity. Given their values, η_0 and κ_0, say, at a prescribed temperature T_0, then at another temperature T one has

$$\eta = \eta_0(T/T_0)^\nu \tag{3.54}$$

and

$$\kappa = \kappa_0(T/T_0)^\nu \tag{3.55}$$

where

$$\nu = \tfrac{1}{2} + 2/\delta \tag{3.56}$$

Equations (3.54) and (3.55) give useful representations of experimental data for reasonable values of δ over limited ranges of temperature.

TABLE 3.7
Sutherland's Constant for Several Gases[a]

Gas	Sutherland's constant (K)	Gas	Sutherland's constant (K)	Gas	Sutherland's constant (K)
He	80	N_2	104	CH_4	164
Ne	56	O_2	125	C_2H_6	252
Ar	142	Air	112	$n\text{-}C_4H_{10}$	358
Kr	188	CO	254	$n\text{-}C_6H_{14}$	436
Xe	252	H_2O	650	HCl	362
H_2	84				

[a]Data from W. Kauzmann (1966).

(c) Chapman–Enskog Theory

The so-called Chapman–Enskog theory is better based than the treatments (a) and (b) summarized above [see Hirschfelder *et al.* (1954) for a complete treatment of the Chapman–Enskog theory]. The theory considers directly the scattering process as molecules collide, and in particular it deals with the quantities $Q_1(g)$ and $Q_2(g)$, as defined below, both having dimensions of area. These quantities measure different weighted scattering cross sections in such collisions. If χ represents the deflection in the scattering process, then the quantities Q_1 and Q_2 are defined by

$$Q_1(g) = 2\pi \int_0^\infty (1 - \cos\chi)b \; db \tag{3.57}$$

and

$$Q_2(g) = 2\pi \int_0^\infty (1 - \cos^2\chi)b \; db \tag{3.58}$$

where b is the impact parameter (see Figure 3.14) while g is the relative velocity of a pair of molecules before they have come into close interaction.

Details of this treatment would take us beyond the scope of the present work; however, the results can be given in summary form as follows[*]:

$$\eta = 5k_BT/8\Omega_{22} \tag{3.59}$$

$$\kappa = 25k_B^2T/16\Omega_{22} \tag{3.60}$$

while

$$D = 3(k_BT)^2/8Mp\Omega_{11} \tag{3.61}$$

We note from equations (3.59) and (3.60) that the ratio η/κ is independent of temperature, which is also the result of dividing the formulas (3.54) and (3.55).[†] Equation (3.61), involving pressure p and the quantity Ω_{11} (see below) is less simply related to η and κ. If μ denotes the reduced mass, then generally $\Omega_{mn}(T)$ is defined as

$$\Omega_{mn}(T) = (1/4\pi)^{1/2} \int_0^\infty \exp(-\mu g^2/2k_BT)(\mu g^2/2k_BT)^{2n+3}Q_m(g) \; dg \tag{3.62}$$

It is to be noted here that the scattering angle χ depends on the pair potential. We refer the reader to Kauzmann (1966) for a more detailed discussion of transport theory (see also Berry *et al.* 1980).

---

[*]The way equations (3.59) through (3.61) follow from Chapman–Enskog theory is treated, for example, by F. Mohling (1982).

[†]The formula $\kappa/\eta c_v = 1$ after equation (3.49) needs refinement, the ratio correlating quite well with c_p/c_v, rather than being constant.

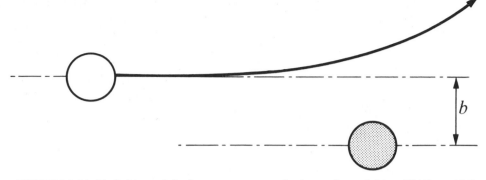

FIGURE 3.14. Definition of the impact parameter [redrawn from Atkins (1982), p. 790].

3.6.4. Molecular Beam Studies

Many types of collisions between molecules are considered in molecular beam studies. Relative kinetic energies can be controlled closely and the result is a beam of monoenergetic molecules which collide with other molecules (i.e., the target molecules) at low pressure, or with a similar beam at right angles. Figure 3.15 is a schematic representation of the setup for a crossed-beam scattering experiment.

We note first the following four points:

1. The scattering that occurs as a result of molecular collision is related to the intermolecular forces.
2. The closely controlled speed (or, equivalently, kinetic energy) of the molecules in the beam enables investigation of the behavior of the primary beam molecules as they collide with the target molecules.
3. The angle of scatter θ, say, is a major piece of information obtained in molecular beam studies.
4. The intensity of the scattered molecules at different angles can be observed by use of devices such as pressure gauges and ionization detectors.

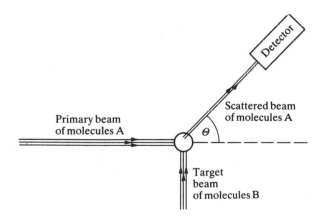

FIGURE 3.15. Schematic diagram of crossed molecular beam scattering experiment [redrawn from Rigby *et al.* (1986), p. 37].

We note next two important quantities in such beam studies. (i) The impact parameter b (see above), which measures the initial vertical separation of the paths of the colliding molecules. Assuming spherical molecules (e.g., Ne atoms or the approximately spherical CH_4 molecule) b is as shown in Figure 3.14. (ii) The differential scattering cross section, which measures the amount of scattering through different angles. For example, let us consider the solid angle $d\omega$ at some angle θ as illustrated in Figure 3.16. The ratio of the number of molecules scattered in the solid angle $d\omega$ to the number in the incident beam can be written

$$I(\theta)\,d\omega = \frac{\text{Number of molecules scattered into } d\omega \text{ at angle } \theta \text{ per unit time}}{\text{Number of molecules per unit area per unit time in incident beam}} \quad (3.63)$$

The points to be stressed here are that the differential cross section is dependent on the impact parameter b introduced above and the form of the intermolecular potential. For example, if the molecules are considered as hard spheres, the collision can be represented by different impact parameters such that $b = 0$ for head-on collisions, or if r_i represents the radius of sphere i then in an A–B collision one can have $0 < b < (r_A + r_B)$. These situations result simply in a different magnitude of b.

The differential scattering cross section $I(\theta)\,d\omega$ can be integrated over all angles to lead to the directly observable integral or total scattering cross section. The measured scattered particle flux at the chosen angle of scatter is a further observable quantity.

Of course, to be fully realistic, one must transcend the hard-sphere model for the scattering, which will in practice depend on the relative speed of approach of the two molecules and the intermolecular forces that come into play.

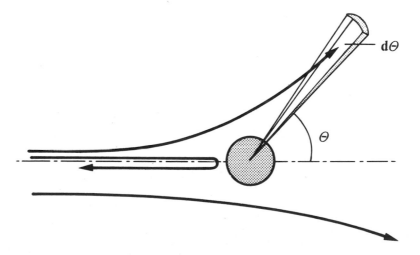

FIGURE 3.16. Definition of the differential scattering cross section [modified from Atkins (1982), p. 791].

It is probably fair to say that at the time of writing, it is still difficult to make unique statements about the form of the pair potential $\phi(r)$ from molecular beam studies. However, it is certainly possible to gain valuable information about the validity of a chosen form for $\phi(r)$. One can expect that molecular beam studies will continue to be important in the investigation of intermolecular forces. For further details of this most useful method the reader is referred to Rigby *et al.* (1986).

3.6.5. Spectra of Molecular Pairs

Owing to molecular interactions, some molecules can form dimers (van der Waals) that have a finite lifetime. For example, two molecules orbiting each other can collide with a third molecule. This third member can carry off some kinetic energy from the pair, which allows a dimer to form. This dimer can then be studied by vibration–rotation spectroscopy as well as by its electronic spectrum (see Chapter 5). Parameters determined from the spectra (force constants, etc.) therefore yield information concerning the molecular interaction energy.

We shall briefly discuss at this point one of the more satisfactory methods for gaining information on intermolecular pair potentials from spectroscopic studies, stemming from the work of Rydberg, Klein, and Rees [the so-called RKR procedure; see Rigby *et al.* (1986)]. The outline of this approach is as follows:

The first step is the determination of the allowed vibration–rotation energy levels, say, of the dimers, which involves assigning vibrational and rotational quantum number changes to each absorption line in the observed spectrum. In this case, the vibrational levels are more easily resolved than the rotational levels; however even information only on the vibrational energies can be utilized to gain information about the intermolecular pair potential.

One then examines a portion of the effective potential energy curve $V(L, r)$; see Figure 3.17 for a pair of structureless particles in the RKR procedure:

$$V(L, r) = U(r) + K/r^2 \tag{3.64}$$

where $K = L^2/2\mu$, L being the initial angular momentum and μ the reduced mass. For convenience the zero of energy is taken at the bottom of the potential well. Further, AB corresponds to the energy of the oscillator with energy V' while r_L and r_R correspond classically to the minimum and maximum separations of the two particles.

The RKR procedure yields

$$r_L^{-1} - r_R^{-1} = \left(\frac{1}{2\pi^2\mu}\right)^{1/2} h \int_x^n \frac{dx}{\bar{r}_v^2(U_n - U_x)^{1/2}} \tag{3.65}$$

where

$$x = v + \tfrac{1}{2} \tag{3.66}$$

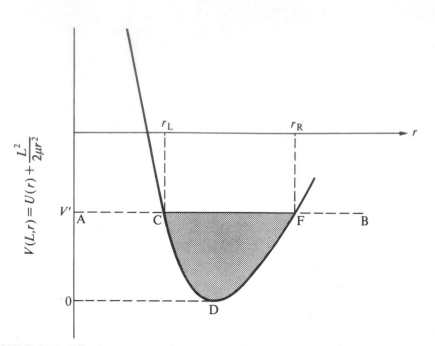

FIGURE 3.17. Effective energy and parameters for RKR analysis [redrawn from Rigby *et al.* (1986), p. 68].

with v the vibrational quantum number; and we find one vibrational level where $x = n$, where the potential energy is U_n. The quantity $\bar{r}_v$ is the effective mean separation of the pair of molecules for the vibrational level with quantum number v.

Of course, the above is simply a statement of the result thus far of the RKR procedure (for details, see Rigby *et al.*, 1986).

Continuing with the RKR summary, the inner and outer branches of the potential energy curve can be obtained by using the expressions shown above up to the dissociation limit of the dimer, where $U_n = \mathcal{E}$. Close to the top of the potential well we have for nonpolar molecules that

$$U(r) = \mathcal{E} - C_6/r^6 \tag{3.67}$$

where C_6 is the coefficient of the dispersion interaction encountered earlier. It is clear therefore that contact with the intermolecular pair potential can be made via spectroscopic analysis [for details see Rigby *et al.* (1986) and Atkins (1991)].

3.6.6. The Inversion of Measured Fluid Structure to Extract Pair Potentials

Apart from the above procedure based on spectroscopic analysis, there is a further fruitful way of using experimental data to extract a pair potential. Johnson and March (1963) pointed out that X-ray and neutron diffraction measurements of fluid structure could be used, in conjunction with classical statistical mechanics, to extract pair potentials when these are appropriate to describe the intermolecular

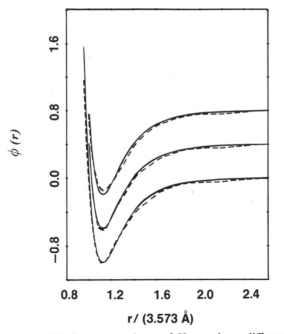

FIGURE 3.18. Pair potential for vapor phase of Kr at three different densities using diffraction data [redrawn from Egelstaff *et al.* (1985)].

force field. While their initial interest was to describe dense liquid interactions, and in particular a dense liquid metal like Na, for which case a full pair potential has been extracted by Reatto *et al.* (1988) by this approach (see also Perrot and March 1990), the method has been implemented for vapor studies on Kr by Egelstaff *et al.* (1985). In the vapor regime, the density can be used as an expansion parameter (cf. Section 3.6.1). Then, from the Mayer function $f(r)$ in equation (3.37), which is directly accessible experimentally from diffraction experiments (in the dilute gas limit), one can extract a pair potential $\phi(r)$.

As an example, we show in Figure 3.18 the results of Egelstaff *et al.* 1985 using diffraction data on the vapor phase of Kr. The three curves shown correspond to different densities, and are displaced vertically from one another in the figure for clarity. Comparison is made in each case with the pair potential of Barker *et al.* (1974) (shown in solid curves), which was obtained in a quite different manner, and the agreement is seen to be excellent. Much progress in extracting interatomic forces from what is, essentially, the pair function of classical fluids, say $g(r)$,* is to be expected in the future. In principle, this technique can also be applied to fluid A–B mixtures, but then three pair functions g_{AA}, $g_{BB}(r)$, and $g_{AB}(r)$ must be determined by experiment (actually diffraction measures the Fourier transform of $[g(r) - 1]$ in a monatomic fluid). To obtain these three $g(r)$'s naturally requires three experiments, say, X-rays and two neutron measurements if different isotopes are available [see, e.g., Enderby (1982)].

---

*If ρ is mean vapor density and if we "sit" on an atom at the origin $r=0$, then $\rho g(r)$ is the density of other atoms at distance r from the atom at $r=0$.

To recapitulate briefly, the basic aspects of the calculation of both long- and short-range intermolecular forces have been reviewed. Especially for the short-range contributions, it is still useful to proceed by empirical or semiempirical procedures, even though quantum-chemical calculations are now becoming tractable on small to medium-size systems. The exp −6 potential seems particularly favorable, as it accounts for the London dispersion force, and its short-range form reflects the properties of wave function and density decay in exponential fashion in an atom.

To complement the above theoretical survey, some of the principal experimental studies by which information can be either extracted or tested relating to intermolecular forces have been reviewed. For details, the reader should consult specialist texts in each area (see Appendixes A3.1 and A3.2).

PROBLEMS

3.1. Show that the minimum of the Lennard-Jones potential energy curve occurs at $r_{\min} = (2)^{1/6}\sigma$.

3.2. Draw a graph of the Lennard-Jones parameter ε against $\frac{3}{2}kT$ (where T is the melting point at 1 atm) for a number of simple substances. What trend do you observe and what is the physical interpretation of this trend?

3.3. Determine the second and third virial coefficients of CO_2 at 273.15 K, 323.15 K, and 373.15 K from tabulated P, V, T data. Hint: Rewrite the virial equation of state in the form

$$V(PV/RT - 1) = B_2 + B_3/V + \cdots$$

Then plot $V(PV/RT - 1)$ *vs.* $1/V$ to determine B_2 and B_3 from the slope and intercept of the line observed at low pressure (V = molar volume).

3.4. Describe theoretical and experimental studies of the forces, other than those leading to the formation of chemical bonds, which may operate between atoms and/or molecules or both. [The interested reader is referred also to G. Robinson *et al.* (1971).]

3.5. Molecular beams and gas imperfections are both important in the study of intermolecular forces. Describe the essentials of each approach.

3.6. Show clearly how one could use the Lennard-Jones potential, say, for a noble gas, and evaluate the second virial coefficient.

3.7. **(a)** The viscosity of Ar at 273 K and $1.01 \times 10^5 P_a$ is 20.99×10^{-6} kg/m · s. Estimate the hard-sphere collision diameter from these data. **(b)** The thermal conductivity for the gas in (a) at 273 K and $1.01 \times 10^5 P_a$ is found to be 9.42×10^{-4} J/m · s · deg K. Estimate the hard-sphere collision diameter also from these data.

3.8. Describe the basis of spectroscopic measurements as a means of studying intermolecular forces.

REFERENCES

P. W. Atkins, *Quanta*, 2nd Ed., Oxford University Press, New York (1991).
P. W. Atkins, *Physical Chemistry, 2nd Ed.*, Freeman and Co., San Francisco (1982).
J. A. Barker, R. O. Watts, J. K. Lee, T. P. Schafer, and Y. T. Lee, *J. Chem. Phys.* **61**, 3081 (1974).
R. S. Berry, S. A. Rice, and J. Ross, *Physical Chemistry*, John Wiley and Sons, New York (1980).
N. F. Carnahan and K. E. Starling, *J. Chem. Phys.* **51**, 635 (1969).
H. B. G. Casimir and D. Polder, *Phys. Rev.* **73**, 360 (1948).

S. Chapman and T. G. Cowling, *The Mathematical Theory of Non-Uniform Gases*, Cambridge University Press (1939).

C. W. N. Cumper, *Wave Mechanics for Chemists*, Academic Press, New York (1966).

J. C. Davis, *Advanced Physical Chemistry*, Ronald Press, New York (1965).

P. A. Egelstaff, F. Barocchi, and M. Zoppi, *Phys. Rev.* **A31**, 2732 (1985).

D. Eisenberg and W. Kauzmann, *The Structure of Properties of Water*, Oxford University Press (1969).

J. E. Enderby, *J. Phys.* **C15**, 4609 (1982).

J. O. Hirschfelder, C. F. Curtiss, and R. B. Bird, *Molecular Theory of Gases and Liquids*, John Wiley and Sons, New York (1954).

W. G. Hoover and M. Ross, *Contemp. Phys.* **12**, 339 (1971)

M. D. Johnson and N. H. March, *Phys. Lett.* **3**, 313 (1963).

W. Kauzmann, *Kinetic Theory of Gases*, W. A. Benjamin, New York (1966).

W. Kauzmann, *Thermodynamics and Statistics: With Applications to Gases*, W. A. Benjamin, Inc., New York (1967).

A. M. Lesk, *Introduction to Physical Chemistry*, Prentice-Hall, Englewood Cliffs, NJ (1982).

J. E. Mayer and M. G. Mayer, *Statistical Mechanics*, Wiley and Sons, New York (1940).

F. Mohling, *Statistical Mechanics*, John Wiley and Sons, New York (1982).

J. N. Murrell, S. F. A. Kettle, and J. M. Tedder, *The Chemical Bond, 2nd Ed.*, John Wiley and Sons, New York (1985).

J. N. Murrell, S. F. A. Kettle, and J. M. Tedder, *Valence Theory, 2nd Ed.*, John Wiley and Sons, New York (1970).

F. Perrot and N. H. March, *Phys. Rev.* **A41**, 4521 (1990).

L. Reatto, *Phil. Mag.* **58**, 37 (1988).

M. Rigby, E. B. Smith, W. A. Wakeham, and G. C. Maitland, *The Forces Between Molecules*, Oxford University Press (1986).

G. Robinson, N. H. March, and R. C. Perrin, *Int. J. Quantum Chem.* **5**, 271 (1971).

M. Ross, *J. Chem. Phys.* **90**, 1209 (1989).

M. Ross, H. K. Mao, P. M. Bell, and J. A. Xu, *J. Chem. Phys.* **85**, 1028 (1986).

D. J. Royer, *Bonding Theory*, McGraw-Hill, New York (1968).

Further Reading

B. I. Blaney and G. E. Ewing, *Ann. Rev. Phys. Chem.* **27**, 553 (1976).

H. Margenau and N. R. Kestner, *Intermolecular Forces*, Pergamon Press, Oxford (1969).

H. Pauly and J. P. Toennies, *Adv. Atomic and Molecular Phys.* **1**, 195 (1965).

J. I. Steinfeld, *Molecules and Radiation; An Introduction to Modern Spectroscopy*, 2nd ed., MIT Press, Cambridge, MA (1985).

ELECTRON DENSITY DESCRIPTION OF MOLECULES

So far, we have focused a lot of attention on one-electron wave functions in molecules. If these wave functions ψ_i are derived from a one-body potential energy $V(\mathbf{r})$ inserted in the Schrödinger equation, then provided the ψ_i's are normalized we can write the electron density $\rho(\mathbf{r})$ as

$$\rho(\mathbf{r}) = \sum_{\substack{\text{occupied} \\ \text{states}}} \psi_i^*(\mathbf{r})\psi_i(\mathbf{r}) \tag{4.1}$$

That it was possible to use $\rho(\mathbf{r})$ to describe the ground state of an atom was clearly recognized by Thomas (1926) and independently by Fermi (1928). As their work forms a basis for the electron density description, we shall outline the Thomas–Fermi (TF) theory in some detail below [see also March (1957) and March (1975)]. The work of Hohenberg and Kohn (1964) formally completed the TF theory and we shall return to their theorem later.

4.1. APPROXIMATE ELECTRON DENSITY–POTENTIAL RELATION

The idea is to write an energy equation for the fastest electron in the molecular electron cloud with density $\rho(\mathbf{r})$. As we shall identify this maximum electronic energy with the chemical potential of the electron distribution, we shall anticipate this by using the symbol μ from the outset. The energy equation for electrons moving in a potential energy $V(\mathbf{r})$ takes the form

$$\mu = (\text{KE})_{\mathbf{r}}^{\max} + V(\mathbf{r}) \tag{4.2}$$

where $(\text{KE})_{\mathbf{r}}^{\max}$ denotes simply the maximum kinetic energy at $\mathbf{r}$. In terms of momentum $p_{\max}(\mathbf{r}) \equiv p_m(\mathbf{r})$ we have

$$\mu = p_m^2(\mathbf{r})/2m + V(\mathbf{r}) \tag{4.3}$$

with m the mass of the electron. Equation (4.3) already has important content. Although each term on the right-hand side depends on position $\mathbf{r}$ in the molecular charge cloud, the sum μ is independent of $\mathbf{r}$. The constancy of μ is already expressing the fact that, in equilibrium, no further charge flow is possible.

But we want now to relate equation (4.3) directly to the electron density $\rho(\mathbf{r})$. To do so, let us regard the electrons around position $\mathbf{r}$ as an electron gas to which we can apply Fermi–Dirac statistics. This means that into a basic cell in phase space* of volume h^3 one can put just two electrons with opposed spin in the ground state. Then, if we consider a volume Ω, and as above take the maximum momentum as p_m, we have for the volume of occupied phase space the result $\frac{4}{3} \pi p_m^3 \Omega$, since a sphere of radius p_m is evidently filled in momentum space. Thus as each cell has volume h^3 the number of cells in phase space is $\frac{4}{3}\pi p_m^3 \Omega / h^3$. With two electrons per cell the number of electrons N in volume Ω is evidently given by

$$N = 2 \cdot \tfrac{4}{3}\pi p_m^3 \Omega / h^3 \qquad (4.4)$$

Hence the number of electrons per unit volume, $N/\Omega \equiv \rho$, is $(8\pi/3h^3)p_m^3$, and applying this result at position $\mathbf{r}$ we have

$$\rho(\mathbf{r}) = (8\pi/3h^3)p_m^3(\mathbf{r}) \qquad (4.5)$$

We can substitute this expression for p_m in terms of ρ into equation (4.3) and obtain

$$\mu = (1/2m)(3h^3/8\pi)^{2/3}[\rho(\mathbf{r})]^{2/3} + V(\mathbf{r}) \qquad (4.6)$$

This is the basic relation for the electron density for a given potential energy $V(\mathbf{r})$, μ being determined from the total number of electrons N by requiring that $\rho(\mathbf{r})$ satisfy

$$\int \rho(\mathbf{r}) \, d\tau = N \qquad (4.7)$$

It is, of course, a great simplification that we have by-passed the calculation of $\rho(\mathbf{r})$ from a one-body wave function ψ_i, the one-body wave-function being derived by inserting $V(\mathbf{r})$ in the Schrödinger equation. However, as is natural, a price has been paid in setting up the relatively simple equation (4.6) to obtain $\rho(\mathbf{r})$ from $V(\mathbf{r})$. It is seen that V at position $\mathbf{r}$ determines ρ at that position. But the $1s$ wave function for the ground state of the hydrogen atom shows that this is an over-simplification. For the wave function $\psi_{1s} = (1/\pi a_0^3)^{1/2} \exp(-r/a_0)$, a_0 being the Bohr radius $\hbar^2/me^2$, is finite at $r=0$ with amplitude $(1/\pi a_0^3)^{1/2}$ while the corresponding density $\rho_{1s}(r) = \psi_{1s}^2(r)$ has the value $(1/\pi a_0^3)$ at $r=0$. However, $V(r) = -e^2/r$ is evidently infinite at $r=0$ and, using this form for $V(r)$, equation (4.6) implies that $\rho(r)$ is also infinite at $r=0$.

The explanation is that equation (4.6) is an approximate equation which is quantitatively valid when $V(r)$ varies only slowly in space. Near an atomic nucleus, the potential varies too rapidly for equation (4.6) to be precise. Nevertheless,

---

*This is a product of position and momentum space. Since the Uncertainty Principle gives $\Delta x \Delta p_x \sim h$, the three-dimensional cell in phase space is $\Delta x \Delta p_x \Delta y \Delta p_y \Delta z \Delta p_z \sim h^3$. For the quantitative detail, reference can be made to Appendix A4.1.

equation (4.6) is already very useful for some purposes,* and also, as we shall see below, it is easy to write an exact, formal generalization.

Before going on to this generalization, we turn to relate the above calculation of $\rho(\mathbf{r})$ from the TF theory to the energy of the electrons in a molecule.

4.2. ENERGY RELATIONS IN MOLECULES AT EQUILIBRIUM

Having established the important, though approximate, relation (4.6) between electron density $\rho(\mathbf{r})$ and potential energy $V(\mathbf{r})$, the latter being the usual input information into a single-electron Schrödinger equation, we can make use of this, as it stands, to establish some relations among the different energy terms in molecules at equilibrium. Although equation (4.6) is approximate, we can still usefully combine it with an exact result from wave mechanics, the quantum-mechanical virial theorem, set out in Appendix A2.6. At equilibrium, this theorem tells us, as with any system in equilibrium under pure Coulomb forces, that

$$2(\text{Kinetic energy}) + \text{Potential energy} = 0 \qquad (4.8)$$

Of course, in a fully wave-mechanical theory, these quantities appearing in equation (4.8) are the usual quantum-mechanical averages. If we denote the total kinetic energy by T, and the total potential energy by U, then we can usefully divide the latter into the sum of three contributions: (i) electron–nuclear potential energy, with $V_N(\mathbf{r})$ denoting the nuclear potential energy felt by an electron at position $\mathbf{r}$ (e.g., $-e^2/r$ in the hydrogen atom), (ii) electron–electron potential energy, denoted by U_{ee}, and (iii) nuclear–nuclear potential energy U_{nn}. Thus we can write equation (4.8) in the equivalent, but more explicit, form

$$2T + U_{en} + U_{ee} + U_{nn} = 0 \qquad (4.9)$$

We now seek to combine the exact consequence of the Schrödinger equation as written in equation (4.9) for molecules at equilibrium, with the TF density-potential relation (4.6). To do so, we multiply both sides of equation (4.6) by the electron density $\rho(\mathbf{r})$, recall the constancy of the chemical potential through the whole of space, and note that the normalization condition on the density written in equation (4.7) leads to

$$N\mu = \tfrac{5}{3}c_k \int [\rho(\mathbf{r})]^{5/3} \, d\tau + \int \rho(\mathbf{r}) V(\mathbf{r}) \, d\tau \qquad (4.10)$$

where we have introduced the constant $c_k = (3h^2/10m)(3/8\pi)^{2/3}$ for convenience later when we identify the form of the total kinetic energy T in the TF theory.

The next step in relating equation (4.10) and the virial equation (4.9) at equilibrium is to divide the potential energy $V(\mathbf{r})$ into two parts, one $V_N(\mathbf{r})$ being due to the electron–nuclear attraction and the other part $V_e(\mathbf{r})$ to the

---

*In particular, it becomes increasingly valid for large numbers of electrons.

electron–electron repulsion energy of the charge cloud. Thus we write, in terms of the nuclear potential energy $V_N(\mathbf{r})$ introduced above:

$$V(\mathbf{r}) = V_N(\mathbf{r}) + V_e(\mathbf{r}) \tag{4.11}$$

and in the simplest (self-consistent field) theory we can regard the electronic contribution in equation (4.11) as being generated by the electronic charge cloud $\rho(\mathbf{r})$ according to classical electrostatics. Thus, the electrostatic potential at position $\mathbf{r}$ due to an element of the charge cloud of volume $d\tau'$ at $\mathbf{r}'$, with electron density of charge $-e\rho(\mathbf{r}')\, d\tau'$, is $-e\rho(\mathbf{r}')\, d\tau'/|\mathbf{r} - \mathbf{r}'|$, the denominator evidently measuring the distance separating points $\mathbf{r}$ and $\mathbf{r}'$. Integrating over the entire charge cloud, and remembering that to go from electrostatic potential, referring to unit positive charge, to potential energy felt by an electron, we must multiply by the electron charge $-e$, we readily find

$$V_e(\mathbf{r}) = e^2 \int \frac{\rho(\mathbf{r}')\, d\tau'}{|\mathbf{r} - \mathbf{r}'|} \tag{4.12}$$

Thus, we can write equation (4.10) in the equivalent forms

$$N\mu = \tfrac{5}{3}c_k \int [\rho(\mathbf{r})]^{5/3}\, d\tau + U_{en} + \int \rho(\mathbf{r}) V_e(\mathbf{r})\, d\tau$$

$$= \tfrac{5}{3}c_k \int [\rho(\mathbf{r})]^{5/3}\, d\tau + U_{en} + e^2 \int \frac{\rho(\mathbf{r})\,\rho(\mathbf{r}')}{|\mathbf{r} - \mathbf{r}'|}\, d\tau\, d\tau' \tag{4.13}$$

The right-hand side of equation (4.13) is seen to be determined once the nuclear framework of the molecule is specified to give $V_N(\mathbf{r})$, by the electron density $\rho(\mathbf{r})$. However, we have yet to relate the term involving $\rho^{5/3}$ directly to the kinetic energy, and this is the next important step to take.

The phase space argument used to relate electron density $\rho(\mathbf{r})$ to maximum momentum through equation (4.5) can be generalized to calculate the kinetic energy of the electron cloud in this theory of the inhomogeneous electron gas, and the proof is set out in Appendix A4.2. Here we shall be content with a simple argument based on dimensional analysis, which will make clear the idea of the way $\rho^{5/3}$ appearing in equations (4.10) and (4.13) relates to the kinetic energy density (i.e., per unit volume) of the electronic charge cloud.

The starting point for dimensional analysis is the recognition that the kinetic energy arises, in quantum mechanics, from the expectation value of the operator $-(\hbar^2/2m)\nabla^2$. Thus, we shall assume that the kinetic energy density $t(\mathbf{r})$ of the electron gas at position $\mathbf{r}$ has the form

$$t(\mathbf{r}) = \text{constant}(h^2/m)\{\rho(\mathbf{r})\}^d \tag{4.14}$$

While the power law assumption made in writing equation (4.14) has not produced the power d, to be found below from dimensional analysis, the assumption is verified by the phase space argument of Appendix A4.2. It is clear now how to determine the power d in equation (4.14) from dimensional analysis. The left-hand side clearly has dimensions of energy per unit volume. In terms of mass M, length L, and time T, we can write that the left-hand side of equation (4.14) has the dimensions

$$\frac{(\text{mass})(\text{velocity})^2}{(\text{length})^3} \equiv ML^{-1}T^{-2}$$

Turning to the right-hand side of equation (4.14), h (Planck's constant) is order of momentum (i.e., MV) multiplied by the distance $L \equiv ML^2T^{-1}$ and hence we have the dimensional equivalence

$$ML^{-1}T^{-2} = M^2L^4T^{-2}M^{-1}L^{-3d}$$

since the electron density is a number per unit volume, i.e., ρ is of the order of L^{-3}. Clearly the powers of $T(-2)$ and $M(1)$ agree on both sides. As for L we evidently must have

$$L^{-1} = L^{4-3d}$$

or $3d = 5$. Of course the constant in equation (4.14) cannot be determined by this dimensional argument. But in Appendix A4.2 it is demonstrated by direct calculation in phase space that

$$t(\mathbf{r}) = c_k [\rho(\mathbf{r})]^{5/3}$$

and hence the kinetic energy in the TF theory is given by integrating this energy density through the whole of space to yield

$$T = c_k \int [\rho(\mathbf{r})]^{5/3} \, d\tau \tag{4.15}$$

This important result shows that, in the approximation implied by the use of this simplest theory of the inhomogeneous electron gas, once the electron density $\rho(\mathbf{r})$ is known, the kinetic energy can be calculated by quadrature from equation (4.15).

For our immediate purposes, however, we can now return to equation (4.13) and rewrite it in the form

$$N\mu = \tfrac{5}{3}T + U_{en} + e^2 \int \frac{\rho(\mathbf{r})\,\rho(\mathbf{r}')\,d\tau\,d\tau'}{|\mathbf{r} - \mathbf{r}'|} \tag{4.16}$$

The final term in equation (4.16) is clearly related to the (electrostatic) self-energy of the charge distribution $\rho(\mathbf{r})$. In fact, it is twice this energy, since we are counting electron–electron interactions twice over. Thus, in the notation introduced earlier in this section we can write equation (4.16) as

$$N\mu = \tfrac{5}{3}T + U_{en} + 2U_{ee} \tag{4.17}$$

In Appendix A4.5, where we discuss the TF theory of the total binding energy of atomic ions, it is shown that, for the neutral TF atom, the chemical potential μ is identically zero. Assuming this to be true for neutral molecules as well (there is no difficulty in establishing this result in the special case of homonuclear diatoms), we obtain from equation (4.17)

$$(U_{en} + 2U_{ee})/T = -\tfrac{5}{3} \tag{4.18}$$

Using the virial theorem at equilibrium in the form $2T + U_{en} + U_{ee} + U_{nn} = 0$, with U_{nn} being as above the nuclear–nuclear potential energy, and eliminating U_{en} between this equation and equation (4.18) gives the further relation

$$(U_{ee} - U_{nn})/T = \tfrac{1}{3} \tag{4.19}$$

and finally, eliminating U_{ee} between this equation and equation (4.18),

$$(U_{en} + 2U_{nn})/T = -\tfrac{7}{3} \tag{4.20}$$

only two of the equations (4.18) through (4.20) being independent.

Equation (4.20) was recognized to be useful for molecules at equilibrium by Politzer (1976). One can term the results in equations (4.18) through (4.20) "energy scaling relations," and these will be tested against self-consistent field (SCF) calculations via wave functions in Section 4.6.

4.3. CAN THE TOTAL ENERGY OF A MOLECULE BE RELATED TO THE SUM OF ORBITAL ENERGIES?

March and Plaskett (1956) showed that for atoms the total energy E is proportional to the sum over the occupied states of the orbital energies, E_s say. That this is so follows readily from the density description given above. Thus, we have the Schrödinger equation

$$\frac{-\hbar^2}{2m}\nabla^2\psi_i + [V(\mathbf{r}) - \varepsilon_i]\psi_i = 0 \tag{4.21}$$

where in the theory $V(\mathbf{r}) = V_N(\mathbf{r}) + V_e(\mathbf{r})$. If we multiply equation (4.21) by ψ_i^* on the left, integrate over all space and then sum over occupied states, we readily find

$$-\frac{\hbar^2}{2m} \sum_{\substack{\text{occupied} \\ \text{states}}} \int \psi_i^* \nabla^2 \psi_i \, d\tau + \int \rho(\mathbf{r}) V(\mathbf{r}) \, d\tau = E_s = \sum_{\substack{\text{occupied} \\ \text{states}}} \varepsilon_i \tag{4.22}$$

where we have used equation (4.1) in the term involving $V(\mathbf{r})$. The first term in equation (4.22) is simply the kinetic energy T while the second, using equations (4.11), (4.10), and (4.14), can be written as the nuclear–electron potential energy U_{en} plus twice the electron–electron potential energy U_{ee}. Thus we have

$$T + U_{en} + 2U_{ee} = E_s \qquad (4.23)$$

By subtracting equations (4.23) and (4.17), we can eliminate $U_{en} + 2U_{ee}$ to find

$$N\mu = \tfrac{2}{3}T + E_s \qquad (4.24)$$

Finally, from the virial theorem for atoms, $2T + U = 0$, or

$$T = -E \qquad (4.25)$$

so that we find

$$E = \tfrac{3}{2}(E_s - N\mu) \qquad (4.26)$$

However, we have noted earlier that the chemical potential μ for neutral atoms in this simplest density description is zero, and hence we arrive at the desired result

$$E = \tfrac{3}{2}E_s \qquad (4.27)$$

Is this result, derived above for (heavy) atoms, still valid for molecules at equilibrium? Interest in this question was revived by Ruedenberg (1977), who demonstrated, from available SCF calculations that we shall discuss in Section 4.5, that relation (4.27) was also well obeyed for molecules at equilibrium. From his semiempirical studies he suggested the coefficient 1.55, which involved the use of relation (4.20) due to Politzer (1976).

A detailed argument shows how the density description can also lead to equation (4.27) for molecules at equilibrium [March (1977)]. The changes in the argument above are straightforward. First, however, we must note that if we press the consequences of the density description given above quite literally, there is a theorem of Teller's (referred to again in Appendix A4.3) which says that there is no molecular binding in the TF electron density theory. Nevertheless, at equilibrium, we have the result again that $T = -E$ from exact considerations (see Appendix A2.6). Then the only change in the above argument is that we must add the nuclear–nuclear energy U_{nn} to the total energy E, but not, of course, to the density–potential relation. Again by an identical argument, we are led to equation (4.27). We shall come back to the scaling relations (4.18), (4.19), and (4.20), and to the predicted relation between E and E_s after we have discussed the SCF theory of molecules in Section 4.5. However, before doing this, we want to consider, on the basis of the electron density theory above, the foundations of Walsh's rules for molecular shape.

4.4. FOUNDATIONS OF WALSH'S RULES FOR MOLECULAR SHAPE

For molecules, the work of Hückel, to be discussed at some length in Chapter 6 for π-electron assemblies, and the discussion of molecular conformation by Walsh (1953), following earlier important work by Mulliken (1942), represented the total energy E by the eigenvalue sum E_s. The density description gives, as we have seen above, a justification for such a representation for atoms and, in the present context, for molecules at equilibrium.

To be more specific about Walsh's work, we show in Figure 4.1 the correlation diagram (compare the detailed discussion in Chapter 2) as we go from a linear HAH molecule to the case when the HAH angle is 90°. In the linear molecule, the classification of the lowest states is quite clear, into two σ-states, even and odd, and into a doubly degenerate π_u nonbonding state, assuming that s and p atomic orbitals are used as building blocks. The degeneracy is due to the possibility of rotating a p-orbital about the HAH axis by 90°. When one considers the 90°

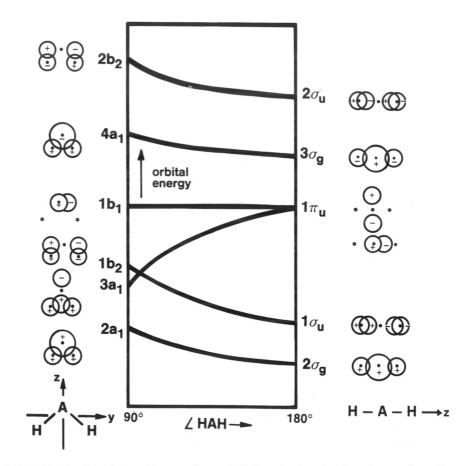

FIGURE 4.1. Correlation diagram for an HAH molecule, showing changes from linear molecule to 90° HAH angle (see also Appendix AIV).

bent molecule (for symmetry classification, see, e.g., Appendix AIV) the lowest
energy states can be viewed as follows:

1. Two orbitals binding the H atoms to the central atom: These can be
   considered as formed from the overlap of a pure p atomic orbital on A
   with the $1s$ on H. Equivalently, one can form in-phase and out-of-phase
   orbitals which are delocalized, denoted by a_1 and b_2, respectively, in
   Figure 4.1.
2. A p-orbital on atom A pointing perpendicular to the plane of the molecule;
   classification b_1.
3. An s orbital on A, which is nonbinding and of type a_1.

While it is not our purpose here to give a full explanation of all the trends
exhibited in Figure 4.1, in fact the variation in the one-electron energies is schematically as shown for purely qualitative reasons. Because of the steep rise in the
eigenvalue curve joining a_2 and π_u, it is argued by Walsh that it is energetically
unfavorable to have a linear molecule when that state is filled. Since four electrons
fill the lowest two levels, it is predicted for five, six, seven, and eight valence
electrons that one will have bent molecules, whereas with BeH_2 and HgH_2, e.g.,
one will have linear geometry.

The electron density theory, as we have seen, yields the correlation between
E and E_s required by Walsh at the equilibrium state. However, an apparent
difficulty arises, e.g., in AH_2, as one changes the angle from its equilibrium value.
Fortunately at the level of approximation corresponding to equation (4.27), one
can now complete the argument pertaining to Walsh's rules as follows: Relations
(4.17) and (4.23) are true for all molecular configurations from the electron density
theory as presented above. For the case when $\mu = 0$, therefore, the equation

$$T = -\tfrac{3}{2}E_s \tag{4.28}$$

is regained for any molecular geometry. The only step that remains is to relate T
to the total energy E. This seems, at first sight, to present a severe difficulty away
from equilibrium. Fortunately, however, the difficulty is only concerned with bond
distances away from equilibrium, not with bond angles. For instance, in the case
of a diatomic molecule, we have, writing U for the total potential energy,

$$2T + U = -R\,dE/dR \tag{4.29}$$

a relation met in Problem (2.1) and Appendix A2.6. The term on the right-hand
side of equation (4.29) is only zero at equilibrium; otherwise there would be a
contribution to the virial from the forces required to hold the nuclei fast at
separation R.

Let us suppose that, in the case of AH_2, we have constructed Figure 4.1 such
that, at each angle, the energy has been minimized with respect to bond length at
the fixed angle under discussion. Then as, e.g., Nelander (1969) has shown, for
every set of angles for which the energy is minimized with respect to bond distances, the virial theorem holds in its simple form $T = -E$. This then enables one
to complete the first-principles argument to provide a basis for Walsh's rules. Of

course, the argument is clearly approximate, as it assumes the semiclassical phase-space form (4.15) for the kinetic energy T, which is rigorously valid in a uniform electron gas, whereas in molecules the electron density is markedly inhomogeneous. Furthermore, the argument for neutral systems has gone through by putting the chemical potential μ equal to zero. For chemical accuracy, this approximation will also need relaxing. Both of these corrections are to be expected to vary somewhat with configuration, but one can anticipate that only in borderline cases will one have to study either carefully. However, this is clearly the direction in which to examine exceptions in the content of Walsh's rules for molecular shape, which indeed have much wider utility than for the AH_2 case discussed by way of example above.

We want to emphasize at this point that we have been able to proceed thus far without actually attempting to solve for the potential $V(\mathbf{r})$ and density distribution $\rho(\mathbf{r})$ in molecules from the density–potential relation (4.6). This is of course fortunate because, even for atoms, such a solution has to be carried out numerically. Although early work was done in solving the density–potential relation for $V(\mathbf{r})$ and $\rho(\mathbf{r})$ for some molecules [N_2 by Hund (1932); benzene by March (1952); the work by Townsend and Handler (1962) is also notable in this context among earlier studies], it is now clear that, though the equations of the density theory are very useful, one should, whenever high-quality densities $\rho(\mathbf{r})$ are available, use these in all applications.

Nevertheless, we have thought it worthwhile, in Appendix A4.4 to present the way in which the potential field and density can be established in the simplest atom He to which the SCF concept is applicable. Once one has grasped that concept from Appendix A4.4 on He, one can then regard $V(\mathbf{r})$ and $\rho(\mathbf{r})$ for heavy atoms as determined from the density–potential equation (4.6) by such a SCF calculation (see Appendix A4.5). This leads to important scaling predictions for atomic ions with many electrons, and these are also set out in Appendix A4.5. However, our focal point here is molecules, and we must now turn to present ways of establishing high-quality electron densities. These, at the time of writing, still require the calculation of one-electron wave functions, namely MOs.

4.5. SELF-CONSISTENT FIELD TREATMENT FOR DIATOMIC MOLECULES AND THE ROOTHAAN–HALL FORMULATION

The MO wave function is expressed as an antisymmetrized product (i.e., Slater determinant) of spin orbitals. Each spin orbital is a product of a spatial orbital and a spin function (α or β). The space orbitals that minimize the energy are obtained by solving the Hartree–Fock (HF) equations for the closed-shell system (see especially Appendix A4.6):

$$\hat{\mathscr{H}}_{\text{eff}}(1)\phi_i(1) = E_i\phi_i(1) \tag{4.30}$$

where E_i is the orbital energy and

$$\hat{\mathscr{H}}_{\text{eff}}(1) = -\tfrac{1}{2}\nabla_1^2 - \sum_\alpha Z_\alpha/r_{1\alpha} + \sum_j [2\hat{J}_j(1) - \hat{K}_j(1)] \tag{4.31}$$

In equation (4.31) the Coulomb and exchange operators $\hat{J}$ and $\hat{K}$ are defined as

$$\hat{J}_j(1)\phi_i(1) = \phi_i(1) \int \phi_j^2(2)(1/r_{12})\, d\tau_2 \qquad (4.32)$$

and

$$\hat{K}_j(1)\phi_i(1) = \phi_i(1) \int \phi_j^*(2)\phi_i(2)(1/r_{12})\, d\tau_2 \qquad (4.33)$$

The operators in equation (4.31) above have the following interpretation: (i) $-\frac{1}{2}\nabla^2$ is the operator for the kinetic energy of one electron; (ii) $-\sum_a Z_a/r_{1a}$ is the potential energy operator for the attractions between one electron and the two nuclei; (iii) $\hat{J}_j(1)$ is the potential energy operator of interaction between electron 1 and the electronic density ϕ_j^2 and, since each space orbital contains two electrons, the factor of 2 appears in equation (4.31); and (iv) $\hat{K}$ appears because the wave function must be antisymmetric with respect to electron exchange. Also the summation in equation (4.31) is over the $N/2$ occupied spatial orbitals of a molecule with N electrons.

The MOs are orthogonal and hence many integrals would vanish (which is not the case in the AO treatment, leading to great complexity).

It will be clear to the reader that the HF Hamiltonian is a one-electron operator as expressed in equation (4.31). The operator $\mathcal{H}_{\text{eff}}$ depends, therefore, on its own eigenfunctions and the HF equations are solved by an iterative process.

The molecular energy, E_{HF} is expressed as

$$E_{\text{HF}} = 2 \sum_i E_i - \sum_i \sum_j (2J_{ij} - K_{ij}) + U_{NN} \qquad (4.34)$$

where again the summations are over the $N/2$ space orbitals and since there are two electrons in each MO we have put the factor 2 in the first sum in equation (4.34).

Roothaan (1951) [see also Hall (1951)] introduced a procedure whereby the space orbitals are expressed as a linear combination of the AOs ψ_k, hence

$$\phi_i = \sum_k c_{ik}\psi_k \qquad (4.35)$$

If equation (4.35) is substituted into equation (4.30) we have

$$\sum_k c_{ik}\mathcal{H}_{\text{eff}}\psi_k = E_i \sum_k c_{ik}\psi_k \qquad (4.36)$$

If equation (4.36) is multiplied by ψ_j^* followed by integration we obtain

$$\sum_k c_{ik}(H_{jk}^{\text{eff}} - E_i S_{jk}) = 0, \qquad j = 1, 2, 3, \ldots \qquad (4.37)$$

where, as usual

$$H_{jk}^{\text{eff}} = \int \psi_j^* \hat{\mathscr{H}}^{\text{eff}} \psi_k \, d\tau \tag{4.38}$$

and the overlap integrals are defined by

$$S_{jk} = \int \psi_j^* \psi_k \, d\tau \tag{4.39}$$

Equations (4.37), as in the simple LCAO–MO treatment, yield a set of linear homogeneous equations with the C_{ik} as the unknown coefficients and hence

$$\det(H_{jk}^{\text{eff}} - E_i S_{jk}) = 0 \tag{4.40}$$

From equation (4.40) we obtain the orbital energies in the same manner as used in the simple LCAO treatment employed earlier (see Chapter 2).

The AO basis functions ψ_k are often taken as Slater-type orbitals (see Appendix A2.4), and frequently only a minimal basis set composed of the inner shell and valence shell AOs is used.

4.6. ROOTHAAN'S APPROACH COMPARED WITH THE ELECTRON DENSITY DESCRIPTION

It is, of course, important to relate the predictions of the Roothaan equations and the electron density description that was emphasized in the first part of this chapter. Though this connection is not, so far, possible by purely analytical procedures, the work of Allan *et al.* (1985) enables numerical comparisons to be made between the two approaches for a variety of properties.

4.6.1. Kinetic Energy

In Roothaan's LCAO method, once the MOs are determined, the kinetic energy can be calculated without further approximations by calculating the expectation value of the kinetic energy operator with respect to each of the occupied MOs. One merely sums the results over all the occupied orbitals. The method of comparison used by Allan *et al.* is then to compute, with the electron density obtained by summing the squares of these MOs over all occupied states, the TF form (4.15) of the kinetic energy. The wave-mechanical kinetic energy is given by the expectation value of the momentum squared, p^2, and the results of Allan *et al.* for this quantity are plotted in the upper part of Figure 4.2 along the ordinate, while the corresponding TF approximation is plotted along the abscissa, with the use of the wave-mechanical molecular density described above. It can be seen that for all the molecules (namely BeH, BH, CH, NH, OH, LiO, HF, N_2, CO, O_2, and F_2) these quantities correlate linearly to graphical accuracy. The slope is not quite unity in fact but the plot leaves no doubt that the density

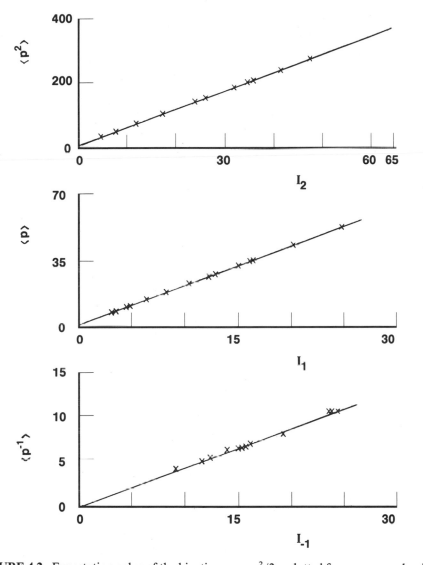

FIGURE 4.2. Expectation value of the kinetic energy $p^2/2m$ plotted from wave mechanical calculations on the ordinate *vs.* the electron density (Thomas–Fermi) result on the abscissa. The same density (of Hartree–Fock–Roothaan quality) was used in all calculations. The middle part of the figure shows a similar plot for mean momentum p. In electron density theory, $\langle p \rangle$ is directly related to exchange energy (see Appendix A4.7) [after Allan *et al.* (1985)]. Note that ordinate $\langle p^m \rangle$ is related to abscissa I_m proportional in turn to $\int \{\rho(\mathbf{r})\}^{1 + (1/3)m} \, d\tau$.

description, used with a good-quality, say, HF–Roothaan, electron density, leads to an excellent representation of the full wave-mechanical kinetic energy. Allan *et al.* also make a similar plot of the expectation value of the momentum itself $\langle p \rangle$, which, as is shown in Appendix A4.7, is directly proportional to the exchange energy in the electron density description. Again the correlation between fully wave-mechanical predictions and the density description is good, but we must refer the reader to the original work for further details.

4.6.2. Test of Energy Scaling Relations

The above test of the kinetic energy expression (4.15) of an inhomogeneous electron gas with the SCF wave function calculations was, of course, absolute. However, we also want to refer to the test of the energy scaling relations (4.18) through (4.20) carried out by Mucci and March (1979), again using SCF calculations on light molecules. In particular, the data given by Synder and Basch (1972) were used to construct Figure 4.3, in which $(U_{nn} - U_{ee})$ is plotted against T. The linearity observed confirms the result (4.19) when it is noted that the slope of the straight line in Figure 4.3 is near to $\frac{1}{3}$. Similarly, their results for $2U_{ee} + U_{en}$ against T for the same set of molecules are displayed in Figure 4.4. Again there is an excellent linear relationship, and the slope of the line is near to $\frac{5}{3}$, in accord with the relation (4.18).

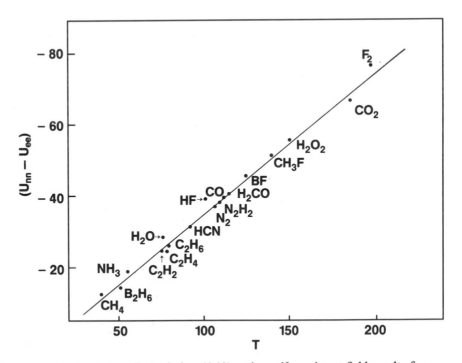

FIGURE 4.3. Test of scaling relation (4.19), using self-consistent field results for energy terms. The solid line has a theoretically predicted slope of $\frac{1}{3}$ [after Mucci and March (1979)].

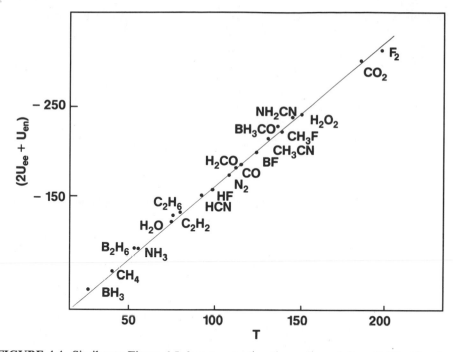

FIGURE 4.4. Similar to Figure 4.3, but now testing the scaling relation (4.18). The slope of the solid line is the predicted value of $\frac{5}{3}$ [after Mucci and March (1979)].

In Appendix A4.7, the exchange energy of an inhomogeneous electron gas is calculated (see also Problem 4.5), the energy density being found to be proportional to the density $\rho(\mathbf{r})$ to the $\frac{4}{3}$ power, analogous to the kinetic energy density given in equation (4.15), proportional to $\rho^{5/3}$, this latter relation being tested against SCF calculations in Figure 4.2. Similarly, Allan *et al.* have related the $\rho^{4/3}$ exchange energy to the appropriate J and K terms in the HF–Roothaan theory (see Appendix A4.6 for more detail). Again the results are encouraging for the electron density description of molecules.

4.7. CHEMICAL POTENTIAL IN RELATION TO ELECTRONEGATIVITY

This is the point at which to return to one of the important concepts on which the electron density description is based, the constancy of the chemical potential μ through the entire charge cloud of a molecule. This is a matter of prime importance in this theory. It is therefore disappointing that the simplest density theory, based directly on the density–potential relation (4.6), leads to the result referred to above that μ equals zero for neutral atoms. This is worth elaborating on before developing the theme of the significance of the chemical potential for molecular physics. For the self-consistent solution of equation (4.6) for such neutral atoms (see Appendix A4.5) both the density $\rho(\mathbf{r})$ and the self-consistent potential energy $V(\mathbf{r})$ tend to zero at infinity. Thus, it follows immediately that the right-hand-side of equation (4.6) tends to zero at infinity. But μ is constant throughout the charge cloud, and can therefore be calculated by substituting any value of $\mathbf{r}$ (and in particular in this example $\mathbf{r}$ tends to infinity) on the right-hand-side. Thus the chemical potential is zero for neutral atoms in the simplest density description based on equation (4.6). The reader should be cautioned that this is not true for positive atomic ions from the density–potential relation (4.6), as they have a finite (semiclassical) radius, and the above argument only applies to the neutral atom, which has an infinite radius.

What is important in what follows is that a formally exact density description does exist. This is clear, starting from the pioneering work led by Thomas (1926), Fermi (1928), and Dirac (1930), who assumed the existence of such a theory and were vindicated by the clear successes of the theory. The theorem was proved later by Hohenberg and Kohn (1964) that the ground-state energy of an atom, molecule, or solid is uniquely determined by the electron density $\rho(\mathbf{r})$. This theorem, proved in Appendix A4.8, can be shown to imply an exact (formal) generalization of equation (4.6) expressing the constancy of the chemical potential through the inhomogeneous electron cloud, and this is the point on which the arguments outlined below rest.

Once the approximations implicit in writing equation (4.6) are removed in this (formal) manner, one can ask what is the value of the chemical potential for an atom, molecule, or solid. We therefore take up the arguments pertaining to the chemical potential with this as a starting point.

In this section, we wish to discuss the chemical potential in relation to electronegativity (see Section 1.4 and also its use in Chapter 2), so we remind the reader that the concept of electronegativity was introduced into chemistry in order to understand the ionic contribution to the chemical binding of molecules (and of solid compounds) [cf. Pauling (1939)]. The electronegativity difference between two atoms A and B is considered to be the driving force for the transfer of electronic charge between them. Because of this transfer of charge, electronegativities become equalized in the AB molecule in particular.

Now let us return to the chemical potential, and in order to be specific to equation (4.6). Consider this equation applied, in turn, to atom A, atom B, and the molecule AB in equilibrium. In principle, the chemical potential takes on values μ_A and μ_B for atoms A and B, but a single value μ_{AB} for the equilibrium molecule. Thus the different* chemical potentials of the atoms A and B are equalized as they are brought into interaction to form the equilibrium AB molecule. The reader will recognize from this that the chemical potential satisfies all the requirements placed on the electronegativity introduced at the beginning of this section.

If one wishes to have a precisely defined quantity, it is clear from the above that one can choose to work with the chemical potential m of the electron density description as the fundamental quantity to replace electronegativity. Having pressed the parallel between the chemical potential and the chemical concept of electronegativity [for more details, see, e.g., Sanderson (1960, 1967, 1978), Ickowski and Margrave (1961), Parr *et al.* (1978), March (1983), and Parr and Yang

---

*We have already remarked that a formally exact generalization of equation (4.6) for the chemical potential exists. This is what we must consider; as $\mu = 0$ for the neutral atom from equation (4.6).

(1989)] we want to reiterate two major points about the electron density treatment, which is the focus of the present chapter.

Equation (4.6) can be derived from the analog of the variation principle using wave functions (see Appendix A2.1) converted into electron density language. Since we evidently minimize the total energy E now with respect to the electron density $\rho(\mathbf{r})$, subject to the normalization condition

$$\int \rho(\mathbf{r}) \, d\tau = N \qquad (4.41)$$

where N as usual is the given total number of electrons in the molecule, we can employ the usual method of Lagrange's undetermined multipliers [see, e.g., Rushbrooke (1944)] to write formally that the variation of the energy with respect to the electron density must be zero, for a stationary principle (see also Appendix A4.6 for a more advanced discussion in wave function language), i.e.,

$$\delta[E - \mu N] = 0 \qquad (4.42)$$

with the chemical potential playing the role of the Lagrange multiplier in this variational approach to equation (4.6).

There is a formally exact generalization of equation (4.6) as discussed above; this can be derived from an (as yet unknown) exact expression for E solely in terms of $\rho(\mathbf{r})$, which the Hohenberg Kohn (1964) theorem of Appendix A4.8 assures us exists in principle.

From equation (4.42), we can follow Hulthen (1935) and write

$$\mu = dE/dN \qquad (4.43)$$

To expand on the meaning of equation (4.43), which is certainly valid in the simplest density description used above, we illustrate in Figure 4.5 the case of an atomic ion, of atomic number Z, the total ground-state energy being plotted against N. We have marked on this plot the energy of the neutral atom corresponding to $N = Z$ along with the values for the singly charged positive and negative ions,[*] $E(Z, Z - 1)$ and $E(Z, Z + 1)$, respectively. We have dotted in an (assumed) smooth curve joining these three energies and, according to equation (4.43), the chemical potential of the neutral atom is the slope of this curve at the point $N = Z$. Also labeled on Figure 4.5 are the ionization potential $I = E(Z, Z - 1) - E(Z, Z)$ and the electron affinity $A = E(Z, Z) - E(Z, Z + 1)$. If one joins the energies of the singly charged positive and negative ions by a straight line, then it is evident from Figure 4.5 that this has slope $(I + A)/2$. Rolle's theorem then ensures that there is a point lying between $Z - 1$ and $Z + 1$ on the dotted curve at which the slope is equal to $(I + A)/2$. This is the value discussed in Chapter 1, which Mulliken (1934) proposed as his electronegativity measure. Of course, if one wished, one could construct, simply, functions $E(N)$ which not only pass through the three energy points marked on Figure 4.5, but also correctly

---

[*]The existence of a stable negative ion would require one to transcend equation (4.6).

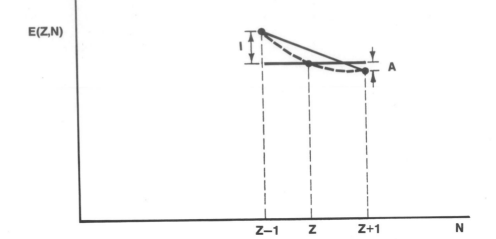

FIGURE 4.5. Schematic illustration of equation (4.43) for chemical potential μ. Relation to Mulliken's approximation to electronegativity $(I + A)/2$ is clearly shown, I being the ionization potential and A the electron affinity. The three energy values are those of a neutral atom under consideration, together with ground-state energies of singly charged positive and negative ions.

incorporate higher ionization potentials [March (1982), Phillips and Davidson (1984)].

Some fundamental attempts have been made to calculate μ directly from the density–potential relation without using equation (4.43). These lead always to a crucial dependence of μ on the ionization potential I as discussed, for example, by March and Bader (1980); see also Alonso and March (1983) in Hartree–Fock theory, where Koopmans' theorem for I is also involved (see Appendix A6.2).

We shall not go into further detail of these studies here. Rather we conclude by noting that provided we are willing to model the chemical potential, useful progress can be made on:

1. The chemical potential in hydrogen halides and mixed halides, which is given as an example of equalizing the chemical potential of the constituent atoms on formation of molecules in Appendix A4.9.
2. The electron density distribution in a series of ten electron molecules; these densities can then be brought into contact with X-ray scattering experiments (Appendix A4.10).
3. Approximate force laws (or potential energy curves) for certain simple types of molecules, described in Appendix A4.11.

In summary, while it is still necessary in molecules to proceed via wave function calculations to obtain high-quality electron densities $\rho(\mathbf{r})$, using the SCF equations for molecules, many regularities from such numerical calculations can be understood simply via the electron density description. A further important bonus of this description is the concept of the (constant) chemical potential at every point in a molecule's electronic charge cloud. This promises to be of considerable interest for chemistry as it resembles, but transcends (because it is, at least in principle, precisely defined) the chemist's valuable concept of electronegativity.

PROBLEMS

4.1. The TF density–potential relation (4.6) for a bare Coulomb potential energy $-Ze^2/r$ leads to the explicit form of the electron density

$$\rho(\mathbf{r}) = (8\pi/3h^3)(2m)^{3/2}(\mu + Ze^2/r)^{3/2} \tag{P4.1}$$

provided the quantity inside the square root (i.e., $\mu + Ze^2/r$) is greater than or equal to zero, and ρ is otherwise zero elsewhere. **(a)** Given that there are N electrons filling up the lowest levels in this bare Coulomb potential show, either by direct calculation of the integral

$$\int \rho(\mathbf{r})\, d\tau = N$$

from equation (P4.1), or using the result from the Bohr formula (A4.5.1) that the energy per closed shell (each holding $2n^2$ electrons for the shell of principal quantum number n) is the constant value $-Z^2e^2/a_0$ independent of n, that the chemical potential μ is given by

$$\mu = -Z^2/(18N^2)^{1/3}\frac{e^2}{a_0}, \qquad a_0 = \hbar^2/me^2 \tag{P4.2}$$

(Hint: if you use the Bohr formula, assume $\mathcal{N}$ closed shells, note that $N = \sum_1^{\mathcal{N}} 2n^2$, and hence relate N and $\mathcal{N}$ as both get large. The reason for dealing with large numbers is to compare with the TF theory, which utilizes statistical mechanics (specifically, it appeals to phase-space arguments). **(b)** Without working out the details in full, generalize the above argument to treat the hydrogen-like molecular ion in which noninteracting electrons are assumed to move in the field created by the presence of two bare point charges Ze separated by distance R. Can you see, in general terms, how the chemical potential depends on Z, N, and R? Will it, in particular, depend on all these three variables? If not, what combination of these variables will enter the final result? (cf. Dreizler and March, 1980).

4.2. In Appendix A1.3, the so-called $1/Z$ expansion for the total energy $E(Z, N)$ of atomic ions with N electrons and atomic number Z is written as

$$E(Z, N) - Z^2[\varepsilon_0(N) + (1/Z)\varepsilon_1(N) + (1/Z^2)\varepsilon_2(N) + \cdots] \tag{P4.3}$$

(a) Given that, for sufficiently large N (and Z), the coefficients $\varepsilon_n(N)$ vary as $N^{n+1/3}$, show that the series will (formally) sum to the form

$$\underset{\text{lim } Z \text{ and } N \text{ large}}{E(Z, N)} = Z^{7/3}f(N/Z) \tag{P4.4}$$

(b) Equation (P4.4) is the result of the TF SCF method, which in fact determines the function $f(N/Z)$. What would be the function f for the simpler case of the (non-self-consistent) bare Coulomb field treated in the first part of Problem (4.1)? [The reader interested in fuller details may consult March (1975).]

4.3. In Appendix A4.3, reference is made to the generalization of the $1/Z$ expansion in equation (P4.3) to the case of homonuclear diatoms, with two nuclear charges Ze separated by internuclear distance R. Writing $X = ZR$, this generalization reads [see, e.g., March and Parr (1980)]:

$$E(Z, N, R) = Z^2[\varepsilon_0(N, X) + (1/Z)\varepsilon_1(N, X) + 1/Z^2\varepsilon_2(N, X) + \cdots] \tag{P4.5}$$

Given that at sufficiently large N (and Z),

$$\varepsilon_n(N, X) = N^{n + (1/3)} A_n(X) \tag{P4.6}$$

one can write, putting $x = N^{2/3}/X$,

$$R^2/N^{5/3}E(Z, N, R) = (A_0(X)/x^2) + (RN^{1/3}/x)A_1(X) + R^2 N^{2/3} A_2(X) + \cdots \tag{P4.7}$$

by truncating the series (P4.7) at A_2. Using the fact that the equilibrium bond length R_e is determined by $(\partial E/\partial R)_{Z,N,R_e} = 0$, show that, at equilibrium, the approximate result

$$(R^2 E/N^{5/3})_{R_e} = F(RN^{1/3})|_{R_e} \tag{P4.8}$$

follows, where F is a universal function. [N.B. Pucci and March (1983) have tested the scaling predicted in equation (P4.8) for neutral diatoms (i.e., $N = 2Z$). They find it is quite well obeyed, testifying to the approximate validity of truncating the $1/Z$ expansion as above. However, when first, second, and third row molecules are plotted together, although the gross scaling predicted in equation (P4.8) is confirmed, there are Periodic Table effects still visible at graphical accuracy.]

4.4. In momentum space formulations of quantum mechanics, one can go from a space wave function $\psi(\mathbf{r})$ to a momentum space wave function $\phi(\mathbf{p})$ using the Dirac–Fourier transform relation

$$\phi(\mathbf{p}) = (1/h^{3/2}) \int \exp(-i\mathbf{p} \cdot \mathbf{r}/\hbar)\psi(\mathbf{r})\, d\tau \tag{P4.9}$$

(a) Take the case of $\psi(\mathbf{r})$ as the $1s$ wave function for the hydrogen atom and, using first the spherical symmetry to integrate over angles in equation (P4.9) one obtains

$$\phi_{1s}(p) = \frac{1}{h^{3/2}} \int_0^\infty \frac{\sin(pr/\hbar)}{(pr/\hbar)} \psi_{1s}(\mathbf{r})4\pi r^2\, dr \tag{P4.10}$$

Writing $\sin(pr/\hbar)$ as the imaginary part of $\exp(ipr/\hbar)$, evaluate from equation (P4.10) the momentum space wave function $\phi_{1s}(p)$ of the ground state of the hydrogen atom. **(b)** Though this method can now be generalized, approximately, to light atoms with more than one electron, it becomes very unwieldy for heavy atoms. However, for these, the electron density description is appropriate. Using the phase-space arguments, show that the momentum density $\rho(p)$ in an atom or molecule with many electrons is given by [see Allan and March (1983)]:

$$\rho(p) = (2/h^3)\Omega(p) \tag{P4.11}$$

In equation (P4.11), $\Omega(p)$ is the volume enclosed by a surface of constant electron density, the value of p being determined from the constant density on that surface by the phase space equation

$$\rho = (8\pi/3h^3)p^3 \tag{P4.12}$$

[A comparison of equation (P4.11), even for light molecules, with the results of the Roothaan SCF approach described in Section 4.5 has been shown by Allan *et al.* (1985) to be quite good.]

4.5. The exchange energy of a uniform electron gas, the starting point for the derivation of equation (P4.14), can be obtained by averaging the electron–electron interaction energy e^2/r_{ij} representing the Coulombic repulsion between electrons i and j at separation r_{ij}, with respect to the unperturbed free electron gas wave function, which is simply the single Slater determinant formed from plane waves with momentum vectors lying within the Fermi sphere in momentum space of radius p_m. This calculation is carried out in Appendix A4.7. However, this problem is based on an analogy with the kinetic energy density derived in Section 4.2. As there, assume that the exchange energy density is also expressible as a power of the electron density. Thus write

$$\text{Exchange energy density} = -\text{constant} \cdot e^2\rho^n \tag{P4.13}$$

The fact that e^2 comes in is evident from the fact that we are averaging e^2/r_{ij} as discussed with respect to a free-particle Slater determinantal wave function. Use dimensional analysis to show that the exponent n in equation (P4.13) is $\frac{4}{3}$. This result is confirmed in Appendix A4.7, where the constant in equation (P4.13) is also determined such that in an inhomogeneous electron gas:

$$\text{Exchange energy density} = -c_e[\,\rho(\mathbf{r})]^{4/3}$$

$$c_e = (3e^2/4)(3/\pi)^{1/3} \tag{P4.14}$$

which is the basis for the local Dirac–Slater exchange potential [Slater (1951)].

4.6. A one-electron system (e.g., the H atom or H_2^+) has a ground-state wave function $\psi(\mathbf{r})$ and an electron density $\rho(\mathbf{r}) = \psi^2(\mathbf{r})$. Show that the kinetic energy is $(\hbar^2/8m) \int (\nabla\rho)^2/\rho \, d\mathbf{r}$. From Teller's theorem it is an essential quantity in molecular binding [March (1992)].

REFERENCES

N. L. Allan, D. L. Cooper, C. G. West, P. J. Grout, and N. H. March, *J. Chem. Phys.* **83**, 239 (1985).

N. L. Allan and N. H. March, *Int. J. Quantum Chemistry Symp.* **17**, 227 (1983).

J. A. Alonso and N. H. March, *J.Chem. Phys.* **78**, 1382 (1983).

J. A. Alonso and N. H. March, *Chemical Physics* (North Holland–Elsevier) **76**, 121 (1983).

P. A. M. Dirac, *Proc. Camb. Phil. Soc.* **26**, 376 (1930).

R. M. Dreizler and N. H. March, *Z. für Physik* **A294**, 203 (1980).

E. Fermi, *Z. Physik* **48**, 73 (1928).

G. G. Hall, *Proc. Roy. Soc.* (*London*) **A205**, 541 (1951).

P. C. Hohenberg and W. Kohn, *Phys. Rev.* **B136**, 864 (1964).

E. Hückel, *Z. Physik* **70**, 204 (1931).

E. Hückel, *Z. Physik* **76**, 628 (1932).

L. Hulthen, *Z. für Physik* **95**, 789 (1935).

F. Hund, *Z. für Physik* **77**, 12 (1932).

R. P. Ickowski and J. L. Margrave, *J. Am. Chem. Soc.* **83**, 3547 (1961).

N. H. March, *Acta Cryst.* **5**, 187 (1952).

N. H. March, *Advances in Physics* **6**, 1 (1957).

N. H. March, *Self-Consistent Fields in Atoms*, Pergamon Press, Oxford (1975).

N. H. March, *J. Chem. Phys.* **67**, 4618 (1977)

N. H. March, *J. Chem. Phys.* **86**, 2262 (1982).

N. H. March, *Contemp. Phys.* **24**, 373 (1983).

N. H. March, *Electron Density Theory of Atoms and Molecules*, Academic Press, New York (1992).

N. H. March and R. F. W. Bader, *Phys. Lett.* **A78**, 242 (1980).

N. H. March and R. G. Parr, *Proc. Nat. Acad. Sci. USA* **77**, 6285 (1980)

N. H. March and J. S. Plaskett, *Proc. Roy. Soc.* **A235**, 419 (1956).

J. F. Mucci and N. H. March, *J. Chem. Phys.* **71**, 5270 (1979).

R. S. Mulliken, *J. Chem. Phys.* **2**, 782 (1934).

R. S. Mulliken, *J. Chem. Phys.* **3**, 573 (1935).

R. S. Mulliken, *Revs. Modern Phys.* **14**, 204 (1942).

B. Nelander, *J. Chem. Phys.* **51**, 469 (1969).

R. G. Parr, R. A. Donnelly, M. Levy, and W. E. Palke, *J. Chem. Phys.* **68**, 3801 (1978).

R. G. Parr and W. Yang, *Density Functional Theory of Atoms and Molecules*, Clarendon Press, Oxford (1989).

L. Pauling, *The Nature of Chemical Bond*, Cornell University Press, Ithaca, NY (1939) [3rd Ed. (1960)].

P. Phillips and E. R. Davidson in: *Local Density Approximations in Quantum Chemistry and Solid State Physics* (J. P. Dahl and J. Avery, eds.), Plenum Press, New York (1984).

P. Politzer, *J. Chem. Phys.* **64**, 4239 (1976).

R. Pucci and N. H. March, *J. Chem. Phys.* **78**, 2466 (1983).

C. C. J. Roothaan, *J. Chem. Phys.* **19**, 1445 (1951); *Rev. Mod. Phys.* **23**, 69 (1951).

K. Ruedenberg, *J. Chem. Phys.* **66**, 375 (1977).

G. S. Rushbrooke, *Statistical Mechanics*, Oxford University Press, Oxford (1944).

R. T. Sanderson, *Science* **121**, 207 (1955).

R. T. Sanderson, *Chemical Periodicity*, Reinhold, New York (1960).

R. T. Sanderson, *Inorganic Chemistry*, Reinhold, New York (1967).

R. T. Sanderson, *Chemical Bonds and Bond Energy*, Academic Press, New York (1978).

J. C. Slater, *Phys. Rev.* **81**, 385 (1951).

L. C. Snyder and H. Basch, *Molecular Wave Functions and Properties*, John Wiley and Sons, New York (1972).

E. Teller, *Rev. Mod. Phys.* **34**, 627 (1962).

L. H. Thomas, *Proc. Camb. Phil. Soc.* **23**, 542 (1926).

J. R. Townsend and G. S. Handler, *J. Chem. Phys.* **36**, 3325 (1962).

A. D. Walsh, *J. Chem. Soc.* 2260–2331 (1953).

MOLECULAR PARAMETERS DETERMINED BY SPECTROSCOPIC METHODS

So far, we have been concerned primarily with electronic properties of molecules when the nuclei are held fixed. However, other properties of molecules are of considerable interest; e.g., vibrational and rotational degrees of freedom are unique to molecules as opposed to atoms. In the present chapter, we shall begin with a discussion of rotational energy levels of the simplest case, namely diatomic and triatomic linear molecules. We shall see that molecular moments of inertia are important parameters, and this topic will be discussed in relation to pure rotational spectra as observed in the infrared, millimeter wave, and microwave regions. Even from this simplest case, it will be seen as important for the interpretation of such data that one must also consider the effect of molecular vibration on the rotational spectra. This leads into a fuller discussion of vibrational spectra of polyatomic molecules, and again, briefly, the relation to observation. In each case selection rules are central and are given some prominence in this chapter.

We then consider electronic spectroscopy and it will emerge that in this area extremely close contact can be established with the concepts which arise naturally from MO theory. The chapter concludes with a short discussion of nuclear magnetic resonance and electron spin resonance, the latter being treated with the specific aim of providing background and motivation for the ensuing discussion of the MO theory of π-electron assemblies.

It is instructive here to examine the way in which changes in molecules are reflected in the observed spectra, and to have an idea of the energies involved in these changes. Figure 5.1 has been constructed, following Banwell (1966), with this in mind.

The discussion below deals primarily with the essentials of spectroscopy. While many of the important aspects relevant to chemical physics are treated, for the reader interested in further detail, specialist texts must be used to supplemement this account.

5.1. ROTATIONAL SPECTROSCOPY

Rotational and vibrational spectroscopy relate to characteristics of systems composed of more than one atom, and are of prime interest in both free molecules and in solids. The former is our focal point here.

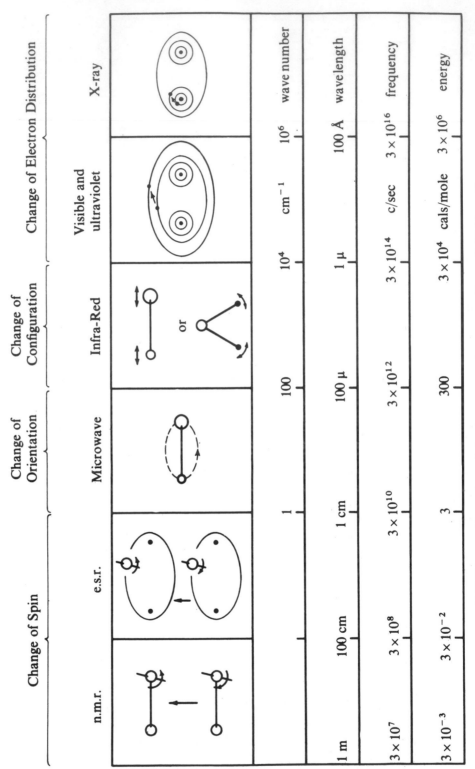

			Change of Spin		Change of Orientation	Change of Configuration		Change of Electron Distribution	
			n.m.r.	e.s.r.	Microwave	Infra-Red		Visible and ultraviolet	X-ray
wave number		10^6				10^4			cm^{-1}
wavelength	100 Å					$1\,\mu$		cm^{-1}	
frequency	3×10^{16}					3×10^{14}		c/sec	
energy	3×10^6					3×10^4		cals/mole	

FIGURE 5.1. The regions of the electromagnetic spectrum [Banwell (1966), p. 8].

125

*Molecular
Parameters
Determined by
Spectroscopic
Methods*

Molecules, in the context of their rotational, vibrational, and electronic spectra can be usefully divided into classes according to their principal moments of inertia.

5.1.1. Principal Moments of Inertia of Molecules

The molecule under discussion may contain a one-, two-, or three-dimensional array of nuclei, each one of which is to be treated as a point mass m_i with coordinates x_i, y_i, and z_i with respect to any set of Cartesian axes whose origin is at the center of mass of the nuclear assembly. The axes are fixed relative to the nuclei so that when the molecule rotates the axes rotate with it. These are termed molecule-fixed axes, in contrast to space-fixed axes, which remain fixed in space while the molecule rotates about them. The moments of inertia about x, y, and z molecule-fixed axes are given by

$$I_x = \sum_i m_i(y_i^2 + z_i^2)$$

$$I_y = \sum_i m_i(x_i^2 + z_i^2) \tag{5.1}$$

$$I_z = \sum_i m_i(x_i^2 + y_i^2)$$

The products of the inertia I_{xy}, I_{yz}, and I_{xz} relating to the same axes are defined as

$$I_{xy} = -\sum_i m_i x_i y_i$$

$$I_{yz} = -\sum_i m_i y_i z_i \tag{5.2}$$

$$I_{xz} = -\sum_i m_i x_i z_i$$

Then the three-dimensional surface defined by the equation

$$I_x x^2 + I_y y^2 + I_z z^2 - 2I_{xy} xy - 2I_{yz} yz - 2I_{xz} xz = 1 \tag{5.3}$$

represents an ellipsoid, technically termed the momental ellipsoid, and shown in Figure 5.2. The Cartesian axes in this diagram represent a particular set of axes for which the products of inertia are zero. Then equation (5.3) evidently reduces to

$$I_a x^2 + I_b y^2 + I_c z^2 = 1 \tag{5.4}$$

where a and b are the major and minor axes of the ellipse in the xy-plane and b and c, respectively, the same quantities in the yz-plane. These axes are termed the

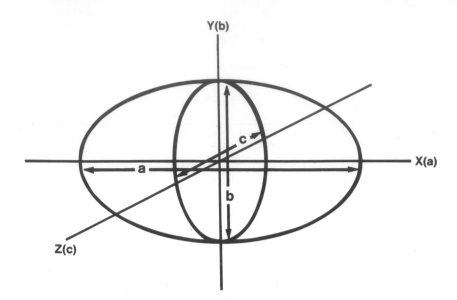

FIGURE 5.2. Momental ellipsoid.

principal inertial axes and I_a, I_b, and I_c are the principal moments of inertia. One of the principal moments of inertia, I_a in this case, is the minimum moment of inertia about any axis; another (I_c here) is the maximum moment of inertia and its axis must always be perpendicular to that of I_a; the third I_b is intermediate in magnitude and about an axis perpendicular to those of both I_a and I_c. In fact, it is helpful to maintain the convention

$$I_c \geq I_b \geq I_a \tag{5.5}$$

and the x-, y-, and z-axes corresponding to these principal moments of inertia are labeled a, b, and c.

Each molecule possesses a characteristic momental ellipsoid which exhibits the property that the distance between any point on the surface and the center of mass is $1/I^{1/2}$, where I is the moment of inertia of the molecule about that line.

For the linear molecule (Figure 5.3), the cross section of the momental ellipsoid in the bc-plane is reduced to a circle.

Let us note immediately some examples. Thus Figure 5.3 shows the principal axes of inertia of the HCN molecule, labeled a, b, and c. Evidently, from Figure 5.3 other principal moments of inertia are such that

$$I_c = I_b > I_a = 0 \tag{5.6}$$

In this case one has explicitly

$$I_c = m_H r_H^2 + m_C r_C^2 + m_N r_N^2 \tag{5.7}$$

where m_X and r_X, respectively, denote the mass of the X atom and the distance along the a-axis of atom X from the center of mass.

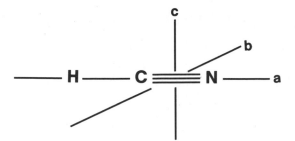

FIGURE 5.3. Principal axes of inertia, labeled a, b, and c, of HCN molecule.

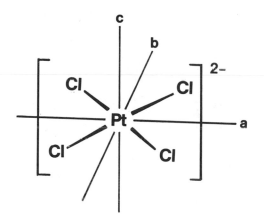

FIGURE 5.4. Principal axes of inertia of oblate symmetric rotor $[PtCl_4]^{2-}$.

5.1.2. The Symmetric Rotor or Symmetric Top

Such a molecule has two equal moments of inertia and the third may be less or greater than the other two. If it is less than the other two, then $I_c = I_b > I_a$ and the molecule is called a prolate symmetric rotor; the linear molecule is embraced as the special case $I_a = 0$. If $I_c > I_b = I_a$ the converse applies and one has the case of an oblate symmetric rotor. An example of this case, for the square planar ion $[PtCl_4]^{2-}$, is shown in Figure 5.4.

5.1.3. Spherical Rotors

Another important case is that of highly symmetrical molecules, for which

$$I_c = I_b = I_a \tag{5.8}$$

These are termed spherical rotors. Tetrahedral molecules such as methane fall into this category, as do regular octahedral molecules such as sulfur hexafluoride.

5.1.4. Rotational Spectra of Diatomic and Linear Polyatomic Molecules

The discussion of rotational spectra below presupposes the sample to be in the gaseous phase at low pressure, enabling one to ignore broadening of rotational transitions by collisions. The energy levels of a rigid rotator are

$$E_J = h^2 J(J+1)/8\pi^2 I \tag{5.9}$$

I being the moment of inertia [see equation (5.6)]. For a diatomic molecule, this is given by μr^2, where μ is the reduced mass and r the internuclear distance. Equation (5.9) applies also to the rotational energy levels of a linear polyatomic molecule such as acetylene C_2H_2 or $O=C=S$.

From the standpoint of observation, a frequency or wave number is observed rather than the energy ΔE of a transition; these are related to one another by $\Delta E = h\nu = hc\tilde{\nu}$. Because of this it is frequently convenient to change from energy level formulas into term value expressions having dimensions of frequency or wave number rather than energy.

Thus equation (5.9) is equivalent to

$$F(J) = E_J/h = hJ(J+1)/8\pi^2 I \tag{5.10}$$

where $F(J)$ has the dimensions of frequency; or, alternatively, with $F(J)$ as a wave number we can write

$$F(J) = E_J/hc = hJ(J+1)/8\pi^2 cI \tag{5.11}$$

Figure 5.5 shows a set of term valaues $F(J)$ in units of cm^{-1} for the CO molecule.

Pure rotational transitions may be observed in the microwave, millimeter wave, or far-infrared regions. They are electric dipole transitions (cf. Appendix A5.5) and the transition moments are vector quantities given by

$$\mathbf{R}^{J'M_J'J''M_J''} = \int \psi_r^{J'M_J'*} \boldsymbol{\mu} \psi_r^{J''M_J''} \, d\tau \tag{5.12}$$

The lower vibrational state involved has quantum numbers J'' and M_J'' and the upper state is characterized by J' and M_J'.

5.1.5. Rotational Selection Rules

The selection rules are obtained from the requirement that the transition moment (5.12) shall be nonzero. This can be shown to be the case in the following:

1. The molecule has a permanent dipole moment, i.e., $\boldsymbol{\mu} \neq 0$.
2. The change in the J quantum number, ΔJ, say, is such that $\Delta J = \pm 1$.
3. $\Delta M_J = 0, \pm 1$, a rule which is of the greatest importance if the molecule is in an electric or magnetic field.

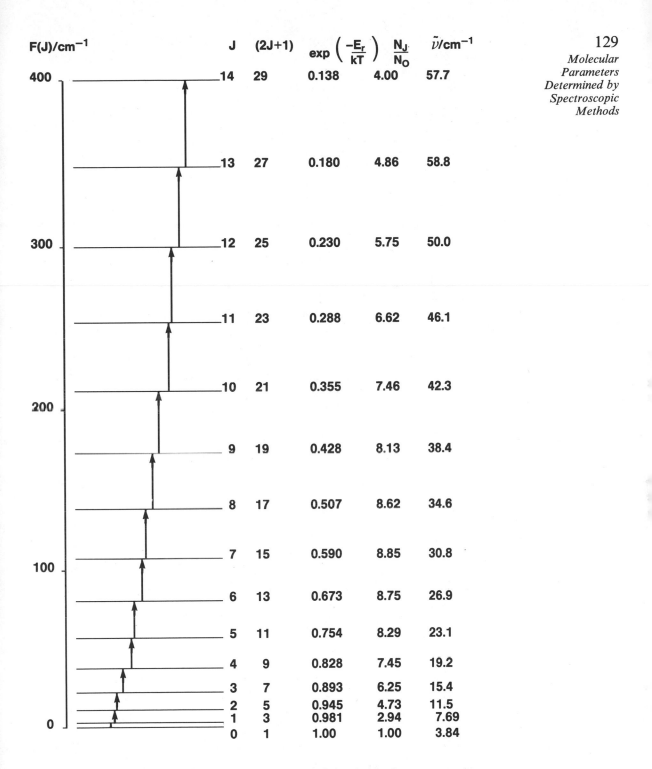

129

*Molecular
Parameters
Determined by
Spectroscopic
Methods*

FIGURE 5.5. Rotational properties of CO molecule. Left-hand side shows term values. Relative populations N_J/N_0 are also shown [modified from Hollas (1987)].

An obvious consequence of (1) is that transitions are allowed in heteronuclear diatoms such as CO, NO, HF, etc. but not in homonuclear molecules such as Cl_2 and N_2.

For linear polyatomic molecules lacking a center of symmetry, e.g., O=C=S, H—C≡N, etc., rotational transitions are allowed, but in molecules like S=C=S or H—C≡C—H they are not.

Since ΔJ is taken conventionally to refer to $J' - J''$, where J' is the quantum number of the upper state and J'' that of the lower level of the transition, selection rule (2) is effectively $\Delta J = +1$. The transitions have frequencies given by

$$v = F(J') - F(J'') = F(J + 1) - F(J) = 2B(J + 1) \tag{5.13}$$

from equation (5.10), B being equal to $h^2/8\pi^2 I$. The transitions observed in absorption are shown in Figure 5.5.

Whether the spectrum lies in the microwave, millimeter wave, or far-infrared regions depends on the values of B and J. The observed spectrum of CO, in fact, lies partly in the millimeter-wave and partly in the far-infrared region. Part of the far-infrared spectrum of CO, from 15 to 40 cm^{-1}, is reproduced in Figure 5.6. The transitions are those with quantum number J'' in the range 3 to 9.

The transitions with $J'' = 0$ to $J'' = 6$ have been observed with high accuracy in the millimeter wave region. The frequencies and wave numbers are listed in Table 5.1. The separations between adjacent transitions are given in the final column of the table. They are seen to be very nearly constant, and equal therefore,

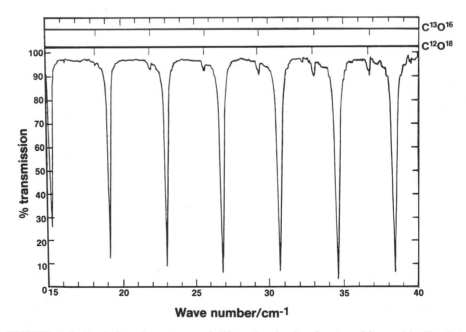

FIGURE 5.6. Far-infrared spectrum of CO molecule showing transitions with $J'' = 3$ to $J'' = 9$ in wave-number range 15 to 40 cm^{-1} [redrawn from Hollas (1982)].

from theory, to $2B_0$, the subscript zero referring to the zero-point vibrational level. B varies with the vibrational quantum number and this point will be discussed further below.

Linear polyatomic molecules generally have larger moments of inertia and, hence, smaller B-values than diatoms. Therefore, it follows that more transitions in their pure rotational spectra occur in the microwave or millimeter-wave regions.

The populations of rotational energy levels, relative to $J = 0$, are obtained from the Boltzmann distribution and are given by

$$N_J/N_0 = (2J + 1) \exp(-E_r/k_B T) \tag{5.14}$$

E_r being the rotational energy and the factor $(2J + 1)$ representing the fact that in the absence of an electric or magnetic field, M_J can take on $(2J + 1)$ values; i.e., the degeneracy of level J is $(2J + 1)$. The values of N_J/N_0 for CO, calculated from equations (5.9) and (5.14), assuming $B = 1.922 \text{ cm}^{-1}$ and $T = 293$ K, are given in Figure 5.5: they show an increase up to $J = 7$ and a subsequent decrease. Such behavior of rotational energy level populations is typical, and can be attributed to the fact that at low J, the increase in N_J/N_0 due to the $(2J + 1)$ degeneracy factor in equation (5.14) dominates the decrease due to the exponential factor; but at high J the opposite situation obtains. The value of J_{max}, the J-value of the energy level with the highest population, is obtained when

$$d(N_J/N_0)/dJ = 0 \tag{5.15}$$

which yields

$$J_{max} \simeq (k_B T/2hB)^{1/2} - \tfrac{1}{2} \tag{5.16}$$

written in terms of B with dimensions of frequency. The approximate sign is because J_{max} must be an integer.

When considering rotational transitions accompanying either vibrational transitions, or electronic transitions, the intensity variation with J is given quite

131
*Molecular
Parameters
Determined by
Spectroscopic
Methods*

TABLE 5.1

Frequencies and Wave Numbers of Pure Rotational
Transitions of the CO Molecule as Observed in
the Millimeter Wave Region

Wave number (cm^{-1})	J''	J'	Frequency (GHz)	Separation between adjacent transitions
3.84503	0	1	115.271	115.271
7.68992	1	2	230.538	115.267
11.53451	2	3	345.796	115.258
15.37866	3	4	461.041	115.245
19.22222	4	5	576.268	115.227
23.06504	5	6	691.473	115.205

accurately by the population variation represented in equation (5.14) applied to the rotational levels of the initial state of the transition. For pure rotational transitions, this is far from the case, although the population variation is an important factor. The intensities are discussed more fully in Appendix A5.2, the treatment there including the rotational partition function.

5.2. CENTRIFUGAL DISTORTION

Although they are almost constant, as predicted by simple theory, close inspection of the final column of Table 5.1 shows that the separations between adjacent lines in the millimeter-wave spectrum of CO, decrease regularly as J increases. This effect is due to centrifugal distortion, as will now be explained.

The rigid rotor approximation has been the basis of the theory presented so far. However, since the bond in a diatomic molecule, say, can elongate and contract in a vibrational motion, it evidently is not totally rigid and the model adopted is therefore oversimplified. The effect of the bond being flexible in this way is that owing to centrifugal acceleration, the rotational motion tends to throw the nuclei outward from the center of mass. Classically the effect increases with the speed of rotation and, therefore, in quantum-mechanical language now with the value of J. This centrifugal distortion of the molecule increases the internuclear distance r and, as B is proportional to I^{-1}, it decreases B. This effect can be allowed for by replacing the term values of equation (5.10) by [see Hollas (1982)]

$$F(J) = B[1 - uJ(J + 1)]J(J + 1) \tag{5.17}$$

where u is a constant. Equation (5.17) is quite equivalent to

$$F(J) = BJ(J + 1) - DJ^2(J + 1)^2 \tag{5.18}$$

D being termed the centrifugal distortion constant, which is always greater than zero for diatomic molecules. D can be related to B and to the vibrational wave number $\bar{v}$ by

$$D = 4B^3/\bar{v}^2$$

Even though the gist of the variation in the last column of Table 5.1 can be explained in this manner, in fact when very accurate experimental data are employed, even equation (5.18) is not wholly satisfactory. Owing to the fact that molecular vibration is not accurately described as harmonic, but needs correcting for anharmonic interactions, there are additional terms in higher powers of $J(J + 1)$ in the rotational term value expression. Thus one has to write

$$F(J) = BJ(J + 1) - DJ^2(J + 1)^2 + HJ^3(J + 1)^3 + \cdots \tag{5.19}$$

However, for many purposes, the essential interpretation of the main features of rotational spectra can be given without the term involving the new constant H shown in equation (5.19).

5.3. VIBRATIONAL SPECTROSCOPY OF DIATOMIC MOLECULES: HARMONIC APPROXIMATION

Consider a classical particle of mass M tied to an origin 0, say, with a restoring force $-kx$, proportional to the displacement x from 0, the negative sign indicating that the force is pulling the particle back toward the origin.

The Newtonian equation of motion is evidently

$$M\ddot{x} = -kx \tag{5.20}$$

and this has a solution, given that $x = 0$ at time $t = 0$,

$$x = A \sin \omega t \tag{5.21}$$

where the angular frequency ω is readily verified to be

$$\omega = (k/M)^{1/2} \tag{5.22}$$

which is clearly determined by the force constant k and the mass.

Now take this treatment over to quantum mechanics of vibrational motion of a diatomic molecule; M now becomes the reduced mass μ, and in any textbook on quantum mechanics it is shown that the energy levels are

$$E_v - (v + \tfrac{1}{2})h\nu, \qquad v = 0, 1, 2, \text{ etc.} \tag{5.23}$$

where ν is the classical frequency related to angular frequency ω in equation (5.22) by

$$\nu = \omega/2\pi = (1/2\pi)(k/\mu)^{1/2} \tag{5.24}$$

In this case, the force or "spring" constant k of the diatomic molecule can be regarded as a measure of the bond strength connecting the two vibrating nuclei in the molecule. To show the correlation of k with bond order, Table 5.2 lists data for force constants. Though in many problems in chemical physics at a microscopic

TABLE 5.2
Force Constants k of Some Selected Diatoms[a]

Molecule	Force constant k in SI units $(aJ \cdot Å^{-2})$	Molecule	Force constant k
HCl	5.16	O_2	11.41
HF	9.64	NO	15.48
Cl_2	3.20	CO	18.55
F_2	4.45	N_2	22.41

[a]N.B. (i) The increase of k with bond order is the main point to be noted. (ii) Compare the data in this table with the trend shown in the lower part of Figure 1.2 (units are different in that figure).

level, the atomic units $e = m = \hbar = 1$ are very useful, the most favored unit for force constants in the SI system seems to be aJ Å^{-2} (attojoule per Ångström squared) and these are in fact employed in this table.

5.4. ANHARMONICITY

If one looks at a potential energy curve as in Figure 5.7 (or for C_2 in Figure 5.11), one can see that it can be expanded around the equilibrium bond length, corresponding to the minimum of this curve. It can also be seen that for small displacements from the minimum the curve is parabolic (potential energy is related to force by $F = -\partial V / \partial x$ and for the force in equation (5.20) the potential energy can be taken to be $V = \frac{1}{2}kx^2$). However, such a representation of the potential energy curve, fitted to have the correct curvature at the minimum (this curvature essentially determines the force constant k), soon departs from the correct potential energy curve for larger distances from the minimum and we say that anharmonic corrections are required.

One example of an anharmonic effect of immediate chemical interest is that in both heteronuclear and homonuclear diatomics, the polarizability (measuring the response of a molecule to the application of an electric field) varies during vibrational motion (see Section 1.8.5(b): the so-called vibrational Raman effect). This change in polarizability during vibrational motion may be represented by a Taylor series in the vibrational displacement x: namely the polarizability α, say, can be expressed as

$$\alpha = \alpha_e + (d\alpha/dx)_e x + \tfrac{1}{2}!(d^2\alpha/dx^2)_e x^2 + \cdots \tag{5.25}$$

where the derivatives are to be calculated at equilibrium. The fact that there is a term in equation (5.25) of order x^2, plus higher powers of x not explicitly shown,

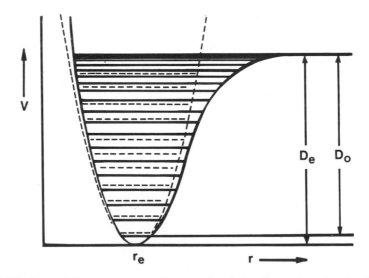

FIGURE 5.7. Potential energy curve and energy levels for diatomic molecule with anharmonic behavior, compared with harmonic approximation shown with dashed lines.

135
*Molecular
Parameters
Determined by
Spectroscopic
Methods*

is what causes one to refer to this as anharmonicity. This particular anharmonicity expressed through equation (5.25) is termed electrical anharmonicity.

The example above, which shows that the electrical behavior of a diatomic molecule is not accurately harmonic, has its counterpart in the mechanical behavior discussed above with reference to the expansion of a potential energy curve as in Figure 5.7 around its minimum. At large values of the internuclear distance r, well beyond the equilibrium distance r_e, we know that the molecule dissociates. Therefore the potential energy function flattens out at $V = D_e$, where D_e is the dissociation energy (cf. Chapter 1) measured relative to the equilibrium potential energy. As dissociation is approached, the force constant tends to zero, which means that the bond gets weaker. The effect is to make the potential energy curve more shallow than that for a harmonic oscillator with the same curvature at the minimum, for $r = r_e$. At small values of r, there is strong mutual repulsion of the two atoms in the molecule which, of course, opposes their approach to one another. This causes the curve to be steeper than for an harmonic oscillator. The deviations found between the realistic potential energy curve of a molecule and the parabolic harmonic oscillator curve are due to the mechanical anharmonicity under discussion.

As already noted, a molecule may exhibit both electrical and mechanical anharmonicity, but it is conventional to define a harmonic oscillator as one which is harmonic in the mechanical sense. It follows that a harmonic oscillator, in that sense, may show electrical anharmonicity. The energy of the harmonic oscillator, and its term values $G(v)$, therefore, defined as F_v/h, are modified from the result (5.23) and can be expressed as a power series in $(v + \frac{1}{2})$, namely,*

$$G(v) = v(v + \tfrac{1}{2}) - vx_e(v + \tfrac{1}{2})^2 + vy_e(v + \tfrac{1}{2})^3 + \cdots \tag{5.26}$$

where v as before is the fundamental vibrational frequency while vx_e, vy_e, etc. are anharmonic constants. The term vx_e has the same sign in all diatomic molecules and in order to make it positive the convention is to put a minus sign in front of it in equation (5.26). However, the succeeding terms vy_e, etc. may be either positive or negative [see Hollas (1982)].

Below, we shall consider the conditions under which transitions between the vibrational levels under discussion are accessible to experimental observation.

5.5. VIBRATIONAL SELECTION RULES

Paralleling to the rotational transitions discussed in Section 5.1, we must now discuss the appropriate vibrational selection rules.

In the process of absorption or emission of radiation involving transitions between a pair of vibrational states of a molecule, the interaction is normally between the molecule and the electric field component of the electromagnetic radiation. For this reason, the selection rules which obtain are termed electric dipole selection rules.

---

*Equation (5.26) represents the modification of vibrational term values due to mechanical anharmonicity.

Transitions between vibrational states involving absorption or emission (as opposed to scattering) occur mainly, but not without exception, in the infrared region of the spectrum, and the selection rules are therefore often called infrared selection rules.

In diatomic molecules, by arguments paralleling those for rotational selection rules in Section 5.1, it can be shown that the anharmonic oscillator selection rule is $\Delta v = \pm 1, \pm 2, \pm 3, \ldots$, with the $\Delta v = \pm 1$ transitions, obeying the harmonic oscillator selection rule, being much the strongest. In addition there must be a change of dipole moment accompanying the vibration, which means allowed transitions in heteronuclear, but not homonuclear, molecules.

Taking the case of polyatomic molecules, the selection rule for each of the normal vibrations (cf. Appendix A5.3) is the same as in a diatomic molecule

$$\Delta v_i = \pm 1, \pm 2, \pm 3, \ldots \tag{5.27}$$

and there must be a change of dipole moment during the vibration.

5.6. VIBRATION–ROTATION SPECTRUM

There is a collection of rotational energy levels associated with all vibrational levels. In pure rotational spectroscopy, what is observed corresponds to transitions between rotational energy levels associated with the same vibrational level.

In vibration–rotation spectroscopy, transitions are measured between the stacks of rotational energy levels associated with two different vibrational levels. The transitions accompany all vibrational transitions but, whereas vibrational transitions can be studied even when the sample is in a condensed phase, the rotational transitions can only be detected in the gaseous phase at low pressure and usually in an absorption process.

When a molecule has both vibrational and rotational energy, the total term values S are represented by the sum of the rotational term values $F_v(J)$, described by equation (5.19), and the vibrational term values $G(v)$ as given in equation (5.26). This sum yields

$$S = G(v) + F_v(J) = v_e(v + \tfrac{1}{2}) - v_e x_e(v + \tfrac{1}{2})^2 + \cdots + B_v J(J + 1)$$
$$- D_v J^2 (J + 1)^2 + \cdots \tag{5.28}$$

Let us next consider the rotational levels associated with two vibrational levels v' (upper) and v'' (lower) between which the $\Delta v = \pm 1$ selection rule allows a vibrational transition. The corresponding rotational selection rule governing transitions between the two stacks of levels is $\Delta J = \pm 1$, resulting in an R branch ($\Delta J = +1$) and a P branch ($\Delta J = -1$). Each transition is labeled by $R(J)$ or $P(J)$, where J is understood to represent J'', the J value of the lower state.* The fact that $\Delta J = 0$ is forbidden means that the pure vibrational transition is not observed. The position at which it would occur is known as the band center. Exceptions to this

---

*For a specific example, the reader may refer to Figure 5.8, p. 158 of Hollas (1982). A Raman spectrum for which $\Delta J = 0, \pm 2$ is also shown there.

137

*Molecular
Parameters
Determined by
Spectroscopic
Methods*

selection rule are molecules which have an electronic angular momentum in the ground electronic state, e.g., nitric oxide. The rotational selection rule for such a molecule is

$$\Delta J = 0, \pm 1 \qquad (5.29)$$

and the $(J' = 0) \rightarrow (J'' = 0)$ transition (the first line of the Q-branch) marks the band center.

More customary is the sort of vibration–rotation band shown in Figure 5.8. This spectrum shows the $v = 1 \rightarrow 0$ transition in $^1H^{35}Cl$ and $^1H^{37}Cl$. The ^{35}Cl and ^{37}Cl isotopes occur naturally with a 3:1 abundance ratio. The band due to $^1H^{37}Cl$ is displaced toward a lower wave number relative to that due to $^1H^{35}Cl$ because of the larger reduced mass.

It is clear from Figure 5.8 that the band for each isotope is fairly symmetrical about the corresponding band center and that there is approximately equal spacing between adjacent R-branch lines and between adjacent P-branch lines, with twice the spacing between the first R- and P-branch lines, $R(0)$ and $P(1)$.

The approximate symmetry of the band is due to the fact that $B_1 \simeq B_0$, which in turn tells us that the vibration–rotation interaction is weak in this example. If it is assumed that $B_1 = B_0 = B$, and that centrifugal distortion can be neglected, then the wave numbers of the R-branch transition $\tilde{v}[R(J)]$, say, are given by

$$\tilde{v}[R(J)] = \tilde{v}_0 + B(J + 1)(J + 2) - BJ(J + 1) = \tilde{v}_0 + 2BJ + 2B \qquad (5.30)$$

where $\tilde{v}_0$ is the wave number of the pure vibrational transition. Similarly, the wave numbers of the P-branch transitions, $\tilde{v}[P(J)]$, are found as

$$\tilde{v}[P(J)] = \tilde{v}_0 + B(J - 1)J - BJ(J + 1) = \tilde{v}_0 - 2BJ \qquad (5.31)$$

It is easily verified from equations (5.30) and (5.31) that the "zero gap", given by $\tilde{v}[R(0)] - \tilde{v}[P(1)]$, is $4B$. Furthermore, the spacing between adjacent R-branch lines is $2B$, this being the same as that between adjacent P-branch lines as well; hence the approximate symmetry of the band, as is observed. For a more detailed interpretation of the experimental results in Figure 5.8, the reader is referred to the book by Hollas (1982).

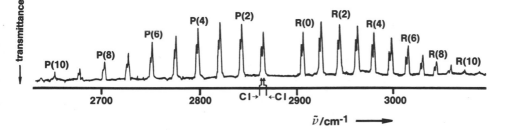

FIGURE 5.8. The $v = 1$–0 infrared spectrum of $^1H^{35}Cl$ and $^1H^{37}Cl$ molecules. P- and R-branch rotational structure is labeled [redrawn from Hollas (1982)].

5.7. ELECTRONIC SPECTROSCOPY

We have discussed the correlation diagram of molecules in Chapter 2 in some detail, so we need not go over this ground again. However, there are a number of other matters meriting discussion in the area of electronic spectroscopy, one such being Rydberg orbitals.

5.7.1. Rydberg Orbitals

Experimentally, as in atoms, it is found that the higher-energy MOs converge to a limit which represents the removal of an electron in an ionization process. The analogy with atoms extends to the MO energies E_n being given by

$$E_n/h = -R/(n - \eta)^2 \tag{5.32}$$

for high values of n. In equation (5.32), R is the Rydberg constant for the molecule concerned, n is an integer, and η is the quantum defect. These term values are quite analogous, say, to the atomic case of the alkali metals.

Any MO resembles an AO when the MO has a large extent in comparison with the size of the molecular core. As an example, an electron in a high-energy orbital in O_2 sees the O_2^+ core as though it were a point. The united atom limit is then appropriate, the resulting field being like that of S^+, and the Rydberg MOs of the O_2 molecule resemble the high-energy AOs of the S atom. It seems clear then that one should be able to correlate the MOs of the O_2 molecule with the AOs of the S atom, and similarly for any other diatomic molecule.

In alkali metal atoms, the quantum defect η of equation (5.32) is a measure of the degree of penetration of the core by an orbital and decreases in the order $s > p > d > f$. This is also the case for the Rydberg MO quantum defect η of equation (5.32), which decreases in the order $s > p > d > f$ corresponding to the united atom orbital.

5.7.2. Classification of Electronic States

When a molecule is undergoing end-over-end rotation, there is evidently an angular momentum associated with that motion. The coupling of this angular momentum to those associated with electronic motion will come up later in this chapter. Here, briefly, we consider the vector picture of momenta due to the orbital and spin motions of the electrons when the molecule is not rotating.

For all multielectron diatomic molecules, the coupling approximation which best describes electronic states is like Russell–Saunders coupling in an atom. The orbital momenta are coupled to give a resultant **L** and the electron spin momenta to give a resultant **S**.

The vector **L** is so strongly coupled to the electrostatic field and, consequently, the precession frequency is so high, that the magnitude of **L** is not defined, which is equivalent to the statement that **L** is not a good quantum number. As already discussed in Chapter 2, only the component $\Lambda\hbar$ along the z-axis is well defined, Λ taking the values

$$\Lambda = 0, 1, 2, \ldots, L \qquad (5.33)$$

In an atom, states with the same magnitudes but different sign of M_L remain degenerate in an electric field. Since Λ is equivalent to $|M_L|$, all states in a diatomic molecule with $\Lambda > 0$ are also doubly degenerate. Classically, this degeneracy can be thought of as being due to electrons with nonzero orbital angular momenta, orbiting clockwise or anticlockwise around the internuclear axis, the energy being equal in the two cases.

We recall here, from Chapter 2, that analogous to the labeling of one-electron orbitals, electronic states are denoted by Σ, Π, Δ, Φ, Γ, ... corresponding to $\Lambda = 0, 1, 2, 3, 4, \ldots$. The dual interpretation of Σ, Π, etc., as symmetry species (of a point group) as well as implying orbital angular momentum quantum numbers, is analogous to the interpretation of S, P, etc. in atoms [see Hollas (1982)].

In Figure 5.9, it is shown that the component of **S** along the internuclear axis is a good quantum number, denoted by $\Sigma\hbar$. The coupling of **S** to the internuclear axis is due, in fact, to the internal magnetic field along the axis owing to the orbital motion of the electrons. For this reason, Σ is analogous to M_s in an atom and can take the values

$$\Sigma - S, S - 1, \quad -S \qquad (5.34)$$

As for atoms, as in, say, Appendix 5.6, there are $(2S + 1)$ components corresponding to the multiplicity of the state with given S; such a situation is depicted again by the notation $^3\Pi$, for instance, for $\Lambda = 1$ and $S = 1$.

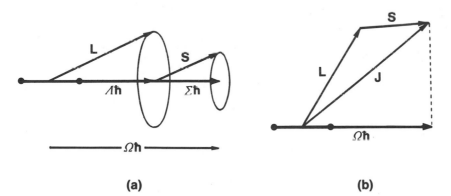

(a)   **(b)**

FIGURE 5.9. (a) Uncoupling of orbital angular momentum vector **L** and spin angular momentum vector **S** in a diatomic molecule in which spin–orbit coupling is weak, as in Hund's case (a) (discussed in Section 5.7). (b) The same in the case when spin–orbit coupling is strong, as in Hund's case (c), **L** and **S** being sufficiently strongly coupled that the field of nuclei can no longer decouple them.

The component of the total angular momentum of the electrons along the internuclear axis is $\Omega\hbar$, indicated in Figure 5.9, where the quantum number Ω is given by

$$\Omega = |\Lambda + \Sigma| \qquad (5.35)$$

The value of $\Lambda + \Sigma$, not $|\Lambda + \Sigma|$, is attached to the term symbol as a subscript. For example, for a $^4\Pi$ term, $\Lambda = 1$, $S = \frac{3}{2}$, $\Sigma = \frac{3}{2}$, $\frac{1}{2}$, $-\frac{1}{2}$, $-\frac{3}{2}$ but the components are labeled $^4\Pi_{5/2}$, $^4\Pi_{3/2}$, $^4\Pi_{1/2}$, $^4\Pi_{-1/2}$, even though $\Omega = \frac{5}{2}, \frac{3}{2}, \frac{1}{2}, \frac{1}{2}$.

Spin–orbit interaction, discussed briefly in Appendix A5.6, splits the component of a multiplet so that the energy level before interaction is shifted by the interaction energy E', where

$$E' = A\Lambda\Sigma \qquad (5.36)$$

in which A is the spin–orbit coupling constant (cf. Appendix A5.6). For a given Λ, which is within a given multiplet, the splitting between adjacent components is constant, unlike the case of atoms. The sign of A determines whether the multiplet is normal or inverted, i.e., whether the energy increases or decreases, respectively, with Ω.

For the case of Σ states, there is no orbital angular momentum and therefore no internal magnetic field to couple S to the axis. For this reason, when spin–spin coupling is neglected, the quantum number Σ is not defined and a Σ state, whatever its multiplicity, has only one component. This situation no longer obtains when spin–spin coupling is included.

Referring again to Figure 5.9, we note that the situation depicted there, in which L and S are completely uncoupled, is the case most frequently met and is an example of what is termed Hund's case (a) coupling. This case is discussed in general terms when the rotational angular momentum is incorporated by Hollas (1982).

As is true for all assumed coupling of angular momenta, Hund's case (a) represents an approximation, though one that is frequently valid. Another extreme coupling approximation is Hund's case (c) illustrated in Figure 5.9b. Here the spin–orbit coupling is sufficiently large that L and S are not uncoupled and Λ is no longer a good quantum number. Nevertheless, the Σ, Π, Δ, . . . labels for states are still employed. This case (c) occurs, for example, if a molecule has at least one highly charged nucleus [considered further by Hollas (1982)].

5.7.3. Electric Dipole Selection Rules for Transitions with $\Delta S = 0$

For Hund's case (a) coupling treated above, the quantum numbers Λ, S and Σ, the last only in states other than those with $\Lambda = 0$, are insufficient to specify selection rules. In addition, two symmetry properties of ψ_e have to be used for homonuclear and one for heteronuclear diatomic molecules.

In the case of homonuclear diatoms, ψ_e may be symmetric or antisymmetric to inversion, and this is shown by a subscript g or u, respectively (cf. Chapter 2).

The electronic selection rules for case (a) coupling, assuming electric dipole transitions, are:

141
Molecular
Parameters
Determined by
Spectroscopic
Methods

1. $\Delta\Lambda = 0, \pm1$: for example $\Sigma \to \Sigma$, $\Pi \to \Sigma$, $\Delta \to \Pi$ transitions are allowed but not $\Delta \to \Sigma$ or $\Phi \to \Pi$.
2. $\Delta S = 0$ in molecules with no highly charged nuclei, analogous to the atomic case: For instance, triplet–singlet transitions are highly forbidden in H_2, but in CO the $a\,^3\Pi \to X\,^1\Sigma^+$ transition is (weakly) observed.
3. $\Delta\Sigma = 0$: this selection rule and the following one are concerned with transitions between multiplet components.
4. $\Delta\Omega = 0, \pm1$.
5. $+ \nleftrightarrow -$, $+ \leftrightarrow +$, $- \leftrightarrow -$: this is relevant only for $\Sigma \to \Sigma$ transitions. It follows that only $\Sigma^+ \to \Sigma^+$ and $\Sigma^- \to \Sigma^-$ transitions are allowed.
6. $g \leftrightarrow u$, $g \nleftrightarrow g$, $u \nleftrightarrow u$: for example, a $\Sigma_g^+ \to \Sigma_g^+$ transition is forbidden.

5.7.4. Transitions with $\Delta S \neq 0$

If one or both of the nuclei of diatomic molecules have high charges, the selection rule $\Delta S = 0$ may break down. Many transitions with $\Delta S = \pm1$ have been observed and these are mainly between singlet and triplet states, as in the $a\,^3\Pi \to X\,^1\Sigma^+$ transition of CO referred to above. A few other types of $\Delta S = \pm1$ transitions such as the $a\,^4\Sigma^- \to X\,^2\Pi$ transition of the short-lived molecule SiF have been observed, but these will not be treated further here. Nor will transitions with $|\Delta S| > 1$ be dealt with: for example, between a singlet and a quintet state.

The $\Delta S = 0$ selection rule is broken down by spin–orbit interaction. For a molecule which, in the absence of spin–orbit coupling, has manifold singlet and triplet states, shown in Figure 5.10, the presence of such coupling leads to a mixing

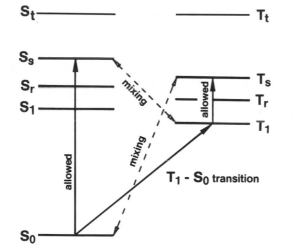

FIGURE 5.10. Processes by which the triplet–singlet transition $T_1 \to S_0$ can gain intensity by spin–orbit coupling, which has the effect of mixing singlet and triplet states [redrawn from Hollas (1982)].

of singlet and triplet states, the admixture being the greater the stronger the coupling.

Let us consider the transition from the ground singlet state S_0 to the first excited triplet state T_1. This $T_1 \rightarrow S_0$ transition may gain intensity either by S_0 being mixed with triplet states by spin–orbit interaction or by T_1 being mixed with singlet states, or both.

While in principle S_0 may mix with several triplet states, in practice, because of restrictions of symmetry and energy separation, it usually mixes preferentially with only one, say, T_s. Similarly, T_1 may mix with several singlet states but usually it mixes preferentially with one, say, S_s. If, as in Figure 5.10, transitions $S_s \rightarrow S_0$ and $T_s \rightarrow T_1$ are allowed, the $T_1 \rightarrow S_0$ transition may gain intensity by intensity-stealing from either $S_s \rightarrow S_0$ or $T_s \rightarrow T_1$. It might steal intensity from both, but one mechanism normally dominates and this is often the one resulting from the S_s and T_1 type of mixing.

The concept of intensity-stealing by one transition from another has arisen solely because of the initial neglect of spin–orbit coupling, and in that sense is a little artificial.

5.8. EXCITED-STATE POTENTIAL ENERGY CURVES

The form of the potential energy curve of a diatomic molecule in its electronic ground state was discussed in some detail in Chapter 2. We now note that the total term value S for a molecule in an electronic state with electronic term value T, corresponding to the equilibrium configuration, and with vibrational and rotational term values of, say, $G(v)$ and $F(J)$ is given by

$$S = T + G(v) + F(J) \tag{5.37}$$

In each excited electronic state of a diatomic molecule there is a corresponding potential energy curve. Most of these are qualitatively similar in appearance to that for the ground electronic state. As an example, Figure 5.11 shows potential energy curves for the ground state $X\,^1\Sigma_g^+$ and several excited states of the short-lived molecule C_2.

It is of interest to note here that the information contained in Figure 5.11 has been obtained using various experimental techniques to observe absorption and emission spectra. The mixtures of techniques for observing the spectra, including a high-temperature furnace, flames, arcs, discharges, and, for a short-lived species, flash photolysis, is in no way untypical of molecular spectroscopy.

Apart from the laboratory-based experiments, bands of C_2 are important in astrophysical problems. The so-called Swan bands in the spectral region 785–340 nm have been observed in the emission spectra of comets and also in the absorption spectra of stellar atmospheres, including that of the sun.

The transitions involved in this example are allowed by the electric dipole selection rules of Hund's case (a) detailed above. The products of dissociation are indicated on the right of Figure 5.11. Most of the potential energy curves of C_2,

143
*Molecular
Parameters
Determined by
Spectroscopic
Methods*

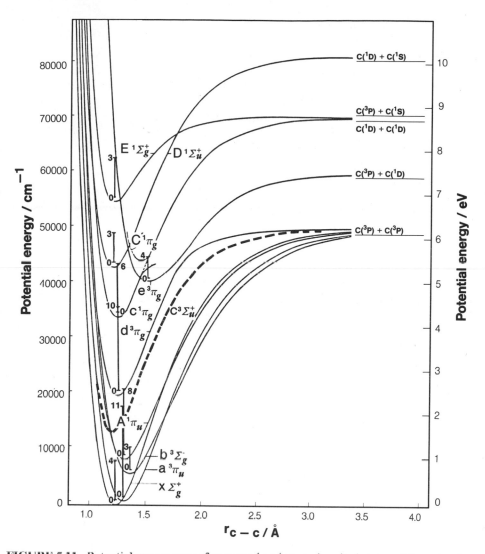

FIGURE 5.11. Potential energy curve for ground and several excited states of C_2 molecule [reproduced with permission from Ballik and Ramsay (1963)].

and of other diatomic molecules, have vibrational energy levels associated with them which converge smoothly toward the dissociation limit. The vibrational term values $G(v)$ relevant to any electronic state can be expressed by

$$G(v) = v(v + \tfrac{1}{2}) - vx_e(v + \tfrac{1}{2})^2 + vy_e(v + \tfrac{1}{2})^3 + \cdots \tag{5.38}$$

The vibrational wave number v, the anharmonic constants vx_e, vy_e, ..., and the equilibrium internuclear distance r_e are all dependent on the electronic structure and therefore are constant only for a particular electronic state.

5.8.1. Progressions and Sequences

Figure 5.12 shows sets of vibrational levels associated with two electronic states between which we shall assume that an electronic transition is allowed. The vibrational levels of the upper and lower states are labeled by quantum numbers v' and v'', respectively. We shall be dealing with absorption as well as emission processes and it will be assumed, unless otherwise stated, that the lower state is the ground state.

In electronic spectra there is no restriction on the values Δv can take but, as can be seen from Appendix A5.4, the Franck–Condon principle imposes restrictions on the intensities of the transitions.

5.8.2. Vibronic Transitions

Vibrational transitions accompanying an electronic transition are termed vibronic transitions. These vibronic transitions, with their accompanying rotational or, strictly, rovibronic transitions, give rise to bands in the spectrum, and the set of bands associated with a single electronic transition is referred to as an electronic band system. Vibronic transitions are conveniently divided into progressions and sequences.

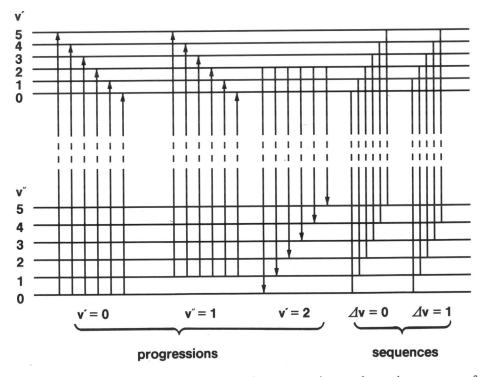

FIGURE 5.12. Vibrational progressions and sequences in an electronic spectrum of diatomic molecules.

A progression, as Figure 5.12 shows, involves a series of vibronic transitions with a common lower or upper level. For example, the $v'' = 0$ progression members all have the $v'' = 0$ level in common.

Apart from the necessity for Franck–Condon intensities of vibronic transitions to be appreciable, it is essential for the initial state of a transition to be sufficiently highly populated for a transition to be observed. Under equilibrium conditions, the population N of the v''th level is related to that of the $v'' = 0$ level by

$$N_{v''}/N_0 = \exp\{-[G(v'') - G(0)]h/k_B T\} \tag{5.39}$$

according to the Boltzmann law.

Owing to the relatively high population of the $v'' = 0$ level, the $v'' = 0$ progression is likely to be prominent in the absorption spectrum. In emission, the relative populations of the v' levels depend on the method used for excitation. In a low-pressure discharge in which there are not many collisions to provide a channel for vibrational deactivation, the populations may be somewhat random. On the other hand, higher pressure may result in most of the molecules being in the $v' = 0$ state with the $v' = 0$ progression being prominent.

A group of transitions with the same value of Δv is referred to as a sequence. Because of the population requirements, long sequences are mostly observed in emission. For instance, sequences of five or six members are observed in the $C\,^3\Pi_u \to B\,^3\Pi_g$ band system in the emission spectrum in the visible and the near-ultraviolet from a low-pressure discharge in N_2 gas. The vibrational wave number is high (2035.1 cm^{-1}) in the C state, and equilibrium population of the vibrational levels is not achieved before emission.

It is clear from Figure 5.12 that progressions and sequences are not mutually exclusive. Each member of a sequence is also a member of two progressions. But the distinction is useful because of the nature of typical patterns of bands found in a band system. Members of a progression are generally widely spaced with approximate separations of v'_e in absorption and v''_e in emission. On the other hand, sequence members are closely spaced, with approximate separations of $v'_e - v''_e$.

The normal notation for indicating a vibronic transition between an upper and lower level with vibrational quantum numbers v' and v'', respectively, is $v' - v''$, consistent with the general spectroscopic convention. Thus the electronic transition is labeled $0 - 0$.

5.9. DETERMINATION OF IONIZATION POTENTIALS OF POLYATOMIC MOLECULES

Prior to the invention of photoelectron spectroscopy, data on valence electron ionization potentials of polyatomic molecules were hard to come by. We now consider this technique, but shall have to be content with discussing in some detail the results for one polyatomic molecule. We have chosen benzene, the HeI UPS spectrum taken from the work of Karlsson et al. (1976) being reproduced in

Figure 5.13. This spectrum, in fact, poses interpretational problems typical of large molecules. These concern the difficulty of identifying the onset of a band system where there is overlapping with neighboring systems and, generally, a lack of resolved vibrational structure.

One needs, usually, a combination of theory and experiment to unravel such a spectrum successfully. For benzene, we shall obtain in Chapter 6 the π-electron MOs by the Hückel method using only the $2p_z$ AOs on the C atoms. In the ground state, it is then found that four of the π-electrons are in the outer $1e_{1g}$ and two in the $1a_{2u}$ MO. More precisely, an SCF calculation following the method described in the last chapter has been carried out by Schulman and Moskowitz (1967) using as a basis set the $1s$, $2s$, and $2p$ AOs on C and the $1s$ AO on H. This leads to the results collected in Table 5.3, where a comparison is made with the ionization energies obtained from the UPS spectra. Koopmans' theorem, discussed in Chapter 4 (see also Appendix A6.2), has been used in making the comparison. The calculation does not take account of electron correlation, and this is at least part of the reason that the calculated values are considerably higher than the observed ionization energies. However, the ordering of the orbital energies is more reliable than the quantitative values. The calculation shows clearly that the UPS spectrum in the 11.5–12.5 eV region consists of two band systems, and the 14–16 eV region, of three. By including electron correlation, following von Niessen *et al.* (1976), the disagreement with experiment is significantly reduced, especially for the low ionization energies, as Table 5.3 makes plain.

A feature of the calculations which will be taken up briefly in Chapter 6 is that the $3e_{2g}$ orbital, which is of σ-type, has a somewhat lower ionization energy than the $1a_{2u}$ π-type orbital. This must cast some doubt on MO calculations which take only the π-electrons into account.

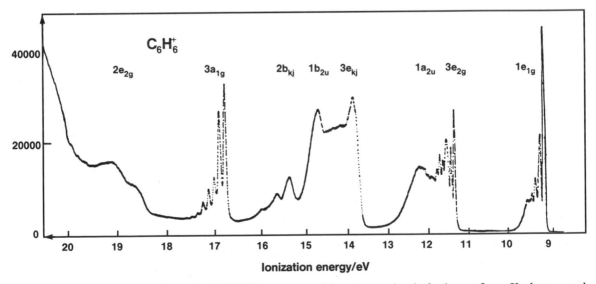

FIGURE 5.13. HeI UPS spectrum of benzene molecule [redrawn from Karlsson *et al.* (1976)].

147

*Molecular
Parameters
Determined by
Spectroscopic
Methods*

TABLE 5.3
Comparison of Experimental and Calculated Ionization Potentials
of Benzene Molecule[a]

Orbital	σ or type	Ionization potential (eV)		
		Experiment	Theory (MO)	Theory with electron–electron correlation
$1e_{1g}$	$\pi 2p$	9.3	10.1	9.31
$3e_{2g}$	$\sigma 2p$	11.4	14.3	13.5
$1a_{2u}$	$\pi 2p$	12.1	14.6	13.8
$3e_{1u}$	$\sigma 2p$	13.8	16.9	16.1
$1b_{2u}$	$\sigma 2p$	14.7	17.8	17.0
$2b_{1u}$	$\sigma 2s$	15.4	18.0	17.4
$3a_{1g}$	$\sigma 2p$	16.9	20.1	19.5
$2e_{2g}$	$\sigma 2s$	19.2	23.0	22.4
$2e_{1u}$	$\sigma 2s$	22.5	28.2	27.7
$2a_{1g}$	$\sigma 2s$	25.9	31.8	31.4

[a]N.B. Vertical ionization potentials are recorded in the third column labeled experiment.

5.10. NUCLEAR MAGNETIC RESONANCE SPECTRA

The nuclear magnetic resonance spectrum arises from transitions between different energy states of atomic nuclei. We will now consider the proton alone, both for convenience and, more importantly, because the proton spectrum is by far the most commonly studied, but the procedure outlined can be applied to other nuclei provided the appropriate value for the nuclear spin is employed. In part, the treatment presented here follows that of Wiberg (1964).

Without any external magnetic field, a proton will have a single definite energy state. On application of a magnetic field, it can assume two states: one in which its angular momentum vector lies in the direction of the applied field and the other in which it points opposite to the field. In the case of the proton, and other nuclei having a spin quantum number of $\frac{1}{2}$, the magnetic quantum number may be $+\frac{1}{2}$ or $-\frac{1}{2}$. Similarly, with nuclei with spin unity, the projection on the z-axis may be +1, 0, and −1.

We now treat the case of a single proton in an applied magnetic field along the z-axis. The spin wave function for the spin pointing along the field will be denoted by α and that for the antiparallel case by β, with α and β being as usual normalized spin functions that are orthogonal to one another. The energy of either state is given by the expectation value of the Hamiltonian $\mathcal{H}$ with respect to the appropriate wave function. In our case, the Hamiltonian has the form

$$\mathcal{H} = \frac{1}{2\pi} \sum_i \gamma H_i I_z(i) \tag{5.40}$$

where γ is the magnetogyric ratio of a proton, H_i is the magnetic field at the ith proton and I_z is the component of the spin operator. I_z operating on α gives $\frac{1}{2}\alpha$ while on β it yields $-\frac{1}{2}\beta$.

It is now straightforward to take the expectation value $\langle \mathscr{H} \rangle$ of $\mathscr{H}$ with respect to α and β, and we find

$$\langle \mathscr{H} \rangle_\alpha = (1/4\pi)\gamma H \equiv \tfrac{1}{2}\nu \qquad (5.41)$$

and

$$\langle \mathscr{H} \rangle_\beta = -(1/4\pi)\gamma H \equiv -\tfrac{1}{2}\nu \qquad (5.42)$$

where we have introduced the frequency ν such that it measures the energy separation between the two levels.

To cause transitions between the two states in NMR experiments one supplies the appropriate energy by a radio frequency generator via a coil mounted perpendicular to the applied field in which the sample under study is placed. Evidently rf absorption will only occur when the input frequency equals ν.

The argument for the case of two protons, in which the interaction between them is neglected at first, is set out briefly below. The possible wave functions are evidently

$$\left.\begin{aligned}
\psi_1 &= \alpha(1)\alpha(2) \equiv \alpha\alpha \\[4pt]
\psi_2 &= \alpha(1)\beta(2) \equiv \alpha\beta \\[4pt]
\psi_3 &= \beta(1)\alpha(2) \equiv \beta\alpha \\[4pt]
\psi_4 &= \beta(1)\beta(2) \equiv \beta\beta
\end{aligned}\right\} \qquad (5.43)$$

the corresponding energies being

$$\left.\begin{aligned}
E_1 &= \tfrac{1}{2}(\nu_1 + \nu_2) \\[4pt]
E_2 &= \tfrac{1}{2}(\nu_1 - \nu_2) \\[4pt]
E_3 &= \tfrac{1}{2}(-\nu_1 + \nu_2) \\[4pt]
E_4 &= \tfrac{1}{2}(-\nu_1 - \nu_2)
\end{aligned}\right\} \qquad (5.44)$$

where $\nu_i = (1/2\pi)\gamma H_i$.

The selection rules for transitions to occur [Whiffen (1966)] are that there must be a net change of 1 in spin.* The allowed transitions are between the states listed below, and the allowed frequencies are also shown:

$$2 \rightarrow 1 \quad \Delta E = \nu_2 \qquad\qquad 4 \rightarrow 2 \quad \Delta E = \nu_1$$

$$3 \rightarrow 1 \quad \Delta E = \nu_1 \qquad\qquad 4 \rightarrow 3 \quad \Delta E = \nu_2$$

---

*Intrinsic nuclear angular momenta are quantized, the spin quantum number I being an integer or half-integer. The angular momentum vector can take up $(2I+1)$ directions in space, with resolved momenta m_I along a specified direction of values $I, I-1 \ldots -I$.

149

*Molecular
Parameters
Determined by
Spectroscopic
Methods*

Evidently, two resonances should occur, at frequencies v_1 and v_2. Now we must discuss what determines the difference between v_1 and v_2.

5.10.1. Chemical Shifts

From their definitions, $v_1 = v_2$ if $H_1 = H_2$. If one were dealing with an assembly of bare protons, then the field would simply be the applied field, and would be the same at each proton, leading to the equality of v_1 and v_2. However, in practice the protons in molecules are magnetically shielded by the electrons and thus the local field at a proton is not the same as the applied field. Further, this shielding varies from one proton to another if these are differently situated in the molecules. The difference between v_1 and v_2 is therefore referred to as a chemical shift.

The internal field caused by the electronic motion is oriented in the opposite direction to the external field H_0, say, and is proportional to this field. Therefore the strength of the magnetic field at the nucleus, H_n, say, is

$$H_n = H_0(1 - \sigma_n) \tag{5.45}$$

where σ_n is the shielding constant of the nucleus. For the transition frequency, the relationship

$$v = (\gamma/2\pi)H_0(1 - \sigma_n) \tag{5.46}$$

is the appropriate one, with $\gamma = g_N \beta_N / \hbar$ where γ is the gyromagnetic ratio, g_N the nuclear g factor, and β_N the nuclear magneton [Whiffen (1966)]. Since v and H_0 cannot be determined independently with sufficient accuracy, absolute values of σ_n cannot be obtained. Thus, only relative measurements can be performed and for this purpose the chemical shift δ_n, say, is defined as

$$\delta_n = [(H_m - H_r)/H_r] \times 10^6 \text{ ppm} \tag{5.47}$$

where H_m (H_r) is the induction of the magnetic field corresponding to the peak of the measured (reference) proton (see Figure 5.14). The differences in the magnetic

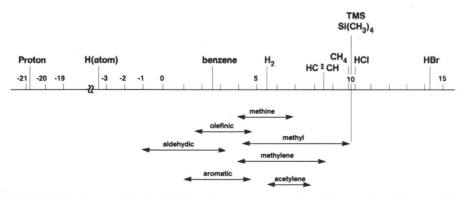

FIGURE 5.14. Proton chemical shifts of many compounds relative to the arbitrary origin (reference). For convenience, the arbitrary reference $Si(CH_3)_4$ is set at $[Si(CH_3)_4] = 10$ ppm (i.e., 10.0×10^{-6}) [redrawn from Flygare (1978), p. 181].

inductions H_m and H_r are very small, being of magnitude 10^{-6} and thus the definition (5.47) is appropriate. Referring to Figure 5.14, we see that the proton chemical shifts for several molecules and molecular systems are displayed relative to the standard, tetramethylsilane, $Si(CH_3)_4$, i.e., TMS. The resonances for protons in most organic compounds correspond to magnetic fields lower than TMS; hence its selection as the reference molecule. For convenience the proton magnetic shielding in TMS is set at $\sigma_H [Si(CH_3)_4] = 10.0 \times 10^{-6}$ or 10 parts per million (ppm).

5.10.2. Chemical Applications

As an instructive example of the chemical use of NMR spectroscopy, we show in Figure 5.15 the NMR spectrum of annulene. Here, the NMR spectrum has two groups of bands with an intensity ratio of 2:1. The first corresponds to the external and the second to the internal protons. The internal protons are strongly shielded, leading to the very considerable difference between the two groups. This strong shielding is attributed to the diamagnetic effect of the π-electrons circulating in response to the external field. The aromaticity of the annulene molecule can be estimated from the position of the peripheral proton signal.

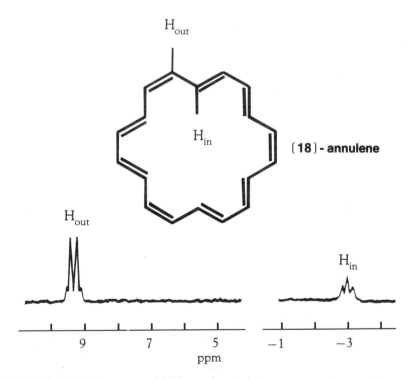

FIGURE 5.15. NMR Spectrum of [18]-annulene. The low temperature spectrum reveals that the "inside" (H_{in}) protons resonate at -3 ppm; and the "outside" (H_{out}) protons resonate at 9 ppm. This large chemical shift difference arises from the ring current associated with the 18π-electron system. When warmed the molecule turns inside out rapidly and repeatedly, which averages the chemical shift.

151

Molecular
Parameters
Determined by
Spectroscopic
Methods

5.10.3. Interaction Constants: Spin–Spin Coupling

One complication which arises in NMR is the interaction between the magnetic moments of active nuclei, but it also has the effect of enriching the phenomena, and makes the technique especially valuable when studying small differences in the structure of substances. In analyzing the spin–spin coupling peaks, it is necessary to start from the fact that the splitting of the peak of the protons under discussion depends on the number of equivalent protons in neighboring positions; n protons bonded in the position next to the proton being studied lead to splitting of the line of this proton into $(n + 1)$ lines.

We consider the NMR spectrum of acetaldehyde as an example. Figure 5.16 shows the spectrum, with, low (a), and high (b) resolving power.* It is clear from the diagram that H atoms neighboring the H atom in the CHO group (i.e., the three Hs of the methyl group) lead to splitting of the original peak of the H atom in the CHO group into a quadruplet; the H atom of the aldehyde group, in turn, causes splitting of the CH_3 band into a doublet. The chemical shift of the multiplet is given by the position of its center; the distance of the multiplet lines measures the interaction constant J, say, whose value does not depend on the strength of the external field. The lines corresponding to different multiplets can be distinguished by recording the spectrum at two different strengths of the magnetic field.

A group of equivalent protons creates a magnetic field proportional to $\sum_i I_{z,i}$, where $I_{z,i}$ is the zth component of the spin angular momentum of the ith nucleus. The proton spin orientation indicated in Figure 5.17 for the proton in the CHO group, and is realistic for the protons of the CH_3 group. It is clear that there are two fields by which the proton of the CHO group can influence those in the CH_3 group, these two fields corresponding to two kinds of possible orientations of the nuclear spin. In the substance under investigation, these two forms are present in a ratio of 1:1, which leads to the splitting of the methyl peak into a doublet, both peaks of the doublet, of course, having the same intensity. The formation of the quadruplet and the ratio of the intensities of its peaks, 1:3:3:1, can be understood in a similar fashion.

5.10.4. Comments on Analyses of NMR Spectra

Spin–spin coupling, as already noted, adds richness to NMR spectra, but makes the spectra difficult to interpret under certain circumstances, with the degree of difficulty depending on the ratio of the chemical shift δ and the interaction constant J. If the chemical shift is large compared with the interaction constant, then the (first-order) spectrum can be interpreted. This is true when the inequality

$$\delta/J \geq 10 \tag{5.48}$$

is fulfilled. When the reverse inequality holds, leading to high-order spectra, the situation is much less favorable for interpretation.

---

*A standard reference here is J. A. Pople *et al.* (1959).

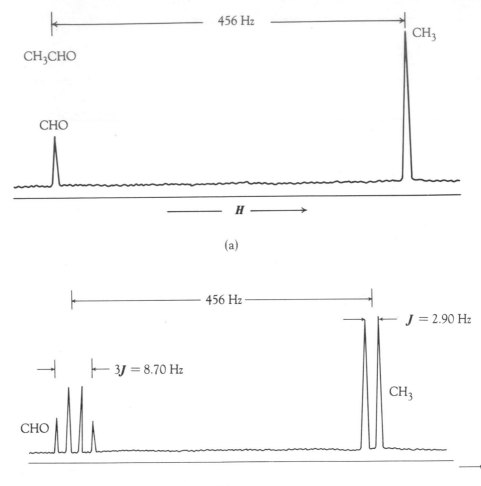

FIGURE 5.16. NMR spectrum of acetaldehyde for low (a) and high (b) resolution. The high-resolution spectrum provides a classic example of the magnetic interaction that exists between chemically nonequivalent neighboring nuclear spins. Note that the aldehydic proton (CHO) is split into four peaks with intensity ratios of 1:3:3:1, whereas the methyl proton is split into a doublet of equal intensity [see Figure 5.17 from Dixon (1965)].

5.10.5. Rules for Interpreting First-Order Spectra

Rules which apply to the interpretation of first-order spectra will be illustrated here with reference to the system $A_m X_n$ which has $m + n$ interacting nuclei [Zahradník and Polák (1980)]. All m nuclei of atom A are magnetically equivalent as are the n nuclei of atom X.

It is important, first of all, that interaction among magnetically equivalent nuclei is not reflected in the NMR spectra. The multiplicity of the peaks of the

153

*Molecular
Parameters
Determined by
Spectroscopic
Methods*

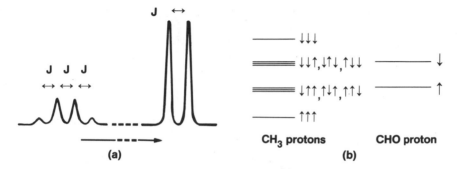

FIGURE 5.17. NMR spectrum of acetaldehyde showing spin orientations. (a) High resolution as indicated in Figure 5.16. (b) The spin orientation of the CH_3 protons and the CHO protons. The three protons of the CH_3 group will be aligned by the field to give $\sum M_I = \frac{3}{2}, \frac{1}{2}, -\frac{1}{2},$ or $-\frac{3}{2}$ with statistical weights of $1:3:3:1$, where M is the quantum number and I is the spin. The CHO proton has $M = \frac{1}{2}$ or $-\frac{1}{2}$ with equal weights. The CH_3 proton resonance is therefore split by the CHO proton into two lines of equal intensity, and the CHO proton resonance is split by the three CH_3 protons into four lines of relative intensity $1:3:3:1$ [redrawn from Dixon (1965)].

nuclei of the A atoms depends on both the number of nuclei of atoms X, of which there are n, and on the nuclear spin quantum number I_A, this multiplicity being given by $2nI_A + 1$. Therefore, assuming that the nuclei of H^1, C^{13} or F^{19} are responsible for the splitting, each having spin $\frac{1}{2}$, the number of lines in the multiplet equals $n + 1$. If the nuclei of atoms A interact with another group of p nuclei Y_p, say, then the multiplicity of the peak of nuclei A is given by the product $(2nI_X + 1)(2pI_Y + 1)$.

The multiplet corresponding to the A atoms is symmetrical and is formed by a series of equidistant peaks, the intensity of which is given by the coefficients of the binomial series, provided $I_X = \frac{1}{2}$.

This simple situation is complicated when the multiplets lie relatively close together, i.e., when the first-order spectrum changes into one of higher order.

5.10.6. The Relation between Chemical Shift and Electronegativity

One approach to obtaining changes in electron density is through the electronegativity difference between two atoms discussed in Chapter 1. Therefore a comparison of the chemical shift for a proton with the electronegativity difference between the C to which it is attached and the group that is bonded to the C might give a linear relation, as proposed by Bothner-By and Naar-Colin (1958). Figure 5.18 shows that this correlation indeed occurs for a series of methyl derivatives. However, there is a whole body of data which demonstrates that the electron density is not the only factor involved, as the difference in slope between the two lines in Figure 5.18 is but one piece of evidence. Bothner-By and Naar-Colin have studied this effect.

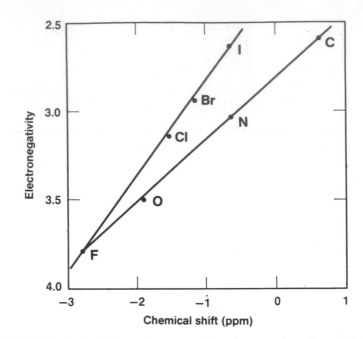

FIGURE 5.18. Chemical shift *vs.* electronegativity for a series of methyl derivatives [redrawn from Bothner-By and Naar-Colin (1958)].

5.11. ELECTRON SPIN RESONANCE AND π-ELECTRON DENSITIES: HYPERFINE STRUCTURE

We shall now discuss experimental information that can be obtained on π-electron densities. In principle, we can obtain the total π-electron density ρ_π, say, from X-ray scattering. We can also write ρ_π as the sum of the densities for $\uparrow$ and $\downarrow$ spins, namely, at each point in space:

$$\rho_\pi = \rho_\pi^\uparrow + \rho_\pi^\downarrow \tag{5.49}$$

The spin density in a crystal is accessible from neutron scattering experiments. Below, we shall be concerned with a limited study about unpaired spins from electron spin resonance (ESR) experiments. It should be noted that while many

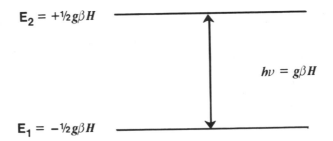

FIGURE 5.19. Energy change for ESR absorption or emission.

155

*Molecular
Parameters
Determined by
Spectroscopic
Methods*

of the principles of NMR apply also to ESR, the most important difference derives from the fact that in molecules electrons regularly are in pairs with opposed spins, in accord with the Pauli principle. Only when an unpaired electron is present can a transition be induced by microwave radiation.

The hyperfine structure of ESR spectra results from the interaction between the magnetic moment due to an unpaired electron spin and the magnetic moments of certain nuclei (usually protons) in the molecule.

It is possible to reorient the spin of an electron only if the electron is unpaired. Hence ESR deals exclusively with ions and molecules with odd electrons; closed-shell molecular species do not give an ESR signal. ESR transitions are observed in the microwave region of the spectrum.

In ESR spectroscopy microwave radiation of constant frequency v is used, and the external field strength H is varied, with the high-frequency magnetic field perpendicular to H. Absorption or emission takes place only when the energy change hv has a definite value. Figure 5.19 depicts this change in energy, where g

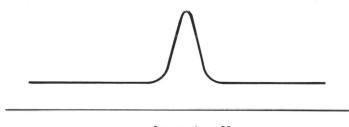

Increasing H ⟶

FIGURE 5.20. Peak *vs.* field strength in an ESR spectrum. [The derivative curve is often plotted (see Figures 5.21 and 5.23).]

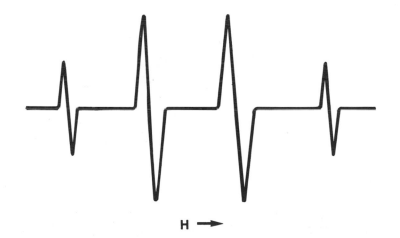

H ⟶

FIGURE 5.21. ESR spectrum of CH$_3$ radical (schematic derivative curve). The electron interacts with the three equivalent protons of CH$_3$ and so four peaks are seen with intensities in the ratio $1:3:3:1$. The separation of the peaks is a measure of the splitting constant a.

is the spectroscopic splitting factor and is a constant (2.0023 for a free electron) and β is the Bohr magneton. A peak is observed as in Figure 5.20 at the value of the field strength where the transition between the two spin states takes place.

If an electron interacts with just one nucleus of spin-$\frac{1}{2}$, its ESR signal consists of two peaks of equal intensity which correspond to the two possible orientations of the nuclear spin. An electron which interacts with two spin $\frac{1}{2}$ nuclei yields an ESR signal of three peaks whose intensities are in the ratio 1:2:1, i.e., those would correspond to the nuclear orientations ↑↑, ↑↓, and ↓↑, and ↓↓.

In the case of the methyl radical (CH$_3^\cdot$), the unpaired electron interacts with three equivalent protons and we see four peaks as depicted in Figure 5.21 with intensities in the ratio 1:3:3:1. We now turn to π-systems of ring hydrocarbons.

The amount of spin polarization at a given H will depend on the time the unpaired π-electron spends on the C atom to which that particular H is bonded.

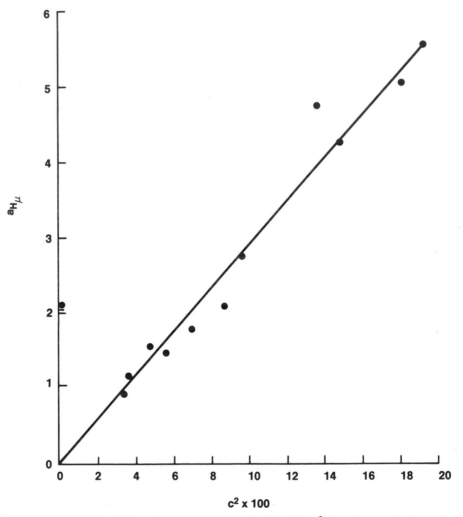

FIGURE 5.22. ESR splitting constants a_{H_μ} *vs.* spin densities c^2 for alternate hydrocarbon radicals [redrawn from Lowe (1978)].

157

*Molecular
Parameters
Determined by
Spectroscopic
Methods*

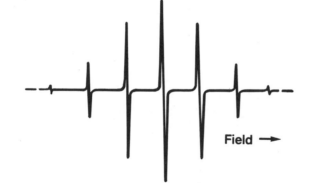

Field →

FIGURE 5.23. The ESR spectrum of $C_6H_6^-$ (schematic derivative curve). The unpaired electron interacts with six equivalent protons and yields a spectrum of seven lines. The intensity ratio is $1:6:15:20:15:6:1$, each line arising from a molecule with different numbers of $+\frac{1}{2}$ and $-\frac{1}{2}$ proton spins [see also Problem 5.11].

McConnell therefore proposed a relationship between the distribution of π-spin density in a radical and the hyperfine coupling constants characterizing its ESR spectrum. He made the simplest assumption possible, namely that the hyperfine splitting constant, a_{H_μ}, say, for a proton bonded directly to the μth C, is proportional to the net spin density ρ_μ on that C:

$$a_{H_\mu} = Q\rho_\mu \qquad (5.50)$$

We shall see in Chapter 6 that the Hückel theory predicts that the spin density* at the μth C due to an unpaired electron in the mth MO is simply $|c_{\mu m}|^2$. Figure 5.22 shows the observed hyperfine splitting constants a_{H_μ} plotted as a function of c^2 for fused ring alternate hydrocarbon radical anions, and there is evidently a correlation. The proportionality constant Q, given by the slope of the line in Figure 5.22, depends somewhat on the type of system and the charge of the radical.

The interaction between an unpaired π-electron and a proton falls off rapidly with increasing separation. Therefore, the hyperfine structure may be ascribed to interactions involving protons directly bonded to carbons in the π-system (α-protons) or else separated from the π-system by two σ-bonds (β-protons).

The equilibrium position of an α-proton is in the nodal plane of the π-system and it is clear that any net spin density at the proton results only indirectly from the presence of an unpaired π-electron. This indirect effect arises because the unpaired π-electron interacts slightly differently with α- and β-spin σ-electrons. Therefore the spatial distributions of these σ-electrons become slightly different, producing net spin density at the proton. This effect is referred to as the spin polarization of the σ-electrons by the π-electron.

Figure 5.23 is a schematic representation of the ESR spectrum of the benzene radical anion, $C_6H_6^-$. The unpaired electron interacts with six equivalent protons and yields a spectrum of seven lines.

*With respect to negative spin densities we refer the reader to Lowe (1978).

The above discussion of ESR in relation to π-electron densities leads naturally into a treatment by the MO method of the π-electron assemblies, the focus of the following chapter.

PROBLEMS

5.1. Consider a one-electron Hamiltonian H given by

$$H = -(\hbar^2/2m)\nabla^2 + V(\mathbf{r}) \tag{P5.1}$$

with normalized wave functions $\psi_i(\mathbf{r})$ and corresponding eigenvalues ε_i. Form the generalization

$$S(\mathbf{r}, \beta) = \sum_{\text{all } i} \psi_i(\mathbf{r})\psi_i^*(\mathbf{r}) \exp(-\beta\varepsilon_i) \tag{P5.2}$$

that corresponds to the partition function $\sum_{\text{all } i} \exp(-\beta\varepsilon_i)$, with $\beta = 1/k_B T$. **(a)** Prove that this quantity $S(\mathbf{r}, \beta)$, known as the Slater sum, is such that the partition function given above, denoted by $Z(\beta)$ can be calculated as

$$Z(\beta) = \int S(\mathbf{r}, \beta) \, d\mathbf{r} \tag{P5.3}$$

(b) Use the Schrödinger equation $H\psi_i = \varepsilon_i\psi_i$ to prove that in a one-dimensional problem, with $\mathbf{r}$ replaced by x, the Slater sum $S(x, \beta)$ satisfies the (partial) differential equation ($\hbar = 1$, $m = 1$) [March and Murray (1960)].

$$\frac{1}{8}\frac{\partial^3 S}{\partial x^3} = V(x)\frac{\partial S}{\partial x} + \tfrac{1}{2}S\frac{\partial V}{\partial x} + \frac{\partial^2 S}{\partial x \, \partial \beta} \tag{P5.4}$$

(c) Verify that for the harmonic oscillator, with $V = \tfrac{1}{2}kx^2$, the Slater sum is given by

$$S(x, \beta) = \left(\frac{v}{\sinh \omega\beta}\right)^{1/2} \exp[-x^2 k^{1/2} \tanh(\tfrac{1}{2}k^{1/2}\beta)] \tag{P5.5}$$

where $\omega = k^{1/2} = 2\pi v$ is the angular frequency of the oscillator.

5.2. Using the same Hamiltonian H as in the previous problem, the TF approximation (cf. Chapter 4) to the Slater sum is related to the free electron value $S_0(\beta)$ by

$$S(\mathbf{r}, \beta) = S_0(\beta) \exp[-\beta V(\mathbf{r})] \tag{P5.6}$$

Evidently this is exact when $V(\mathbf{r}) = $ constant, since then the wave functions $\psi(\mathbf{r})$ are unchanged by V and the eigenvalues are shifted by a constant. **(a)** What would be the Slater sum in the TF approximation for a constant electric field along the x-axis? (Hint: the potential energy corresponding to this constant electric field E, say, is $-Ex$.) **(b)** What is the quantity $S_0(\beta)$ in equation (P5.6) for the Slater sum S when one is dealing with (i) one-dimensional free particles and (ii) three-dimensional free particles?

5.3. Describe the expected proton NMR spectrum for CH_3CH_2Cl at 10,000 G. What changes would occur if the field were increased to 14,000 G.

5.4. Explain why the chemical shift of the acidic proton of a carboxylic acid dissolved in a nonpolar solvent like carbon tetrachloride varies less with concentration than that of the OH proton of an alcohol under the same conditions.

5.5. The values of the chemical shift δ for protons in methyl halides are CH_3F: 4.26 ppm; CH_3Cl: 3.05 ppm; CH_3Br: 2.68 ppm; and CH_3I: 2.16 ppm. Explain what might account for this trend.

159

*Molecular
Parameters
Determined by
Spectroscopic
Methods*

5.6. The microwave spectrum of CN contains a series of lines spaced equally by 3.7978 cm^{-1}. **(a)** What is the rotational constant B? **(b)** What is the reduced mass of the molecule? **(c)** What is the bond length?

5.7. The first excited vibrational level of CO is 2143.3 cm^{-1} above the ground vibrational level, and the second excited vibrational level is 2116.4 cm^{-1} above the first excited level. Estimate the frequency v and D_e, the dissociation energy, using the Morse potential.

5.8. How would the pure rotational spectrum, the vibrational–rotation spectrum at high resolution, and the electronic absorption and emission spectrum of a heteronuclear diatomic molecule change if the temperature of the sample were raised from 150 K to 600 K? Assume that the sample is in the ideal state at both temperatures.

5.9. Fluorine has a nuclear spin of $\frac{1}{2}$. Describe the expected proton and fluorine NMR spectra of CH_3F. What about the NMR spectrum of CH_2F_2?

5.10. Both NMR and ESR spectroscopy differ from other types of spectroscopy in one important aspect. Explain.

5.11. The ESR spectrum of the benzyl anion $C_6H_5CH_2^-$ is found to have three lines with relative intensities $1:2:1$. Explain.

5.12. Sketch the predicted ESR hyperfine patterns for an electron coupled to: **(a)** two equivalent spin-1 nuclei; **(b)** three equivalent spin-$\frac{3}{2}$ nuclei; **(c)** three equivalent protons with coupling constant a and two other equivalent protons with coupling constant $2a$.

REFERENCES

E. A. Ballik and D. A. Ramsay, *Astrophys. J.* **137**, 84 (1963).

C. N. Banwell, *Fundamentals of Molecular Spectroscopy*, McGraw-Hill, New York (1966).

A. A. Bothner-By and C. Naar-Colin, *Ann. N.Y. Acad. Sci.* **70**, 833 (1958).

R. N. Dixon, *Spectroscopy and Structure*, John Wiley and Sons, New York (1965).

W. H. Flygare, *Molecular Structure and Dynamics*, Prentice-Hall, Englewood Cliffs, NJ (1978).

J. M. Hollas, *High Resolution Spectroscopy*, Butterworths, London (1982).

J. M. Hollas, *Modern Spectroscopy*, John Wiley and Sons, New York (1987).

L. Karlsson, L. Mattson, R. Jadony, T. Bergmark, and K. Siegbahn, *Physica Scripta* **14**, 230 (1976).

J. P. Lowe, *Quantum Chemistry*, Academic Press, New York (1978).

N. H. March and A. M. Murray, *Phys. Rev.* **120**, 830 (1960).

J. A. Pople, W. G. Schneider, and H. J. Bernstein, *High-Resolution Nuclear Magnetic Resonance*, McGraw-Hill, New York (1959).

J. M. Schulman and J. W. Moskowitz, *J. Chem. Phys.* **47**, 3491 (1967).

A. Streitwieser, Jr., *Molecular Orbital Theory*, John Wiley and Sons, New York (1961).

Von Niessen, L. S. Cederbaum and W. P. Kraemer, *J. Chem. Phys.* **65**, 1378 (1976).

D. H. Whiffen, *Spectroscopy*, Longmans, London (1966).

R. Zahradnik and E. Polak, *J. Phys. Chem.* **84**, 3312 (1980).

Further Reading

P. W. Atkins, *Molecular Quantum Mechanics*, 2nd Ed., Oxford University Press, Oxford (1983).

P. W. Atkins, *Quanta*, 2nd Ed., Oxford University Press, Oxford (1991).

I. N. Levine, *Molecular Spectroscopy*, John Wiley and Sons, New York (1975).

J. I. Steinfeld, *Molecules and Radiation: An Introduction to Modern Molecular Spectroscopy*, 2nd Ed., MIT Press, Cambridge, MA (1985).

MOLECULAR ORBITAL METHODS AND POLYATOMIC MOLECULES

In the previous chapter we were largely concerned with those properties of molecules that are associated with their rotational and vibrational degrees of freedom. In the present chapter, we again take up electronic structure, discussed earlier in Chapter 2 for diatomic molecules and in Chapter 4 via electron density theory. Though polyatomic molecules were embraced by the electron density treatment, it is still of importance to describe orbital theories of such molecules, which is the main purpose of the present chapter.

After a discussion of conformation in H_2O and NH_3, and their UPS spectra, the important concept of hybridization is set out primarily in relation to C compounds. This approach is valuable not only in free molecules, but also in solids, e.g., diamond as well as amorphous C, and similarly for crystalline and amorphous Si.

This discussion is followed by a treatment of π-electrons, and associated properties, in organic molecules. While the treatment is mainly at the level of Hückel MO theory, some detailed calculations transcending this approximation will be discussed, with particular reference to the interpretation of the HeI UPS spectra of benzene. Important motivation, in fact, for making accurate MO calculations on large molecules is the usefulness of such treatments in aiding the interpretation of photoelectron spectra.

In addition to the treatment of electronic states of polyatomic molecules, a brief discussion is included of the so-called Renner–Teller effect, which can be thought of as a breakdown of the Born–Oppenheimer approximation (see Appendix A2.3) on which most of the electronic structure theory presented in this volume is based.

6.1. DIRECTED BONDS: CONFORMATION OF H$_2$O AND NH$_3$

As a preliminary to an examination of the UPS spectrum of H_2O in Section 6.2, let us tackle by elementary methods, first of all, the problem of the conformation of polyatomic molecules such as H_2O and NH_3.

The starting point of the treatment based on directed bonds, say, for H_2O, is to note the electron configuration of the O atom (cf. Appendix A1.2):

$$1s^2 2s^2 2p_x^1 2p_y^2 2p_z^1 \qquad (6.1)$$

161

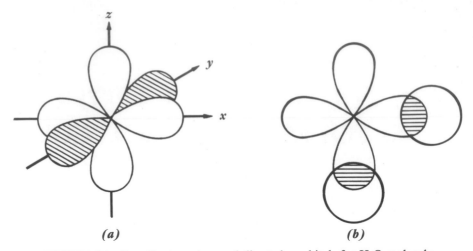

FIGURE 6.1. Coordinate system and directed *p*-orbitals for H_2O molecule.

with the consequence that the $2p_x$ and $2p_z$ orbitals are available for bond formation with the two H $1s$-orbitals. Referring to Figure 6.1, we note that the two unpaired orbitals of O (i.e., p_x and p_z) are at right angles to each other and are available to accept one H atom each, resulting in a molecule with a shape of the two O—H bonds at 90° to each other; experimentally the angle between the two O—H bonds is approximately 105°. This deviation from 90° will be considered later. The same situation is found, with ns, np orbitals in the H_2S and H_2Se molecules, with bond angles quite close to 90°.

A similar analysis can be applied to the series of molecules NH_3, PH_3, AsH_3, and SbH_3. For example, in considering NH_3, we have the electronic configuration of N as $1s^2 2s^2 2p_x^1 2p_y^1 2p_z^1$, with each of the three *p*-orbitals available to accept one electron from the H atoms. By analogy with the H_2O molecule we would expect three mutually perpendicular bonds. Experimentally, the angles are 108°, 93°, 92°, and 91° for NH_3, PH_3, AsH_3, and SbH_3, respectively, approaching the expected 90°. Having established, in a simple fashion, the semiquantitative usefulness of the concept of directed bonds, let us now turn to the interpretation of the UPS spectrum of the bent molecules H_2O and H_2S, together with a discussion of the conformation of their positive ions.

6.2. BENT MOLECULES: WALSH DIAGRAMS AND RULES

Before turning to the Hückel theory of π-electrons in organic molecules, we shall relate some of the previous discussion of ground-state properties to the preceding work in Chapters 2 and 4 on correlation diagrams and Walsh's rules for molecular shape. We shall be concerned therefore in this section with bent molecules, a focal point being the way in which, in molecules like H_2O and H_2S, there can be a large change in angle from the neutral molecule to the ion. This will be discussed in relation to the UPS spectra.

6.2.1. HeI UPS Spectrum of the H_2O Molecule

The notation that will be employed below can be established by referring to the Walsh MO diagram for AH_2 molecules in Figure 6.2. The σ and π classification, with even (g) and odd (u) symmetry in the linear form HAH gives way to the classification at the 90° HAH angle shown in Figure 6.2. Specifically for H_2O, reference to the figure allows the ground MO configuration to be written as $(2a_1)^2(1b_2)^2(3a_1)^2(1b_1)^2$ and the HOH angle in this ($\tilde{X}\,^1A_1$) state is 104.5°.

The HeI UPS spectrum is shown in Figure 6.3. The combination of this spectrum and other experimental evidence, supplemented by various types of calculations, allows some interesting conclusions to be drawn about the effect of removing an electron from one of the above MOs.

(a) Removal of Electron from the $1b_1$ Orbital

Removal of an electron from the $1b_1$ orbital (to give actually the $\tilde{X}\,^2B_1$ state of H_2O^+) turns out not to affect the HOH angle very much. Both symmetric

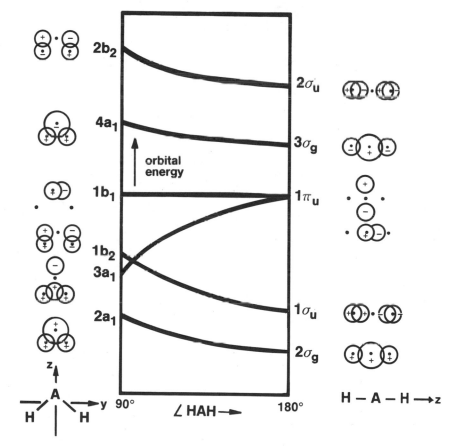

FIGURE 6.2. Walsh MO diagram for AH_2 molecules. (For the convenience of the reader, this is reproduced here from Chapter 4.)

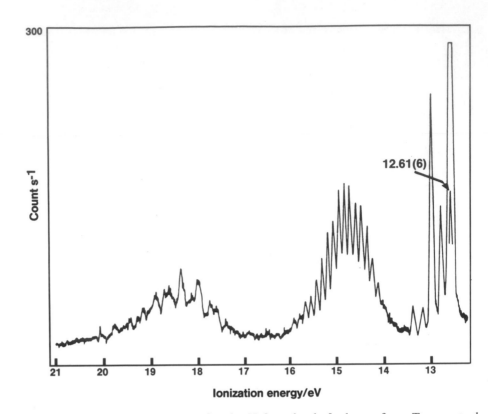

FIGURE 6.3. HeI UPS spectrum for the H_2O molecule [redrawn from Turner *et al.* (1970)].

stretching (v_1) and angle bending (v_2) vibrations show only short progressions. Franck–Condon factor calculations (see Appendix A5.4), including the effects of vibrational anharmonicity (see Section 5.4), indicate that the OH bond length $r(OH)$ increases from 0.9572 Å in the $\tilde{X}\,^1A_1$ state of H_2O to 0.995 ± 0.005 Å in the $\tilde{X}\,^2B_1$ state of H_2O^+. It was estimated originally that the HOH angle increases from 104.5° to about 109°, but high-resolution emission spectrum measurements on H_2O^+ have demonstrated that the angle in its ground state is 110.5°.

(b) Removal of an Electron from the $3a_1$ Orbital

Removal of an electron from the $3a_1$ MO in H_2O turns out to lead to a large increase in angle in the $\tilde{A}\,^2A_1$ state of H_2O^+. This was the conclusion from the emission spectrum of H_2O^+, although the angle was not determined (the angle is 144° in the corresponding state of NH_2). A long progression (see Section 5.5) in the bending vibration v_2 in the second band system of the UPS spectrum is consistent with such a large increase in angle.

Some irregularity in the first few members of the progression is consistent with the ion being not quite linear in the $\tilde{A}\,^2A_1$ state. The $\tilde{X}\,^2B_1$ and $\tilde{A}\,^2A_1$ states of bent H_2O^+ correlate with a $^2\pi_u$ state in the linear ion and vibrational–electronic interaction between the $\tilde{X}$ and $\tilde{A}$ states results in a Renner–Teller effect (see Section 6.12) and splitting of vibronic levels. The potential functions for these two states turn out to resemble those in Figure 6.30d.

If the HOH angle in the $\tilde{A}$ state of H_2O^+ is about 144°, the barrier to linearity is low and only the first few vibrational levels of v_2 will be below it. These levels will all be close to the barrier and (it turns out) will be irregularly spaced, consistent with the observations in the second band system. Above the barrier there is much greater regularity of levels (particularly those with $\sum$ vibronic symmetry in the linear molecule).

The third band system, arising by removal of an electron from the $1b_2$ orbital, is vibrationally complex, consistent with the orbital being strongly bonding and the orbital energy having a minimum value when HOH = 180°. Presumably both v_1 and v_2 are excited but the bands in this system are considerably broadened, making analysis difficult.

6.2.2. HeI UPS Spectrum of H_2S

From the admittedly elementary discussion of Section 6.1, it should come as no surprise that the HeI UPS spectrum of H_2S bears a strong resemblance to that of H_2O. Of the three band systems, the first is typical of the removal of an electron from a nonbonding orbital, analogous to the $1b_1$ orbital of H_2O (and involves the $\tilde{X}\,^2B_1$ state of H_2S^+). The third involves removal of an electron from a b_2 bonding orbital to give the $\tilde{B}\,^2B_2$ state and shows progressions in v_1 and v_2.

Figure 6.4 displays the second band system in H_2S. As with H_2O, it is dominated by a long progression in the bending vibration v_2 because an electron has

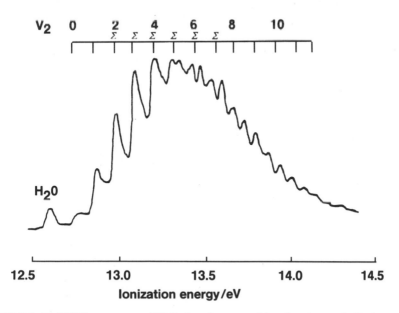

FIGURE 6.4. HeI UPS spectrum of H_2S, showing second band system only [redrawn from Hollas (1982)].

been removed from an orbital whose energy falls sharply as the bond angle decreases. The system of H_2S shows vibrational intervals of v_2 decreasing smoothly as far as $v_2' = 4$, beyond which is a small, somewhat complex, region, followed by a change to approximately half the vibrational spacing which then smoothly increases. This behavior is expected for the Σ vibronic levels, with $K = 0$. The discontinuity between $v_2' = 4$ and $v_2' = 6$ corresponds to the region at the top of the barrier to linearity. Both an energy barrier and a Renner–Teller effect must be invoked to explain the observed band system (see Section 6.12 below).

6.3. UPS OF AH₃ MOLECULES

In view of the brief discussion of the conformation of AH₃ molecules in Section 6.1, it is worth summarizing briefly here the main information that emerges from a study of the UPS of such polyatomic molecules. It has to be said first that in the UPS spectra of larger molecules there is a tendency toward greater, and often unresolvable, vibrational complexity.

The first band system of NH₃, with an adiabatic ionization potential of 10.16 eV, is an exception, being dominated by a long progression, of some 15 members, in the inversion vibration v_2. The reason for this is that the electron is taken from the outer $3a_1$ orbital whose energy falls steeply as the HNH angle decreases. This can be seen by reference to the Walsh MO diagram for AH₃ molecules shown in Figure 6.5. The result of this is that the ground state of NH₃⁺ is planar, as are Rydberg states of NH₃, and v_2 is strongly excited on

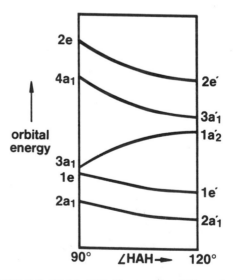

FIGURE 6.5. Walsh MO diagram for AH₃ molecules.

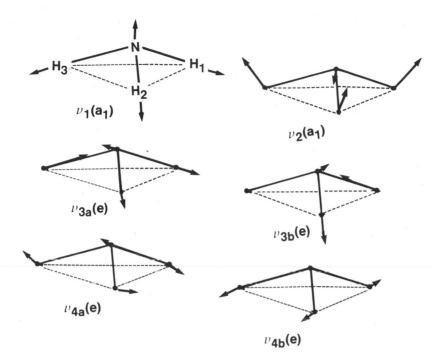

$\nu_1(a_1)$

$\nu_2(a_1)$

$\nu_{3a}(e)$

$\nu_{3b}(e)$

$\nu_{4a}(e)$

$\nu_{4b}(e)$

FIGURE 6.6. Normal vibrations of the NH_3 molecule.

ionization. The first band system of PH_3 shows a similar progression but now much less well resolved.

The second band systems of NH_3 and PH_3, stemming from the removal of an electron from the $1e$ orbital (or its equivalent in the molecule PH_3), do not exhibit such well-resolved vibrational structures as the first systems. Nevertheless, that of NH_3, with an adiabatic ionization potential of 14.7 eV, shows some structure and evidence of a Jahn–Teller effect (see Appendix A6.1).

The Jahn–Teller effect may split the degenerate $\tilde{A}\,^2E$ state of NH_3^+ by distortion in the direction of either component of the e vibrations ν_3 and ν_4, illustrated in Figure 6.6. Distortion in the direction of the normal coordinate Q_{3a} or Q_{4a} causes one NH bond to become different from the other two and thereby lowers the molecular symmetry. This results in the ^2E state splitting into $^2A'$ and $^2A''$ states. Alternatively, distortion in the direction of Q_{3b} or Q_{4b} causes all NH bonds to be different. The ^2E state splits into two ^2A states, again as a consequence of symmetry lowering. The first alternative is thought to be the more likely [see Hollas (1982)].

The second band system of NH_3 shows two broad maxima separated by about 1 eV and it seems that these correspond to the two components of the split $\tilde{A}\,^2E$ state. The vibrational structure is not sufficiently resolved to allow unambiguous interpretation.

To conclude this discussion of AH_3 molecules, it is worthy of note that the UPS spectrum of CH_3 has been recorded, the CH_3 being produced by the pyrolysis of azomethane ($CH_3N{=}NCH_3$) at 1700 K. Removal of an electron from the outer

$1a_2''$ orbital to give the $\tilde{X}\,^1A_1'$ ground state of CH_3^+ produces a band system typical of a nonbonding electron. Evidence from the very short vibrational progressions favors a planar ground state of both CH_3^+ and CH_3.

6.4. THE NEED FOR HYBRID ORBITALS

Having discussed in some detail the conformation of H_2O, NH_3, their ions, and related molecules, by a combination of experiment and theory, let us return to the earlier discussion of directed chemical bonds. We now wish to generalize the argument of Section 6.1 by considering hybrid orbital formation, and the shape of other molecules. Let us turn then, specifically, to a study of C compounds as an example of the concept of mixing pure AOs, i.e., hybridization. Consider the configuration

$$(1s)^2(2s)^2(2p_x)^1(2p_y)^1 \qquad (6.2)$$

of the ground state of the C atom. This shows that C has two unpaired electrons, which would lead to divalent properties with a bond angle of 90°.

However, the C atom is tetravalent and we must now consider how to account for this tetravalent character.

Pauling introduced the concept of hybridization to explain this behavior and represented the electronic configuration of C as

$$(1s)^2(2s)^1(2p_x)^1(2p_y)^1(2p_z)^1 \qquad (6.3)$$

with the corresponding bonding orbitals as

$$\phi(2s),\ \phi(2p_x),\ \phi(2p_y),\ \phi(2p_z)$$

These orbitals are combined to produce a set of four new equivalent orbitals that are directed toward the corners of a tetrahedron. This process is called sp^3-hybridization. In order to obtain the electron configuration shown in equation (6.3), an electron from the s-orbital has to be promoted to the empty p-orbital and an excited state formed. The energy to obtain such an excitation is not large and the energy gained by the formation of extra bonds (through hybridization) more than offsets that required for the excitation to take place. The sp^3-hybridization accounts for the four equivalent bonds in methane, CH_4; furthermore, these bonds are directed to form a tetrahedral molecule with equal angles of 109°28' to each other. We will examine this in detail presently in addition to considering other hybrid orbitals, but, first, a few further points need to be stressed. Hybrid orbitals are directed so that interelectronic repulsion is minimized; also the hybrid orbitals concentrate the wave function in the direction of bonding leading to a greater amount of overlap with the wave function of the other atom, which yields a stronger bond. For example, consider the mixing of an s- and a p-orbital as shown in Figure 6.7 and note how different the resulting hybrid orbitals appear from the mixing component orbitals. It should also be noted that the sp-hybrid in Figure 6.7 has a more pronounced directional property than the *p*-orbitals. Consequently strong bonds will be formed when an H atom orbital overlaps with the positive lobe of such a hybrid orbital.

We now turn to a detailed consideration of the various types of hybrid orbitals that can be formed and their properties, and then apply these concepts to study the characteristics of bonding and shape for some molecules resulting from this hybridization.

(a) sp Hybrids

We consider the formation of the hybrid wave functions composed of one *s*- and one p_z- wave function. The two hybrid wave functions are usually expressed as

$$\phi_1 = c_1\phi(s) + d_1\phi(p_z), \qquad \phi_2 = c_2\phi(s) + d_2\phi(p_z) \qquad (6.4)$$

where c_1, c_2, d_1, and d_2 are coefficients to be determined remembering that ϕ_1 and ϕ_2 are normalized, i.e.,

$$\int \phi_1^2 \, d\tau = 1 \qquad \text{and} \qquad \int \phi_2^2 \, d\tau = 1 \qquad (6.5)$$

and that the orthogonality property (cf. Appendix A1.4) is operative, i.e.,

$$\int \phi_1\phi_2 \, d\tau = 0 \qquad (6.6)$$

$\psi(s)$ contributes equally to the new wave functions, which means that the probability density d_1^2 must contain $\frac{1}{2}s^2$ character; therefore $c_1^2 = c_2^2 = \frac{1}{2}$ and hence

$$c_1 = c_2 = 1/\sqrt{2} \qquad (6.7)$$

Using (6.4), we have

$$\int \phi_1^2 \, d\tau = c_1^2 \int \phi^2(s) \, d\tau + 2c_1d_1 \int \phi(s)\phi(p_z) \, d\tau + d_1^2 \int \phi^2(p_z) \, d\tau \qquad (6.8)$$

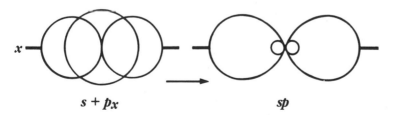

$$s + p_x \qquad\qquad sp$$

FIGURE 6.7. Comparison of hybrid *sp*-orbitals with pure *s*- and *p*-orbitals.

Remembering the normalization stipulation on ϕ_1, and the normalization and orthogonality properties of $\phi(s)$ and $\phi(p_z)$, one obtains, by using equation (6.8),

$$c_1^2 + d_1^2 = 1 \qquad \text{or} \qquad d_1 = 1/\sqrt{2} \qquad (6.9)$$

Continuing by using the orthogonality condition (6.6) applied to ϕ_1 and ϕ_2, we have from equation (6.4) that

$$\int \phi_1 \phi_2 \, d\tau = c_1 c_2 + d_1 d_2 = \tfrac{1}{2} + d_2/\sqrt{2} = 0 \qquad (6.10)$$

and

$$d_2 = -1/\sqrt{2} \qquad (6.11)$$

Consequently equations (6.4) become explicitly

$$\phi_1 = \frac{1}{\sqrt{2}} [\phi(s) + \phi(p_z)]$$

$$\phi_2 = \frac{1}{\sqrt{2}} [\phi(s) - \phi(p_z)] \qquad (6.12)$$

If we use the angular parts of the wave functions* (see Appendix A1.1) corresponding to $\phi(s)$ and $\phi(p_z)$, equation (6.12) becomes

$$\phi_1 = \frac{1}{2\sqrt{2\pi}} (1 + \sqrt{3} \cos \theta)$$

$$\phi_2 = \frac{1}{2\sqrt{2\pi}} (1 - \sqrt{3} \cos \theta) \qquad (6.13)$$

Since ϕ_1 and ϕ_2 must be equivalent orbitals, we can solve for $\cos \theta$ in equation (6.13), where θ is the valence angle. We find $\cos \theta = -1$, which means that $\theta = 180°$ and thus ϕ_1 and ϕ_2 are pointing in opposite directions, i.e., at 180° from each other as shown in Figure 6.8.

The *sp*-hybrid, therefore, leads to a linear molecule and we shall have an opportunity to give examples of molecules displaying this type of hybridization later in this section. We now turn to *sp²*-hybrids and the determination of their bonding angle.

---

*Below, following Pauling, we ignore the difference in radial parts of s and p orbitals for simplicity of presentation.

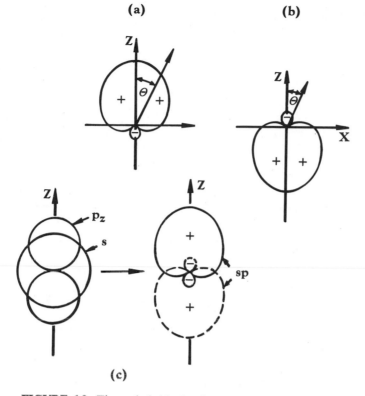

(a)   (b)

(c)

FIGURE 6.8. The *sp*-hybrids, leading to a linear molecule.

(b) sp^2 Hybrids

In sp^2-hybridization, the three new wave functions are composed of one s-, and two p-orbitals (p_x and p_y), and we can write

$$\phi_1 = c_1\phi(s) + d_1\phi(p_x) + f_1\phi(p_y)$$

$$\phi_2 = c_2\phi(s) + d_2\phi(p_x) + f_2\phi(p_z) \qquad (6.14)$$

$$\psi_3 = c_3\phi(s) + d_3\phi(p_x) + f_3\phi(p_y)$$

Employing the technique used in *sp*-hybridization, we find that the equivalent sp^2-hybrid orbitals are

$$\phi_1 = \frac{1}{\sqrt{3}}\,\phi(s) + \sqrt{\tfrac{2}{3}}\phi(p_x)$$

$$\phi_2 = \frac{1}{\sqrt{3}}\,\phi(s) - \frac{1}{\sqrt{6}}\,\phi(p_x) + \frac{1}{\sqrt{2}}\,\phi(p_y) \qquad (6.15)$$

$$\phi_3 = \frac{1}{\sqrt{3}}\,\phi(s) - \frac{1}{\sqrt{6}}\,\phi(p_x) - \frac{1}{\sqrt{2}}\,\phi(p_y)$$

and using the angular parts of the wave functions* (see Appendix A1.1):

$$\phi_1 = \frac{1}{2\sqrt{\pi}}\left(\frac{1}{\sqrt{3}} + \sqrt{2}\cos\phi\right)$$

$$\phi_2 = \frac{1}{2\sqrt{\pi}}\left(\frac{1}{\sqrt{3}} - \frac{1}{\sqrt{2}}\cos\phi + \sqrt{\tfrac{3}{2}}\sin\phi\right) \tag{6.16}$$

$$\phi_3 = \frac{1}{2\sqrt{\pi}}\left(\frac{1}{\sqrt{3}} - \frac{1}{\sqrt{2}}\cos\phi - \sqrt{\tfrac{3}{2}}\sin\phi\right)$$

The bond angle ϕ in this case turns out to be 120°, and therefore the sp^2-hybrids are directed as shown in Figure 6.9. The sp^2-hybrids, therefore, lie in a plane with a valence angle of 120°. Above, and in Figure 6.9, the hybrid orbitals are given in the xy-plane, i.e., $\theta = \pi/2$.

(c) sp^3 Hybrids

In this case, we would have four equivalent hybrid orbitals made up of one s- and three p-orbitals (p_x, p_y, p_z) with

$$\phi_1 = \tfrac{1}{2}\phi(s) + \frac{\sqrt{3}}{2}\phi(p_z)$$

$$\phi_2 = \tfrac{1}{2}\phi(s) + \frac{\sqrt{2}}{\sqrt{3}}\phi(p_z) - \frac{1}{2\sqrt{3}}\phi(p_z)$$

$$\phi_3 = \tfrac{1}{2}\phi(s) - \frac{1}{\sqrt{6}}\phi(p_x) + \frac{1}{\sqrt{2}}\phi(p_y) - \frac{1}{2\sqrt{3}}\phi(p_z) \tag{6.17}$$

$$\phi_4 = \tfrac{1}{2}\phi(s) - \frac{1}{\sqrt{6}}\phi(p_x) - \frac{1}{\sqrt{2}}\phi(p_y) - \frac{1}{2\sqrt{3}}\phi(p_z)$$

The bonding angle here turns out to be the one appropriate to a tetrahedron (i.e., 109°28′). Hence, sp^3-hybrids are directed to the corners of a tetrahedron, which is exactly what one finds experimentally for the molecule CH_4 (methane).

Other hybrid orbitals can be formed (using in addition to the s- and p-orbitals the d-orbitals) such as dsp^2, dsp^3, and d^2sp^3. Table 6.1 lists some of the properties of these hybrids.

We noted earlier that stronger bonds resulted from hybridization. It is, therefore, of interest to compare the strengths of bonds formed using the various types of hybrid orbitals. Figure 6.10 shows how the overlap integral between two atoms

---

*Again common radial wave functions are assumed. Angular wave functions p_x and p_y are $(3/4\pi)^{1/2}$ times $\sin\theta\cos\phi$ and $\sin\theta\sin\phi$.

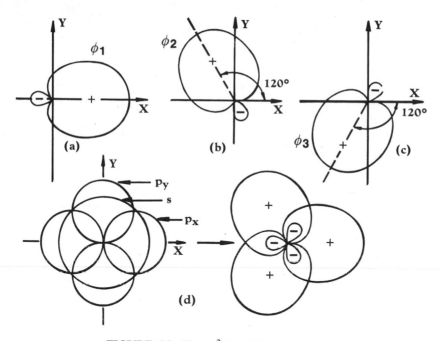

FIGURE 6.9. The sp^2-hybridized orbitals.

TABLE 6.1
Shapes of Hybrid Orbitals

Hybrid orbital	Number of orbitals	Shape
sp	2	Linear
sp^2	3	Planar, 120°
sp^3	4	Tetrahedral
dsp^2	4	Square planar
dsp^3	5	Trigonal bipyramidal
d^2sp^3	6	Octahedral

varies with the percentage of s character in the hybrid bond, and we see that the overlap integral is largest for sp, followed by sp^2 and sp^3. This is substantiated by experimental results.

We now turn to some specific C compounds where sp^3-, sp^2-, or sp-hybridization plays a major role.

6.5. SOME CARBON COMPOUNDS WITH sp^n HYBRIDIZATION ($n = 1, 2, 3$)

Since the main principles should now be clear, we shall proceed to a short summary of the salient points for a set of molecules taken in turn below.

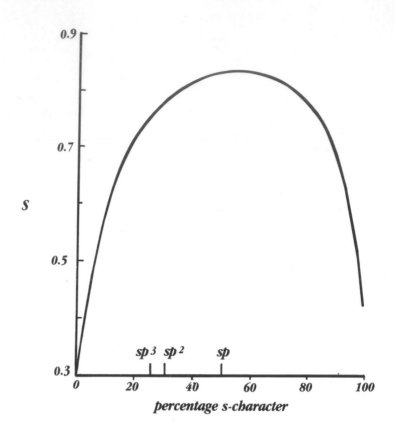

FIGURE 6.10. The overlap integrals between two atoms as a function of the amount of *s*-character in the hybrid bond [redrawn from Coulson (1961)].

(a) *sp³ Hybridization*

Methane (CH_4). One forms four equivalent sp^3-hybrid bonds for C. The resulting shape is plainly tetrahedral. Localized C—H bonds can be constructed. These are four single bonds formed by overlap with *s*-orbitals of H. The resulting bonding is of σ-character.

Ethane (C_2H_6). In this molecule one forms four equivalent sp^3-hybrid orbitals from each C atom. Three of these overlap with *s*-orbitals of three H atoms while the fourth overlaps with its equivalent hybrid on the second C atom. All bonding is of single-bond character. The H—C—H angle is 109°28′ (cf. the tetrahedral angles in CH_4). The bonds throughout are of σ-character.

(b) *sp² Hybridization*

Ethylene (C_2H_4). The bond angles, H—C—H, are close to the predicted 120° for sp^2 hybrid orbitals. The *s*-orbital and two *p*-orbitals of the C atoms hybridize to form three coplanar hybrid orbitals: two of which form localized MOs of σ-type by overlapping with the *s*-orbital of H, while the third overlaps with its counterpart of the other C to form another localized MO. Therefore all σ-types lie in a plane. Each C atom is left with a single occupied p-orbital having its axis perpendicular to the plane of the atoms. The lateral overlap of these two p-orbitals gives rise to a localized MO (bond) between the two carbons—this π-bond makes up the second part of the C—C double bond. The π-bond is weaker than the σ-bond.

To illustrate the above, Figure 6.11 shows how the atomic orbitals, namely the H $1s$ and the C sp^2 hybrids (a) are combined to form localized bonds. This is σ-bond formation; π-bonding is depicted in part (b). Combining these, we have the final picture (c) for the bonding in ethylene, showing the double bond as the outcome of σ- and π-bonding.

Benzene (C_6H_6). Briefly, Figure 6.12a depicts the use of sp^2 hybrid orbitals on the C atoms to form σ-bonds in benzene. There is then one p-orbital, which is perpendicular to the plane of the benzene ring (z-axis), on each C atom that is still available for bonding. Linear combinations of these orbitals lead to delocalized MOs and π-bonding, as depicted in Figure 6.12b. We shall return later to a more detailed discussion of the π-level spectrum in C_6H_6.

(c) sp Hybridization

Acetylene (C_2H_2). Using the same approach as with ethylene above, one first forms the σ-bonding pattern shown in Figure 6.13a, from the H $1s$ orbitals and the sp-hybrids on C. The $2p_z$ orbitals on each C are then still available for bonding, namely the formation of π-bonds (see Figure 6.13b). As the final stage, one has the triple bond formed between the C atoms, with two π-bonds and a σ-bond, yielding

$$H-C\equiv C-H$$

as a linear molecule with a triple bond as shown.

In short, many experimental facts on the conformation of polyatomic molecules fit into a clear and simple pattern which emerges from the concept of hybridization of AOs. While many excellent numerical solutions of the Schrödinger wave equation can now be found in the literature, they do not radically change the basic picture presented above, though of course they do refine it quantitatively.

6.6. MOLECULAR ORBITAL THEORY OF π-ELECTRON SYSTEMS

In this section, molecular orbital theory will be applied to π-electron systems in unsaturated C compounds. Because it can be developed analytically and leads also to insight into some general features of π-electronic structure, the simple Hückel treatment will be extended to embrace somewhat more refined methods. These will be compared, toward the end of the chapter, with the chemically interesting results that can be obtained from the so-called $(e, 2e)$ experiment, where the particular example of acetylene has been chosen by way of illustration.

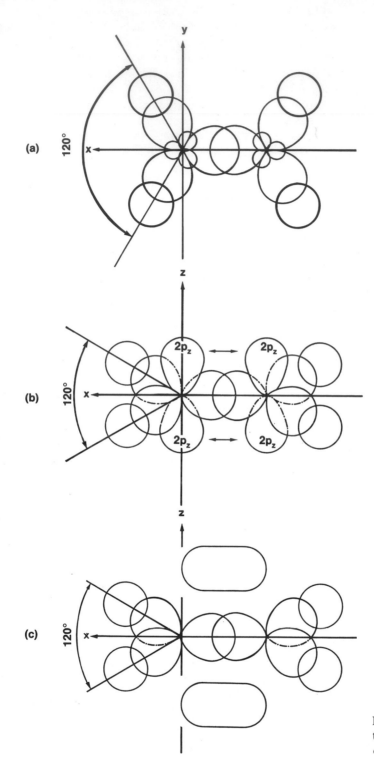

FIGURE 6.11. Formation of the double bond in ethylene via σ- and π-bonding.

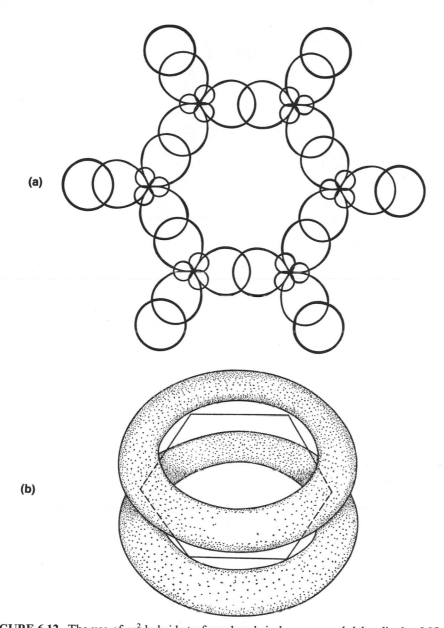

FIGURE 6.12. The use of sp^2-hybrids to form bonds in benzene, and delocalized π-MOs.

6.6.1. The Hückel Molecular Orbital Method

It has long been clear that numerous chemical and spectroscopic properties of conjugated hydrocarbons involve primarily the π-electrons. This has led to the assumption that the energy difference between the π and σ electronic energies is large, with the consequence that the π- and σ-electrons can be treated separately. Hückel, in the early 1930s, was the first to make this assumption of π–σ separability. This turns out to be a somewhat drastic assumption and there is ample

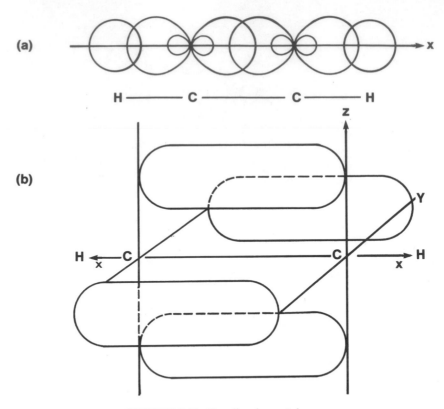

FIGURE 6.13. Bonding in acetylene.

evidence, from both experiment and theory, that there is significant π–σ interaction. Nevertheless, the Hückel molecular orbital (HMO) theory has been extremely fruitful, and we shall therefore present its predictions at length below, before summarizing progress using more refined MO methods for π-electron assemblies.

The assumption of π–σ separability can be used to simplify the Schrödinger equation for a conjugated hydrocarbon:

$$\hat{\mathcal{H}}\Psi_{\text{tot}} = E_{\text{tot}}\Psi_{\text{tot}} \tag{6.18}$$

Within the Born–Oppenheimer approximation (see Appendix A2.3), and using

$$\hat{\mathcal{H}} = \hat{\mathcal{H}}_\sigma + \hat{\mathcal{H}}_\pi \tag{6.19}$$

$$E_{\text{tot}} = E_\sigma + E_\pi \tag{6.20}$$

$$\Psi_{\text{tot}} = \psi_\sigma \psi_\pi \tag{6.21}$$

it follows that

$$\hat{\mathscr{H}}_\sigma \psi_\sigma = E_\sigma \psi_\sigma \qquad (6.22)$$

$$\hat{\mathscr{H}}_\pi \psi_\pi = E_\pi \psi_\pi \qquad (6.23)$$

In the Hückel approximation we shall be concerned only with the π-electronic eigenfunctions and eigenvalues. The π-electrons are assumed to be moving in a potential field due to the nuclei and a σ-core. The Hamiltonian operator for the π-electron system can then be expressed as

$$\hat{\mathscr{H}}_\pi = \sum_{i=1}^{n} \mathscr{H}_i^{\text{core}} + \sum_{i,j}^{n} 1/r_{ij} \qquad (6.24)$$

where $\mathscr{H}_i^{\text{core}}$ represents the kinetic and potential energy terms for the π-electrons in the field of all the cores; and $1/r_{ij}$ is as usual the two-electron interaction term. The electronic interaction terms, as we saw in the case of the H_2 molecule, are difficult to handle; therefore Hückel chose to ignore any explicit treatment of these interactions and used only an "effective" Hamiltonian

$$\hat{\mathscr{H}}_\pi = \sum_{1}^{n} \hat{\mathscr{H}}_i^{\text{eff}} \qquad (6.25)$$

where a chosen π-electron is now moving in an average field generated by all the other σ- and π-electrons.

Using the LCAO method for the π-electron wave functions we have

$$\psi = \sum_i c_i \phi_i \qquad (6.26)$$

where the ϕ_i are Slater-like functions for the AOs occupied by the π-electrons, one per atomic center.

Using the variational method (see Appendix A4.6), Hückel introduced the following notation characteristic of his HMO method:

$$\int \phi_\mu \phi_\nu \, d\tau \equiv S_{\mu\nu} = \begin{cases} 1 & \mu = \nu \\ 0 & \text{otherwise} \end{cases} \qquad (6.27)$$

$$\int \phi_\mu \hat{\mathscr{H}}_\pi \phi_\nu \, d\tau \equiv H_{\mu\nu} = \begin{cases} \alpha_\mu & \mu = \nu \\ \beta_{\mu\nu} & \mu\nu - \text{nearest neighbors} \\ 0 & \text{otherwise} \end{cases} \qquad (6.28)$$

The first integral is the overlap integral (cf. Chapter 2 especially) which is assumed to be either unity, in the case of identical orbitals, or zero. This is equivalent to assuming an orthonormal basis set of orbitals and aids in keeping the calculations simple.

The second integral is an energy integral. If μ and v are identical, it is designated the Coulomb integral and given the symbol α and the subscript μ. In the case or conjugated hydrocarbons, in which all of the π-electrons are associated with C atoms, these must necessarily all be identical, and the subscript is dropped. In the case of heteroatomic molecules other than C, adjustments need to be made [see Lowe (1978)].

When μ and v in the second integral are not identical, the integral is called the resonance integral and is designated by $\beta_{\mu v}$. The resonance integral is a measure of the interaction between a π-electron on core μ with the neighboring core v, and is dependent on the distance between the two nuclei. However, as the HMO method assumes all bond lengths identical in conjugated hydrocarbons, all $\beta_{\mu v}$ are of equal magnitude and subscripts are usually dropped. In addition, as in the case of the overlap integral, the value of $\beta_{\mu v}$ is assigned as zero if the cores μ and v are not directly bonded to one another; this assumes negligibly small interactions between nonnearest neighbors. Therefore in a system of n π-electrons, the linear combination will consist of n basis functions and will lead to n simultaneous linear equations of the form

$$\sum_{\mu=1}^{n} c_{\mu}(H_{\rho\mu} - S_{\rho\mu}E) = 0 \tag{6.29}$$

Solutions to these give n values of the energy and the corresponding n MO wave functions. Let us consider some examples using the HMO.

6.6.2. An Example of π-Level Spectra: Ethylene

The two π-electrons in ethylene yield a linear combination of the form

$$\psi = a_1\phi_1 + a_2\phi_2 \tag{6.30}$$

with the secular equations

$$a_1(\alpha - E) + a_2\beta = 0$$
$$a_1\beta + a_2(\alpha - E) = 0 \tag{6.31}$$

and secular determinant

$$\begin{vmatrix} \alpha - E & \beta \\ \beta & \alpha - E \end{vmatrix} = 0 \tag{6.32}$$

It is useful to divide through the determinant by β and let

$$x = (\alpha - E)/\beta \tag{6.33}$$

The resulting determinant is, then,

$$\begin{vmatrix} x & 1 \\ 1 & x \end{vmatrix} = 0, \qquad x^2 - 1 = 0 \qquad (6.34)$$

with solutions $x = \pm 1$, and hence from equation (6.33) the π energy levels are

$$E = \alpha + \beta, \alpha - \beta \qquad (6.35)$$

where the resonance integral β is negative. Consequently the root $E_1 = \alpha + \beta$ will have a lower (i.e., more negative) value than $E_2 = \alpha - \beta$. Substitution of the energy eigenvalues back into the secular equations leads to values for the coefficients:

$$\text{For } E_1: \quad a_1 = a_2$$

$$(6.36)$$

$$\text{For } E_2: \quad a_1 = -a_2$$

Hence, the normalized wave functions are

$$\psi_1 = \frac{1}{\sqrt{2}}(\phi_1 + \phi_2)$$

$$(6.37)$$

$$\psi_2 = \frac{1}{\sqrt{2}}(\phi_1 - \phi_2)$$

exactly analogous to the case of the H_2 molecule discussed in Chapter 2. In the ground state of ethylene the two π-electrons occupy the lower energy level, with spins paired. The orbital energy level diagram showing the occupancy is depicted in Figure 6.14. The total ground state π-electron energy according to Hückel is simply the sum of the energies of the individual electrons (see Chapter 4 for a relation between total energy and orbital energy sum):

$$E = 2\alpha + 2\beta \qquad (6.38)$$

and the total spatial ground-state wave function is the product function

$$\psi_\pi = \psi_1(1) \cdot \psi_1(2) \qquad (6.39)$$

6.6.3. The HeI UPS Spectrum of Ethylene

The HeI UPS spectrum of ethylene shows five band systems with adiabatic ionization potentials of 10.51, 12.38, 14.47, 15.68, and 18.9 eV. Figure 6.15 illustrates the first, which has the following pronounced features: (1) a progression in v_2, the (a_g) C=C stretching vibration, (ii) $\Delta v_4 = 2$ transitions in v_4, the (a_u) torsional vibration depicted in Figure 6.16, and (iii) a short progression in v_3, the

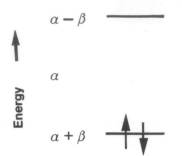

FIGURE 6.14. Hückel level spectrum of ethylene.

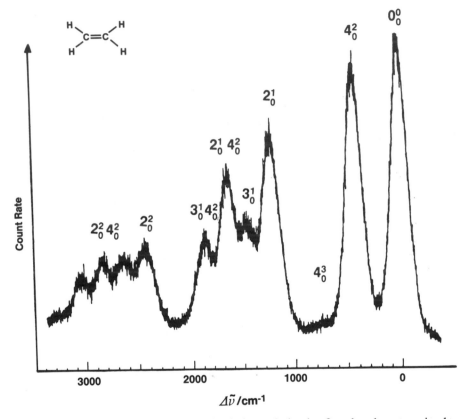

FIGURE 6.15. HeI UPS spectrum of ethylene. Only the first band system is shown [redrawn from Hollas (1982)].

(a_g) CH_2-scissoring vibration. The strong intensity of bands involving $2\nu_4$ parallels similar observations in the first Rydberg transition in ethylene. In the first Rydberg state of C_2H_4 and the ground state of $C_2H_4^+$ the equilibrium configuration is nonplanar, the CH_2 groups being twisted relative to each other by about 30°.

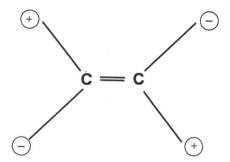

FIGURE 6.16. Torsional vibration in ethylene.

According to the Franck–Condon principle discussed in Appendix A5.4, it is this change in geometry that results in the intense transitions with $\Delta v_4 = 2$.

Returning now to electronic structure, the above discussion can be formalized into the ground configuration of ethylene as

$$(1b_{2u})^2(3a_g)^2(1b_{3g})^2(1b_{3u})^2(1b_{2g})^0$$

where $1b_{3u}$ denotes a π-bonding and $1b_{2g}$ a π^*-antibonding orbital (see Figure 6.17). The first excited singlet state results from promotion of an electron from $1b_{3u}$ to $1b_{2g}$. The molecule is twisted in this state, with an angle of 90° between the two CH_2 planes. The twisting can be rationalized by the supposition that the repulsion in the $1b_{2g}\,\pi^*$ orbital across the xy nodal plane is minimized by this change of shape.

Bearing this explanation in mind, it may appear a little surprising that taking an electron out of the $1b_{3u}$ orbital should also result in a twisted configuration, though now by considerably less than 90° [see Hollas (1982)].

6.6.4. Other Examples of π-Level Spectra

Having considered ethylene at some length, let us continue with some further examples of π-level spectra in organic molecules, according to Hückel theory.

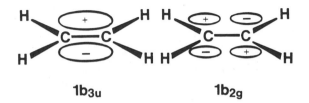

FIGURE 6.17. Bonding and antibonding orbitals in ethylene.

(a) *Butadiene*

$$\begin{array}{cccc}
\text{H} & \text{H} & \text{H} & \text{H} \\
| & | & | & | \\
\text{H}-\text{C}_1 & = \text{C}_2 & -\text{C}_3 & = \text{C}_4-\text{H}
\end{array}$$

The π-electrons are assumed to be delocalized and able to move freely among all four C atoms, so that there is at least a partial double bond between atoms 2 and 3, making β_{23} nonzero.

The linear combination of atomic orbitals is now written

$$\psi = a_1\phi_1 + a_2\phi_2 + a_3\phi_3 + a_4\phi_4 \tag{6.40}$$

with the corresponding secular determinant being

$$\begin{vmatrix}
\alpha - E & \beta & 0 & 0 \\
\beta & \alpha - E & \beta & 0 \\
0 & \beta & \alpha - E & \beta \\
0 & 0 & \beta & \alpha - E
\end{vmatrix} = 0 \tag{6.41}$$

or

$$\begin{vmatrix}
x & 1 & 0 & 0 \\
1 & x & 1 & 0 \\
0 & 1 & x & 1 \\
0 & 0 & 1 & x
\end{vmatrix} = 0 \tag{6.42}$$

which has the roots

$$x = \pm 1.6180, \pm 0.6180 \tag{6.43}$$

These yield the eigenvalues

$$E = \alpha \pm 1.6180\beta, \qquad \alpha \pm 0.6180\beta \tag{6.44}$$

and the set of energy levels shown in Figure 6.18.

In the ground electronic state, the four π-electrons (with spins paired) of butadiene occupy the two lowest energy levels.

The total π-energy is then, following Hückel,

$$E_\pi = 2(\alpha + 1.6180\beta) + 2(\alpha + 0.6180\beta) = 4\alpha + 4.4720\beta \tag{6.45}$$

If we consider butadiene as having localized bonds, β_{23} would be set equal to

$$\alpha - 2\beta$$
$$\alpha - 1.6\beta \quad \underline{\qquad}$$

$$\alpha - \beta$$
$$\alpha - 0.6\beta \quad \underline{\qquad}$$

$$\alpha$$

$$\alpha + 0.6\beta \quad \underline{\uparrow\downarrow}$$

$$\alpha + \beta$$

$$\alpha + 1.6\beta \quad \underline{\uparrow\downarrow}$$

$$\alpha + 2\beta$$

Energy →

FIGURE 6.18. Hückel π-level spectrum of butadiene.

zero, resulting in the Hückel determinant*

$$
\begin{vmatrix}
x & 1 & 0 & 0 \\
1 & x & 0 & 0 \\
- & - & - & - \\
0 & 0 & x & 1 \\
0 & 0 & 1 & x
\end{vmatrix} = 0 \tag{6.46}
$$

The localized butadiene would have the characteristics of two degenerate ethylene molecules joined by the σ-bond between atoms 2 and 3, with energy levels shown in Figure 6.19.

It is now easy to calculate the total π-electron energy for each of the butadiene problems (i.e., delocalized and localized) and obtain the π-delocalization, or resonance, energy.

*In this form, the similarity with equation (6.34) is evident.

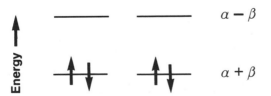

$$\underline{\qquad} \quad \underline{\qquad} \quad \alpha - \beta$$

$$\underline{\uparrow\downarrow} \quad \underline{\uparrow\downarrow} \quad \alpha + \beta$$

Energy →

FIGURE 6.19. Relation of butadiene with localized double bonds to two ethylenic linkages.

Delocalized: $E_\pi = 2(\alpha + 1.6180\beta) + 2(\alpha + 0.6180\beta)$

$$= 4\alpha + 4.4720\beta$$

Localized: $E_\pi = 4(\alpha + \beta) = 4\alpha + 4\beta$

Resonance energy: $= E_\pi(\text{deloc}) - E_\pi(\text{loc}) = 0.472\beta$

(6.47)

Using the same technique, we now consider further examples of calculating electronic energy levels by HMO method.

(b) The Allyl System

With the notation as in Figure 6.20, one writes the π-MOs as

$$\psi = a_1 p_1 + a_2 p_2 + a_3 p_3 \tag{6.48}$$

where p_1, p_2, and p_3 are atomic orbitals. The secular determinant written in full is

$$\begin{vmatrix} H_{11} - ES_{11} & H_{12} - ES_{12} & H_{13} - ES_{13} \\ H_{21} - ES_{21} & H_{22} - ES_{22} & H_{23} - ES_{23} \\ H_{31} - ES_{31} & H_{32} - ES_{32} & H_{33} - ES_{33} \end{vmatrix} = 0 \tag{6.49}$$

With the simplifying assumptions characteristic of the Hückel method:

$$\left. \begin{aligned} H_{11} &= H_{22} = H_{33} = \alpha \\[6pt] H_{12} &= H_{21} = H_{23} = H_{32} = \beta \\[6pt] H_{13} &= H_{31} = 0 \\[6pt] S_{11} &= S_{22} = S_{33} = 1 \\[6pt] S_{12} &= S_{21} = S_{13} = S_{31} = S_{23} = S_{32} = 0 \end{aligned} \right\} \tag{6.50}$$

the determinant (6.49) reduces to

$$\begin{bmatrix} \alpha - E & \beta & 0 \\ \beta & \alpha - E & \beta \\ 0 & \beta & \alpha - E \end{bmatrix} = 0 \tag{6.51}$$

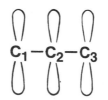

FIGURE 6.20. Notation for the allyl system.

As usual, with $x = (\alpha - E)/\beta$, one finds from equation (6.51) that

$$\begin{vmatrix} x & 1 & 0 \\ 1 & x & 1 \\ 0 & 1 & x \end{vmatrix} = 0 \qquad \text{and} \qquad \begin{gathered} x^3 - 2x = 0 \\ x = -\sqrt{2}, \quad x = 0, \quad x = \sqrt{2} \end{gathered} \qquad (6.52)$$

$$\text{For } x = -\sqrt{2}: \qquad E_1 = \alpha + \sqrt{2}\beta$$

$$\text{For } x = 0: \qquad E_2 = \alpha \qquad\qquad (6.53)$$

$$\text{For } x = \sqrt{2}: \qquad E_3 = \alpha - \sqrt{2}\beta$$

The two π-electrons present in the allyl carbonium ion are both placed in ψ_1 with total energy

$$E_{C_3H_5}(+) = 2(\alpha + \sqrt{2}\beta) = 2\alpha + 2\sqrt{2}\beta \qquad (6.54)$$

The allyl radical has three electrons and the third one must go into ψ_2 (see Figure 6.21); hence the total π-electronic energy of the allyl radical is

$$E_{C_3H_5}(\cdot) = 2(\alpha + \sqrt{2}\beta) + \alpha = 3\alpha + 2\sqrt{2}\beta \qquad (6.55)$$

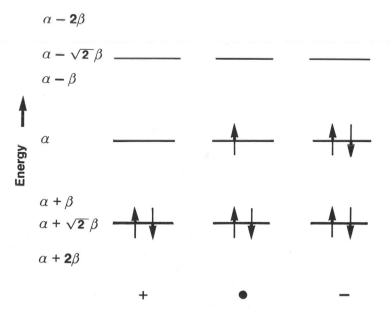

FIGURE 6.21. Energy levels of π-electrons in the allyl system.

In the allyl carbanion still another electron must be considered. This is also put in ψ_2. The anion has two electrons in ψ_1 and two in ψ_2. The total energy is consequently

$$E_{C_3H_5}(-) = 2(\alpha + \sqrt{2}\beta) + 2\alpha = 4\alpha + 2\sqrt{2}\beta \tag{6.56}$$

(c) Cyclopropenyl System:

Referring to Figure 6.22, we get

$$\psi = a_1 p_1 + a_2 p_2 + a_3 p_3 \tag{6.57}$$

The secular determinant is

$$\begin{vmatrix} H_{11} - ES_{11} & H_{12} - S_{12}E & H_{13} - ES_{13} \\ H_{21} - ES_{21} & H_{22} - ES_{22} & H_{23} - ES_{23} \\ H_{31} - ES_{31} & H_{32} - ES_{32} & H_{33} - ES_{33} \end{vmatrix} = 0 \tag{6.58}$$

With the simplifications of the Hückel method:

$$\left. \begin{aligned} H_{11} &= H_{22} = H_{33} = \alpha \\ H_{12} &= H_{21} = H_{13} = H_{31} = H_{23} = H_{32} = \beta \\ S_{11} &= S_{22} = S_{33} = 1 \\ S_{12} &= S_{21} = S_{13} = S_{31} = S_{23} = S_{32} = 0 \end{aligned} \right\} \tag{6.59}$$

equation (6.58) becomes

$$\begin{bmatrix} \alpha - E & \beta & \beta \\ \beta & \alpha - E & \beta \\ \beta & \beta & \alpha - E \end{bmatrix} = 0 \tag{6.60}$$

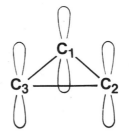

FIGURE 6.22. Notation for the cyclopropenyl system.

With $x = (\alpha - E)/\beta$, equation (6.60) is equivalent to

$$\begin{bmatrix} x & 1 & 1 \\ 1 & x & 1 \\ 1 & 1 & x \end{bmatrix} = 0 \quad \text{and} \quad \begin{array}{l} x^3 - 3x + 2 = 0 \\[4pt] x = -2, \quad x = 1, \quad x = 1 \end{array} \qquad (6.61)$$

$$\text{For } x = -2: \quad E_1 = \alpha + 2\beta$$

$$\text{For } x = 1: \quad E_2 = \alpha - \beta \qquad (6.62)$$

$$E_3 = \alpha - \beta$$

Here we note that there are degenerate orbitals, the energy of ψ_2 and ψ_3 being the same, namely $(\alpha - \beta)$. Hence for the cyclopropenyl carbonium ion:

$$E_{C_3H_3}(+) = 2(\alpha + 2\beta) = 2\alpha + 4\beta \qquad (6.63)$$

For cyclopropenyl carbanion:

$$E_{C_3H_3}(-) = 2\alpha + 4\beta + (\alpha - \beta) + (\alpha - \beta) = 4\alpha + 2\beta \qquad (6.64)$$

The level diagrams are depicted in Figure 6.23.

(d) Cyclobutadiene

Referring to Figure 6.24, we have

$$\psi = a_1 p_1 + a_2 p_2 + a_3 p_3 + a_4 p_4$$

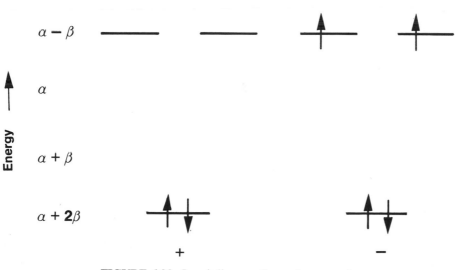

FIGURE 6.23. Level diagram for cyclopropenyl.

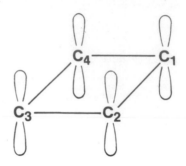

FIGURE 6.24. Notation for cyclobutadiene.

The secular determinant, remembering that $H_{14} = H_{41}$ and putting $x = (\alpha - E)/\beta$, takes the form

$$\begin{vmatrix} x & 1 & 0 & 1 \\ 1 & x & 1 & 0 \\ 0 & 1 & x & 1 \\ 1 & 0 & 1 & x \end{vmatrix} = 0 \tag{6.65}$$

Expanding this determinant immediately yields

$$x\begin{vmatrix} x & 1 & 0 \\ 1 & x & 1 \\ 0 & 1 & x \end{vmatrix} - 1\begin{vmatrix} 1 & 1 & 0 \\ 0 & x & 1 \\ 1 & 1 & x \end{vmatrix} - 1\begin{vmatrix} 1 & x & 1 \\ 0 & 1 & x \\ 1 & 0 & 1 \end{vmatrix} = 0 \tag{6.66}$$

or

$$x^4 - 4x^2 = 0 \tag{6.67}$$

The roots are therefore $x = -2$, $x = 0$, $x = 0$, $x = 2$, corresponding to the following energy levels:

$$\left. \begin{aligned} x = -2: \quad & E_1 = \alpha + 2\beta \\ x = 0: \quad & E_2 = \alpha \\ x = 0: \quad & E_3 = \alpha \\ x = 2: \quad & E_4 = \alpha - 2\beta \end{aligned} \right\} \tag{6.68}$$

and

$$E_\pi = 4\alpha + 4\beta \tag{6.69}$$

(e) Trimethylene Methane

191

*Molecular Orbital
Methods and
Polyatomic
Molecules*

The structure is depicted in Figure 6.25; the π-MOs are written as

$$\psi = a_1 p_1 + a_2 p_2 + a_3 p_3 + a_4 p_4 \tag{6.70}$$

With the simplifications

$$H_{14} = H_{41} = H_{24} = H_{42} = H_{34} = H_{43} = \beta \tag{6.71}$$

and all other H_{AB} with $A \neq B$ taken to be zero, the secular determinant then becomes

$$\begin{vmatrix} x & 0 & 0 & 1 \\ 0 & x & 0 & 1 \\ 0 & 0 & x & 1 \\ 1 & 1 & 1 & x \end{vmatrix} = 0 \tag{6.72}$$

or

$$x \begin{vmatrix} x & 0 & 1 \\ 0 & x & 1 \\ 1 & 1 & x \end{vmatrix} - 1 \begin{vmatrix} 0 & x & 0 \\ 0 & 0 & x \\ 1 & 1 & 1 \end{vmatrix} = 0 \tag{6.73}$$

which yields

$$x^4 - 3x^2 = 0$$

or   (6.74)

$$x = -\sqrt{3}, \quad x = 0, \quad x = 0, \quad x = \sqrt{3}$$

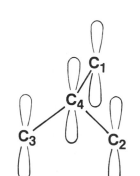

FIGURE 6.25. Notation for trimethylene methane.

The corresponding energies are:

$$x = -\sqrt{3}: \qquad E_1 = \alpha + \sqrt{3}\beta$$

$$x = 0: \qquad E_2 = \alpha$$

$$x = 0: \qquad E_3 = \alpha \qquad\qquad (6.75)$$

$$x = \sqrt{3}: \qquad E_4 = \alpha - \sqrt{3}\beta$$

$$E_\pi = 4\alpha + 2\sqrt{3}\beta$$

6.6.5. The UPS Spectrum of Benzene

Turning to benzene, and with reference to Figure 6.26, the MOs are linear combinations of the $2p_z$ orbitals* denoted by p_i, $i = 1$–6:

$$\psi = a_1 p_1 + a_2 p_2 + a_3 p_3 + a_4 p_4 + a_5 p_5 + a_6 p_6 \qquad (6.76)$$

The secular determinant now reads:

$$\begin{vmatrix} x & 1 & 0 & 0 & 0 & 1 \\ 1 & x & 1 & 0 & 0 & 0 \\ 0 & 1 & x & 1 & 0 & 0 \\ 0 & 0 & 1 & x & 1 & 0 \\ 0 & 0 & 0 & 1 & x & 1 \\ 1 & 0 & 0 & 0 & 1 & x \end{vmatrix} = 0 \qquad (6.77)$$

*Both the total electron density from the Thomas–Fermi method of Chapter 4 and the π-electron density given by these MOs were reported for benzene by March (1952), and compared with available X-ray scattering data (compare also Appendix A4.10).

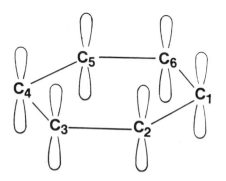

FIGURE 6.26. Notation for the benzene ring.

which yields

$$x^6 - 6x^4 + 9x^2 - 4 = 0 \qquad (6.78)$$

The roots of equation (6.78) are readily verified to be

$$x = -2, \quad x = -1, \quad x = -1, \quad x = 1, \quad x = 1, \quad x = 2$$

and hence the following energy levels are obtained:

$$\left.\begin{array}{ll}
x = -2: & E_1 = \alpha + 2\beta \\[4pt]
x = -1: & E_2 = \alpha + \beta \\[4pt]
x = -1: & E_3 = \alpha + \beta \\[4pt]
x = 1: & E_4 = \alpha - \beta \\[4pt]
x = 1: & E_5 = \alpha - \beta \\[4pt]
x = 2: & E_6 = \alpha - 2\beta
\end{array}\right\} \qquad (6.79)$$

The eigenvalue sum is

$$E = 6\alpha + 8\beta \qquad (6.80)$$

(i.e., two electrons in each of the three lowest levels E_1, E_2, and E_3).

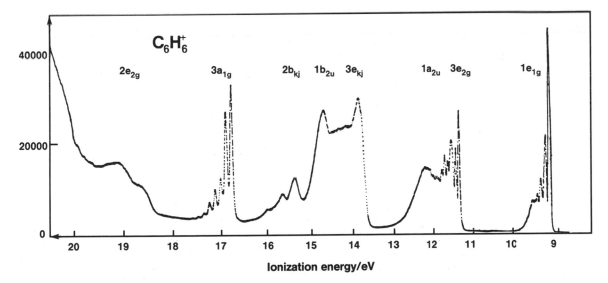

FIGURE 6.27. HeI UPS spectrum of benzene [redrawn from Karlsson *et al.* (1976)]. (For the convenience of the reader this is reproduced here from Chapter 5.)

Figure 6.27 shows the HeI UPS spectrum of benzene (cf. Section 1.8). The problems posed by the interpretation of this spectrum are typical of those encountered for large molecules and concern the difficulty of identifying the onset of a band system when there is overlapping with neighboring systems and, generally, a lack of resolved vibrational structure.

One of the advantages of having reliable MO calculations is that recognition of the onset of a band system is facilitated, and the analysis of such a spectrum of a large molecule is often a result of combining theory with the experiment. In addition, study of the angular dependence of photoelectrons can aid the identification of overlapping systems [see Hollas (1982)].

(a) Use of Hückel* and More Refined MO Calculations on C_6H_6

Returning briefly to the Hückel results discussed above, we recall that the π-electron MOs of benzene were described using only the $2p_z$ AOs, z being perpendicular to the plane of the benzene ring, on the six C atoms. In the ground state, four of the π-electrons are in the outer ($1e_{1g}$) and two in the $1a_{2u}$ MO (Figure 6.27).

As to refinements, Table 6.2 shows the results of a self-consistent field (SCF) calculation of orbital energies for the ground configuration of C_6H_6. Here, the basis set (cf. Appendix A2.4) used comprised the $1s$, $2s$, and $2p$ AOs on C and the $1s$ AO on H, the results being compared in Table 6.2 with the ionization energies obtained from the UPS spectra. Koopmans' theorem (see Appendix A6.2) is assumed in this comparison.

Naturally the calculation does not include electron correlation (see Section 6.11) and, at least partly because of this, the calculated ionization energies are substantially higher than the observed values. However, the order of the calculated energies is considerably more reliable than their absolute values. The above calculation clearly points to the fact that the UPS spectrum in the region between 11.5 and 12.5 eV consists of two band systems and that between 14 and 16 eV of three. Inclusion of correlation reduces the discrepancy between theory and experiment, particularly for low-ionization energies [von Niessen *et al.* (1976)].

A feature of these calculations, as seen from Table 6.2, is that the $3e_{2g}$ orbital (of σ-type) has a rather lower ionization energy than the $1a_{2u}$ (of π-character). This serves as a caution as to the quantitative validity of any kind of MO calculation that takes account only of π-electrons.

At best, vibrational structure is only partially resolved in band systems of large molecules. The structure in the first, denoted here by $(1e_{1g})^{-1}$, which indicates the removal of an electron from an orbital, and the second $(3e_{2g})^{-1}$ system is complex, which could be due, at least in part, to there being a Jahn–Teller effect (see Appendix A6.1) in each of them. The only relatively simple structure is in the seventh $3a_{1g}$ system which is dominated by a progression in the a_{1g} ring-breathing vibration (ν_1). This is consistent with the electron being ejected from a σCC bonding orbital [see Hollas (1982)].

---

*Note added in proof: G. G. Hall [(*Int. J. Quantum Chem.* **45**, 167 (1993)] treats conjugated benzenoid molecules rather generally.

TABLE 6.2
Self-Consistent Field Calculations and Experimental Results for
the Ground State of Benzene[a]

A. Orbital Energies (atomic units) of Benzene Ground and Doublet States

Symmetry	Ground state	E_{1g} cation	A_{2u} cation
$1a_{1g}$	−11.30	−11.57	−11.57
$1e_{1u}$	−11.30	−11.57	−11.56
$1e_{2g}$	−11.29	−11.57	−11.56
$1b_{1u}$	−11.29	−11.57	−11.56
$2a_{1g}$	−1.168	−1.431	−1.427
$2e_{1u}$	−1.037	−1.293	−1.294
$2e_{2g}$	−0.846	−1.096	−1.095
$3a_{1g}$	−0.738	−0.982	−0.981
$2b_{1u}$	−0.662	−0.895	−0.892
$1b_{2u}$	−0.654	−0.914	−0.912
$3e_{1u}$	−0.622	−0.868	−0.868
$1a_{2u}$ (π)	−0.535	−0.757	−0.801
$3e_{2g}$	−0.524	−0.770	−0.769
$1e_{1g}$ (π)	−0.373	−0.623	−0.595
$1e_{2u}$ (π^*)	+0.127	. . .	. . .
$1b_{2g}$ (π^*)	+0.363	. . .	. . .

B. Experimental and Calculated Ionization Potentials of Benzene (eV)

Calculated values		Experimental values		
Orbital energy	Orbital	Rydberg limit	Photoelectron method[b]	Photoelectron method[c]
10.15	$1e_{1g}$	9.24 (π)	9.25	9.25
14.26	$3e_{2g}$		(12.19)	13.88
14.56	$1a_{2u}$	11.48 (π)	11.49	11.51
16.92	$3e_{1u}$		13.67	(14.87)
18.01	$2b_{1u}$		(14.44)	(15.54)
17.80	$1b_{2u}$	16.84 (σ)	16.73	16.84
20.08	$3a_{1g}$		18.75	18.22

[a]Data from J. M. Schulman and J. W. Moskowitz (1967).
[b]Work of M. I. Al-Joburg and D. W. Turner, Jr. (1965).
[c]Work of I. D. Clark and D. C. Frost (1967)

6.7. DETERMINATION OF MOLECULAR ORBITALS AND LCAO COEFFICIENTS

Having considered the calculation of electronic energy values, we can now determine the corresponding MOs. Using the allyl system as an example, we proceed as follows:

1. Normalization requirement on equation (6.48):

$$a_1^2 + a_2^2 + a_3^2 = 1 \qquad (6.81)$$

2. Secular equation (see equation (6.52)):

$$\left.\begin{vmatrix} x & 1 & 0 \\ 1 & x & 1 \\ 0 & 1 & x \end{vmatrix} = 0 \right\} \rightarrow \text{roots} \qquad x = -\sqrt{2}, \quad x = 0, \quad x = \sqrt{2} \tag{6.82}$$

3. Incorporating the coefficients (a_1, a_2, a_3) into a column vector gives

$$\begin{pmatrix} x & 1 & 0 \\ 1 & x & 1 \\ 0 & 1 & x \end{pmatrix} \begin{pmatrix} a_1 \\ a_2 \\ a_3 \end{pmatrix} = 0 \tag{6.83}$$

or

$$a_1 x + a_2 = 0, \qquad a_1 + a_2 x + a_3 = 0, \qquad a_2 + a_3 x = 0 \tag{6.84}$$

For $x = -\sqrt{2}$, one finds immediately from equations (6.84)

$$-\sqrt{2}\, a_1 + a_2 = 0, \qquad a_1 - \sqrt{2}\, a_2 + a_3 = 0, \qquad a_2 - \sqrt{2}\, a_3 = 0 \tag{6.85}$$

which leads to

$$a_1 = a_3 \tag{6.86}$$

Using equations (6.81) and (6.85) yields

$$a_1^2 + 2a_1^2 + a_1^2 = 1 \tag{6.87}$$

and hence

$$a_1 = \tfrac{1}{2} \tag{6.88}$$

Equations (6.86) and (6.85) then give finally

$$a_2 = \sqrt{2}/2, \qquad a_3 = \tfrac{1}{2} \tag{6.89}$$

Therefore the molecular orbital ψ_1 is found explicitly as

$$\psi_1 = \tfrac{1}{2} p_1 + (\sqrt{2}/2) p_2 + \tfrac{1}{2} p_3 \tag{6.90}$$

For $x = 0$

$$a_2 = 0, \qquad a_1 + a_3 = 0, \qquad a_2 = 0 \tag{6.91}$$

and hence

$$a_2 = 0 \quad \text{and} \quad a_3 = -a_1 \tag{6.92}$$

Using the normalization condition

$$a_1^2 + a_1^2 = 1 \tag{6.93}$$

together with equation (6.92) gives

$$a_1 = 1/\sqrt{2}, \quad a_2 = 0, \quad a_3 = -1/\sqrt{2} \tag{6.94}$$

In this case, therefore, the normalized MO is

$$\psi_2 = (1/\sqrt{2})p_1 - (1/\sqrt{2})p_3 \tag{6.95}$$

For $x = \sqrt{2}$:

$$\sqrt{2}a_1 + a_2 = 0, \quad a_1 + \sqrt{2}a_2 + a_3 = 0, \quad a_2 + \sqrt{2}a_3 = 0 \tag{6.96}$$

Thus

$$a_2 = -\sqrt{2}a_1, \quad a_3 = a_1 \tag{6.97}$$

and

$$a_1^2 + 2a_1^2 + a_1^2 = 1$$

or

$$a_1 = \tfrac{1}{2}, \quad a_2 = -\sqrt{2}/2, \quad a_3 = \tfrac{1}{2} \tag{6.98}$$

Hence

$$\psi_3 = \tfrac{1}{2}p_1 - (\sqrt{2}/2)p_2 + \tfrac{1}{2}p_3 \tag{6.99}$$

As another example we can use the same method to find the coefficients and the MOs of (delocalized) butadiene. We shall not give the details here [see Salem (1966)] but this case is set as a problem at the end of the chapter.

6.8. MOLECULAR INDEXES

We are now in a position to use the coefficients of the MOs to calculate some important indexes such as electron density on atoms (ED), charge density (q), bond order (p), and free valence on atoms (F). The following examples will aid in understanding both the meaning of these indexes and the method of calculating them (see also footnote on p. 201).

6.8.1. Electron Density on Atoms

Consider one of the two electrons in the ethylene molecule (i.e., the π-bond). We recall that

$$\psi_1 = a_1 p_1 + a_2 p: \qquad a_1 = a_2 = 1/\sqrt{2} \tag{6.100}$$

and thus

$$a_1^2 + a_2^2 = 1$$

The electron density of electron 1 at C_1 is equal to $a_1^2 = (1/\sqrt{2})^2 = \frac{1}{2}$ and at C_2, $a_2^2 = (1/\sqrt{2})^2 = \frac{1}{2}$. Now there are two π-electrons occupying orbital ψ_1 in ethylene and hence the electron density at C_1 (ED_1) due to both electrons is

$$ED_1 = 2(\tfrac{1}{2}) = 1.00$$

and at C_2

$$ED_2 = 2(\tfrac{1}{2}) = 1.00$$

The conclusion is that the electron density is equal to one at each C atom, both of these atoms being neutral. As a further example, we recall that for the allyl system:

$$\psi_1 = \tfrac{1}{2} p_1 + (\sqrt{2}/2) p_2 + \tfrac{1}{2} p_3 \tag{6.102}$$

The electron density at each position due to one electron in ψ_1 equals

At C_1: $\qquad (\tfrac{1}{2})^2 = \tfrac{1}{4}$

At C_2: $\qquad (\sqrt{2}/2)^2 = \tfrac{1}{2}$

At C_3: $\qquad (\tfrac{1}{2})^2 = \tfrac{1}{4}$

There are two π-electrons in the allyl carbonium ion, and we have for the total electron density at C_1, C_2, and C_3

$$ED_1 = 2(\tfrac{1}{4}) = 0.500$$

$$ED_2 = 2(\tfrac{1}{2}) = 1.00$$

$$ED_3 = 2(\tfrac{1}{4}) = 0.500$$

In the allyl radical we have two electrons in ψ_1 and one electron in ψ_2. Since ψ_1 is given by equation (6.90) and ψ_2 by equation (6.95), it follows that the electron density at C_1 due to the two electrons in ψ_1 is $2(\frac{1}{2})^2$ and that due to the one electron in ψ_2 is $(1/\sqrt{2})^2$. The electron density at C_2 and C_3 can be calculated in a similar fashion. Therefore,

$$ED_1 = 2(\tfrac{1}{2})^2 + (1/\sqrt{2})^2 = 1.00$$

$$ED_2 = 2(\sqrt{2}/2)^2 + 0 = 1.00$$

$$ED_3 = 2(\tfrac{1}{2})^2 + (-1/\sqrt{2})^2 = 1.00$$

The carbanion has two electrons in both ψ_1 and ψ_2 and hence

$$ED_1 = 2(\tfrac{1}{2})^2 + 2(1/\sqrt{2})^2 = 1.50$$

$$ED_2 = 2(\sqrt{2}/2)^2 + 2(0) = 1.00$$

$$ED_3 = 2(\tfrac{1}{2})^2 + 2(-1/\sqrt{2})^2 = 1.50$$

6.8.2. Charge Density on Atoms

Let us consider first ethylene. Prior to the formation of the π-bond, each C atom has associated with it one electron that can be used in π-bond formation. This means that each C atom must have an electron density of 1. We now define the charge density q at any C atom to be 1.00 minus the electron density at that atom:

$$q = 1.00 - ED \qquad \text{(6.103)}$$

For ethylene, the charge density at each position is therefore

$$\text{At } C_1: \quad q_1 = 1.00 - ED_1 = 1.00 - 1.00 = 0$$

$$\text{At } C_2: \quad q_2 = 1.00 - ED_2 = 1.00 - 1.00 = 0$$

We treat next the allyl carbonium ion. It is usually considered as a resonance hybrid of the following structures:

$$\overset{+}{C}=C-C \leftrightarrow C-C=\overset{+}{C}$$

This means that one-half of the positive charge is at each of the terminal positions and that the middle carbon atom is neutral. We find this to be the result using MO theory since:

$$q_1 = 1.00 - 0.500 = 0.500$$

$$q_2 = 1.00 - 1.00 = 0$$

$$q_3 = 1.00 - 0.50 = 0.500$$

In the case of the allyl radical, the charge density at each C is

$$q_1 = 1.00 - 1.00 = 0$$

$$q_2 = 1.00 - 1.00 = 0$$

$$q_3 = 1.00 - 1.00 = 0$$

Hence in the radical, the third π-electron* is in ψ_2; this causes a change in electron density equal to 0.500 at C_1 and C_3, but no change at C_2. Similar results obtain for the allyl carbanion: the minus charge is spread out so that one-half of it resides at C_1 and the other half at C_3:

$$q_1 = 1.00 - 1.500 = -0.500$$

$$q_2 = 1.00 - 1.00 = 0$$

$$q_3 = 1.00 - 1.50 = -0.500$$

The charge density in the cyclopropenyl cation is spread equally over all positions:

$$q_1 = 1 - \tfrac{2}{3} = \tfrac{1}{3}$$

$$q_2 = 1 - \tfrac{2}{3} = \tfrac{1}{3}$$

$$q_3 = 1 - \tfrac{2}{3} = \tfrac{1}{3}$$

For the cyclopropenyl anion:

$$q_1 = 1 - \tfrac{4}{3} = -\tfrac{1}{3}$$

$$q_2 = 1 - \tfrac{4}{3} = -\tfrac{1}{3}$$

$$q_3 = 1 - \tfrac{4}{3} = -\tfrac{1}{3}$$

*The reader can consult equation (6.53) and Figure 6.21.

6.8.3. Bond Order

Bond order* is defined so that the degree of bonding between two C atoms (A and B) is represented by the bond order

$$p_{AB} \equiv \sum a_A a_B \qquad (6.104)$$

where the summation is over all π-electrons. Hence the amount of bonding between two C atoms C_A and C_B due to one electron in a particular MO is equal to the product of the coefficients $a_A a_B$ for that orbital. Some examples follow.†

(a) Ethylene

For the double bond in ethylene, the bond order due to *one* electron is

$$a_1 a_2 = (1/\sqrt{2})(1/\sqrt{2}) = \tfrac{1}{2}$$

With two electrons in this bond, the total π-bond order between C_A and C_B is

$$p_{12} = 2(\tfrac{1}{2}) = 1.00$$

(b) The Allyl System

The π-bond order between C atoms 1 and 2 is the same as that between atoms 2 and 3. For the carbonium ion:

$$p_{12} = p_{23} = 2(\tfrac{1}{2})(\sqrt{2}/2) = 0.707$$

In the radical we recall that there are two electrons in ψ_1 and one in ψ_2. The π-bond order for the radical is therefore

$$p_{12} = 2(\tfrac{1}{2})(\sqrt{2}/2) + (1/\sqrt{2})(0) = 0.707$$

$$p_{23} = 2(\sqrt{2}/2)(\tfrac{1}{2}) + (0)(-1/\sqrt{2}) = 0.707$$

For the carbanion

$$p_{12} = 2(\tfrac{1}{2})(\sqrt{2}/2) + 2(0)(1/\sqrt{2}) = 0.707$$

$$p_{23} = 2(\sqrt{2}/2)(\tfrac{1}{2}) + 2(0)(1/\sqrt{2}) = 0.707$$

These bond orders are useful not only in free space molecules, but also for organic molecules weakly chemisorbed on metal surfaces [see, e.g., Hiett *et al.* (1984)].

---

*Free valence in Table 6.3 below is closely related to bond order. Essentially free valence F measures the difference between maximum bond order C and the total bond order ΣP at a given position. F characterizes reactivity toward attack by free radicals [see McWeeny (1979)].
†A single bond is said to have a bond order of one: a double bond of two.

6.9. SOME QUANTUM-MECHANICAL AND SEMIEMPIRICAL METHODS APPLIED TO POLYATOMIC MOLECULES

As seen already, the HMO method has its value, but the approximations involved are drastic. We therefore consider next the bases of other MO methods applied to polyatomic molecules, all of which depend on the HF equations considered in Section 4.5. This is within the Roothaan–Hall formulation.

It will be recalled from the Hückel method that MO calculations yield molecular indexes of various kinds. These in turn can be employed to calculate properties of molecules that are useful to the chemical physicist. By way of illustration, Table 6.3 summarizes several important properties that one can calculate via MO methods. It is obviously important for reliable calculation of properties to obtain accurate values of the coefficients of the AOs in the MOs, as well as good energies. Let us therefore turn to consider methods less drastic than the Hückel theory.

When the HF equations are combined with the LCAO approximation, it was found in Section 4.5 that the Roothaan–Hall equations follow. Thus the HF SCF equations:

$$H^{\text{SCF}}\phi_i = E_i^{\text{SCF}}\phi_i \tag{6.105}$$

are satisfied by the molecular orbitals ϕ_i. Following Roothaan and Hall, we write

$$\phi_i = \sum_n c_{in}\chi_n \tag{6.106}$$

TABLE 6.3
Some Applications of Molecular Indexes Determined by Molecular Orbital Calculations

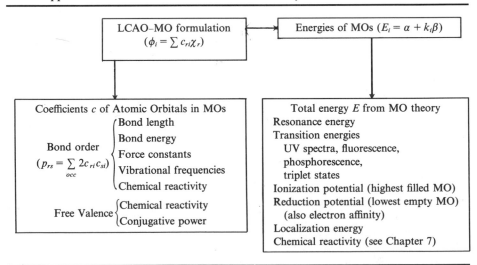

where the c_{in} are LCAO coefficients, with χ_n the atomic orbitals. Employing equation (6.106) in equation (6.105), we can write

$$H^{SCF} \sum_n c_{in}\chi_n = E_i^{SCF} \sum_n c_{in}\chi_n \qquad (6.107)$$

and we are led to the Roothaan–Hall equations:

$$\sum_n H_{mn}^{SCF} c_{in} = \sum_n S_{mn}c_{in}E_i^{SCF} \qquad (6.108)$$

Here S_{mn} is the overlap integral

$$S_{mn} = \int \chi_m^* \chi_n \, d\tau \qquad (6.109)$$

and

$$H_{mn}^{SCF} = H_{mn}^N + G_{mn} \qquad (6.110)$$

where

$$H_{mn}^N = \int \chi_m^* \left[-\tfrac{1}{2}\nabla^2 - \sum_A Z_A/r_A \right] \chi_n \, d\tau \qquad (6.111)$$

The quantity G_{mn} can be expressed in the form

$$G_{mn} = \sum_{ls} [P_{ls}(mn|ls) - \tfrac{1}{2}(ms|ln)] \qquad (6.112)$$

where

$$(mn|ls) = \iint \chi_m^*(1)\chi_n(1/r_{12})\chi_l^*(2)\chi_s(2) \, d\tau_1 \, d\tau_2 \qquad (6.113)$$

while P_{ls} is given in terms of the coefficients c by

$$P_{ls} = 2 \sum_i^{occ} c_{il}^* c_{is} \qquad (6.114)$$

It has now to be recognized that some of these integrals are difficult to evaluate. Hence in approximate MO methods, some are set equal to zero while others are replaced by semiempirical quantities, leading to the class of semiempirical MO methods. Such methods have been useful when performing MO calculations on large, polyatomic molecules in particular.

Below we shall record a variety of such methods but let us start by illustrating the so-called CNDO method (complete neglect of differential overlap). The basic assumption here is that integrals of the type $(mn|ls)$ are zero, unless $m = n$ and $l = s$, in which case one can define an integral

$$\gamma_{ml} = (mm|ll) \qquad (6.115)$$

Furthermore, the integrals as expressed by equation (6.115) are assumed to be dependent only on the atoms m and l and not on the particular type of atomic orbital. With these assumptions, one obtains the following diagonal matrix elements:

$$H_{mm}^{SCF} = H_{mm}^{N} - \tfrac{1}{2} P_{mm} \gamma_{mm} + \sum_{s(\neq m)} P_{ss} \gamma_{ms} \qquad (6.116)$$

and off-diagonal elements

$$H_{mn}^{SCF} = H_{mn}^{N} - \tfrac{1}{2} P_{mn} \gamma_{mn} \qquad (6.117)$$

These are the matrix elements to be employed in this simplification of the Roothaan–Hall formulation. One is led to the secular determinant

$$\det |H_{mn}^{SCF} - E_{i}^{SCF} S_{mn}| = 0 \qquad (6.118)$$

While, at first sight, the result (6.118) seems to have the same form as in the Hückel approach, the matrix elements as expressed in equations (6.116) and (6.117) are now somewhat different.

With this illustration, we list several existing methods*:

1. The CNDO method [Pople and Segal (1965 and 1966)].
2. The partial neglect of differential overlap [PNDO: see Dewar and Klopman (1967)].
3. The intermediate neglect of differential overlap (INDO), and also its modification (MINDO) [see Baird and Dewar (1969), Dewar and Haselbach (1970), and Bingham *et al.* (1975)].
4. The valence electron approximation [see Pople *et al.* (1965)].
5. The extended Hückel method [Hoffmann (1963)].
6. The π-electron approximate methods of Goeppert-Mayer and Sklar [see Parr (1963)].
7. The Pariser–Parr–Pople method [see Parr (1963)].

At this point, we simply outline, as a further example, the Pariser–Parr–Pople (PPP) method which is of continuing importance, having especial value in relation to electronic spectra of polyatomic molecules.

---

*Notable studies by Zerner (1979) and his co-workers should also be consulted in this context.

6.9.1. The Pariser–Parr–Pople Method

In the PPP method, as in the HMO treatment, one assumes the separability of σ- and π-electrons. Molecular orbitals are again formed as a linear combination of atomic orbitals, i.e., the LCAO–MO procedure is adopted and the secular determinant is developed:

$$
\begin{vmatrix}
(\alpha_1 - E) & \beta_{12} & \beta_{13} & \cdots & \beta_{1n} \\
\beta_{21} & (\alpha_2 - E) & \beta_{23} & \cdots & \beta_{2n} \\
\beta_{31} & \beta_{32} & (\alpha_3 - E) & \cdots & \beta_{3n} \\
\vdots & \vdots & \vdots & & \vdots \\
\beta_{n1} & \beta_{n2} & \beta_{n3} & & (\alpha_n - E)
\end{vmatrix} = 0 \qquad (6.119)
$$

This secular determinant appears very similar to the HMO-secular determinant; however, there are important differences, which we now consider.

We recall from the HMO method that

$$
\alpha_p = H_{pp} = \text{Coulomb integral} = \int \chi_p \mathscr{H} \chi_p \, d\tau \qquad (6.120)
$$

and is equal to the energy of an electron localized in atom p.

$$
\beta_{pq} = H_{pq} = \text{Resonance integral} = \int \chi_p \mathscr{H} \chi_q \, d\tau \qquad (6.121)
$$

and is equal to the energy of an electron in the region of overlap of the AOs χ_p and χ_q.

$$
S_{pq} = \text{Overlap integral} = \int \chi_p \chi_q \, d\tau \qquad (6.122)
$$

and is equal to the measure of overlap of the two orbitals of the AOs χ_p and χ_q, and equal to zero [i.e., there is zero differential overlap (ZDO)] in both the HMO and PPP methods; and of course

$$
S_{pp} = \int \chi_p \chi_p \, d\tau = 1 \qquad (6.123)
$$

Furthermore, in the HMO method, the energies α_p and β_{pq} (written simply as α and β in HMO) refer to the energies present in the absence of all other π-electrons. Hence no account is taken of the energies of repulsion owing to the presence of other electrons, i.e., the effects of electronic interactions are ignored. In the PPP method, on the other hand, not only are the energies due to electronic interactions included but molecular geometry is taken into consideration. These important properties, along with others, are incorporated into the improved forms of α_p and β_{pq} which can be expressed in the PPP method as

$$\alpha_p = \alpha_p^{core} + E_p^R \tag{6.124}$$

and

$$\beta_{pq} = \beta_{pq}^{core} + E_{pq}^R \tag{6.125}$$

where E_p^R and E_{pq}^R are electron repulsion energies; and α_p^{core} and β_{pq}^{core} are known as the core energies. Let us pursue the significance and meaning of each term in equations (6.124) and (6.125).

To begin with, E_p^R is the one-center repulsion energy which arises from the interaction of an electron localized on atom p with all the other π-electrons. Similarly, E_{pq}^R in the two-center repulsion energy which arises from the interaction of an electron that is "located" in the overlap regions of atoms p and q with the other π-electrons.

The core energies α_p^{core} and β_{pq}^{core} denote the attractive energies of the electron localized on atom p or between p and q, respectively. The force of attraction arises from the positively charged nuclear framework when one assumes that the other π-electrons are not present.

The repulsion energies and the core energies can be expressed in terms of empirical parameters such as ionization potentials, electron affinities, and expressions taking account of the geometry of the polyatomic molecule. Our aim here has merely been to highlight the differences between the PPP method and the Hückel approach. The interested reader must consult the references for full details. However, we shall point out briefly the way one can proceed to use the expressions thus obtained.

Equations (6.124) and (6.125) allow the secular determinant of the PPP method to be set up. However, in order to accomplish this, we need a set of coefficients* (c's) in order to form the expressions for α_p and β_{pq}. One might well ask, which coefficients do we use? The answer to this question involves an understanding of the iterative procedure that the computer program utilizes to reach what was called in Chapter 4 a self-consistent field. This process requires some explanation. Coefficients (c's) for several molecules resulting for example from HMO calculations are well known or can easily be generated if not available. These HMO coefficients can be used to set up an initial PPP secular determinant. The computer program for the PPP method then generates a set of improved coefficients. These new coefficients are then automatically used again by the computer program to produce still another set of improved coefficients; and the same process is continued through a number of iterations until two successive sets of coefficients yield a consistent set of coefficients. When this consistency is reached, we have a self-consistent field. The program then produces the consistent coefficients and the corresponding MO energies E_i, and these can be used to calculate the desired molecular properties.

---

*For details of LCAO coefficients and the formation of the PPP secular determinant see Parr (1963).

6.9.2. The *Ab Initio* Method

Of particular importance at the time of writing is the *ab initio* method. Here all electrons—bonding, nonbonding, and inner shell—are included in the calculations. With the availability of supercomputers, this method has become much more useful and popular. One effective treatment of this approach is the account in the book by Richards and Cooper (1983). Richards (1983) discusses elsewhere the role of the *ab initio* method in pharmacology.

6.10. MOLECULAR ORBITALS IN ACETYLENE AND THE (e, 2e) EXPERIMENT

In Chapter 5 we noted the importance of MO theory in unraveling photoelectron spectra of molecules. There is a quite different technique, the so-called (e, 2e) experiment, that is also important in a related context, as it has strong connection with the MO description. Though the (e, 2e) method is much more general than the present example, namely the study of MOs in acetylene, we thought it of interest to include in this chapter a description of the principles behind the (e, 2e) experiment and a discussion of the results obtained thereby on acetylene.

It is worth noting here that, in essence, the (e, 2e) experiment measures momenta of electrons. However, as discussed elsewhere in this book, if we know a space wave function $\psi(\mathbf{r})$, then the corresponding wave function in momentum ($\mathbf{p}$) space is obtained by Fourier transforming $\psi(\mathbf{r})$. Another technique that is worth noting in passing which also measures electronic momentum distributions is the Compton line shape experiment, discussed, for example, in the book edited by Williams (1977).

6.10.1. The Technique and Theory of (e, 2e) Studies

The account below represents a summary of the review of McCarthy and Weigold (1983). The essential process involved is to arrange for a beam of electrons, all with equal momenta $\mathbf{p}_0$, say, of some 100 to 10,000 eV energy, to collide with a beam of target atoms (or molecules). Two electron detectors are set up to measure the momenta of the two emitted electrons in such a way that their directions from the collision center are known.

The momentum of a free electron is obtained by measuring its kinetic energy and the direction of its motion.

The electrostatic fields in the two detectors are set up so that the energies of any detected electrons are equal and their sum is less than the energy $p_0^2/2m$ of an incident electron by an energy ε. In addition, the technique of coincidence counting is used to ensure that the two electrons come from the same collision of an incident electron with an atom, say.

For each recorded coincidence, one knows the momenta p_A and p_B recorded in the detectors A and B. If, to illustrate, we take the target as a H atom, one electron is the incident electron while the other will be that originally bound to the nucleus (proton) of the H atom. The coincidence count rate is much higher than the background of random coincidence counts only if the energy difference ε is equal to the energy (13.6 eV) required to separate the electron from the nucleus in the ground state of the H atom. In multielectron atoms, electrons are in various atomic orbitals (cf. Appendix A1.2), some of which have different binding or separation energy. The separation energy ε_i of the orbital i is measured in the same fashion.

In each recorded collision, one has measured the difference between the incident momentum $\mathbf{p}_0$ and the total final momentum $\mathbf{p}_A + \mathbf{p}_B$. This is the recoil momentum $\mathbf{q}$, say, of the ion that is left after the removal of one electron (initially the target atom being at rest). Evidently

$$\mathbf{q} = \mathbf{p}_0 - \mathbf{p}_A - \mathbf{p}_B \tag{6.126}$$

If the incident kinetic energy is high enough, the collision occurs sufficiently rapidly that one has a "clean knockout." Then the recoil momentum is equal and opposite to the momentum of the bound electron at the moment of impact. In that situation it follows that one has a direct measure of the momentum, $\mathbf{p}$ say, of the bound electron:

$$\mathbf{p} = -\mathbf{q} \tag{6.127}$$

If the atomic electron is stationary ($\mathbf{p} = 0$) and unbound ($\varepsilon = 0$) the kinematics of the collision requires that the two electrons emerge at the same angle as

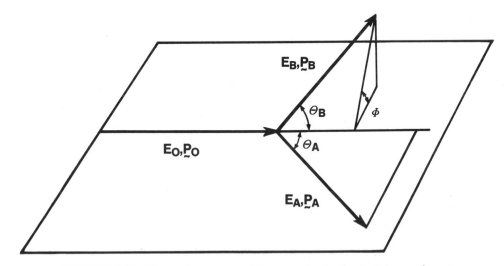

FIGURE 6.28. Noncoplanar symmetric geometry in the (*e*, 2*e*) experiment.

billiard balls after a collision, i.e., at 90° relative to each other. If the electrons have equal energies, this means that the polar angles θ_A and θ_B made with the incident beam direction are each 45°, and the final momentum vectors are coplanar with the incident momentum $\mathbf{p}_0$. If the atomic electron struck is moving, one can only detect it if the detectors are moved from their $\mathbf{p} = 0$ position. One can introduce various shifts of the detectors in order to scan the range of $\mathbf{p}$, but it can be done systematically by keeping the detectors in the plane of the incident beam while holding θ_A fixed and varying θ_B. Another way is to keep $\theta_A = \theta_B$ at approximately 45° and to vary the azimuthal angle ϕ made by B with the OA plane. This is termed noncoplanar symmetric geometry. Figure 6.28 shows a schematic diagram of the kinematics in an $(e, 2e)$ collision.

6.10.2. Results of Measurements on Acetylene

Without more ado, we turn to the results of the $(e, 2e)$ experiment on acetylene (C_2H_2). This is the simplest molecule with a triple bond and serves as a prototype for a large class of more complicated organic molecules. A comparison of precise measurements of the momentum distribution of the valence electrons, carried out by Coplan *et al.* (1978) provides a critical test of the theoretical description of the electronic structure of the molecule.

The molecular orbital valence electron configuration is $(2\sigma_g)^2$-$(2\sigma_u)^2(3\sigma_g)^2(1\pi_u)^4$. The $1\pi_u$ and $3\sigma_g$ electrons are primarily associated with the C—C triple bond while the $2\sigma_u$ and $2\sigma_g$ electrons contribute principally to the C—H bonds.

The momentum distributions, measured as discussed above, are given theoretically by the square of the magnitude of the single-electron wave functions in momentum space. Figure 6.29 displays the relative differential $(e, 2e)$ cross sections for the valence electrons of acetylene. The points represent the measurements while the solid lines are from theory, based on the s and p orbital contributions of the wave functions of McLean and Yoshimine (1967). The vertical error bars on the experimental points represent the degree of precision of the measurements (about 7%) while the horizontal error bars show the resolution in momentum space, about 0.04 atomic units.

From Figure 6.29, it is clear that for the $1\pi_u$ electrons, of particular interest to us in the present context, the measured probability of finding them with low values of momenta is greater than the theory predicts. There is, it is worth adding, a similar discrepancy between the measured and calculated momentum densities for the $2\sigma_u$ electrons.

Coplan *et al.* (1978) point out that the $(e, 2e)$ experiments focus on the fact that the p contribution to a polyatomic wave function results in electron density being more uniformly distributed over a large volume of configuration space than anticipated by molecular orbital theory. Further theoretical work is plainly needed here.

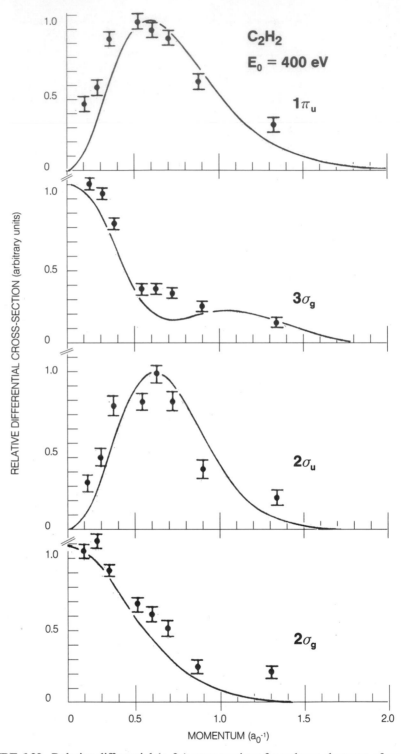

FIGURE 6.29. Relative differential (*e*, 2*e*) cross sections for valence electrons of acetylene [redrawn from Coplan *et al*. (1978)].

6.11. SLATER–KOHN–SHAM ONE-ELECTRON EQUATIONS

To conclude this chapter, we return to the electron density description. There, through equation (4.1), the ground-state electron density of a molecule was written in terms of a one-body potential energy $V(\mathbf{r})$ in which independent electrons were assumed to move. We are afforded a valuable approach in present-day quantum chemistry, following the philosophy of Chapter 4, by adding to the conventional electrostatic contribution to $V(\mathbf{r})$, namely

$$V_{\text{Hartree}}(\mathbf{r}) = V_N(\mathbf{r}) + e^2 \int d\mathbf{r}' \, \rho(\mathbf{r}')/|\mathbf{r} - \mathbf{r}'| \tag{6.128}$$

an exchange-correlation potential $V_{xc}(\mathbf{r})$. As was discussed in Chapter 4, it is often taken to be such that the exchange part is approximated by a constant times $[\rho(\mathbf{r})]^{1/3}$. In the so-called Slater $X\alpha$ method, the constant can be varied, and has been tabulated for numerous atoms by now. A common approximation for the correlation contribution is to use the known result for a uniform interacting electron assembly [cf. Kohn and Sham (1965)], which has been obtained by computer simulation, and fitted by analytic formulas (see Appendix A4.7).

This approach has already been applied to a number of light molecules. We shall not go into further detail here, but the reader is referred to reviews by March (1981) and by Jones and Gunnarsson (1989). It is to be stressed that this method is designed for calculation of the ground-state density as the sum of the squares of the one-electron wave functions of the states occupied by electrons. It can be shown that this can be achieved (in principle exactly) by proper choice of the one-body potential energy [see Slater (1951); Balbás *et al.* (1988)]:

$$V(\mathbf{r}) = V_{\text{Hartree}}(\mathbf{r}) + V_{xc}(\mathbf{r}) \tag{6.129}$$

Unfortunately, exact knowledge of V_{xc} would be equivalent to solving the Schrödinger equation exactly, which has so far not proved feasible for more than one electron.

6.12. RENNER–TELLER EFFECT

Distortion in the direction of a degenerate normal coordinate in a degenerate electronic state of a linear molecule causes a splitting of the electronic state. This is the so-called Renner–Teller effect. This splitting may or may not result in a bent equilibrium nuclear configuration.

To put this into the context of the earlier discussion of Walsh diagrams in Chapters 2 and 4, consider the molecules of types AH_2, AB_2, and HAB. Here the double degeneracy of a π-orbital in the linear configuration of the molecule is split when the molecule bends. In an analogous fashion the degeneracy of a Π (and also a Δ, Φ, . . .) electronic state of a linear molecule is split when the bending vibration is excited.

It is to be noted that in all linear triatomic molecules there is only one bending vibration, denoted as v_2, and a Π electronic state may be split in any of three general ways when the vibration v_2 is excited. Figure 6.30b–d shows the resulting potential functions V plotted as a function of the bending coordinate Q_2, for comparison with that in Figure 6.30a for a nondegenerate electronic state. As shown in Figure 6.30b, the two potential curves, labeled V^+ for the inner and V^- for the outer, have an identical minimum in the linear configuration corresponding to $Q_2 = 0$. In Figure 6.30c V^- has a W shape with minima corresponding to a bent molecule (with Q_2 therefore nonzero). Finally in Figure 6.30d, both V^+ and V^- exhibit W-shaped forms.

One example corresponding to Figure 6.30b is the $\tilde{A}\ ^{}\Pi_u$ excited state of the C_3 molecule in which it is linear; a further, somewhat more complicated, example is provided by a similar $\tilde{G}\ ^1\Pi_u$ state of C_2H_2.

In Figure 6.30d, the two potential curves represent two separate electronic states of a bent molecule, which become degenerate in the nonequilibrium linear configuration. A transition between two such states which both correlate with a $^2\Pi_u$ state of the linear molecule is to be found in NH_2 (and also in PH_2 and AsH_2). This would result from a configuration $(1\sigma_u)^2\ (1\pi_u)^3$ obtained using an MO diagram. An example of Figure 6.30c is to be found in an $\tilde{A}\ ^2\Pi_u - \tilde{X}\ ^2A_1$ transition of BH_2 [see Hollas (1982)].

Some further detail of the Renner–Teller effect is given by Hollas (1982) where a useful potential function for the bending coordinate Q_2 is recorded.

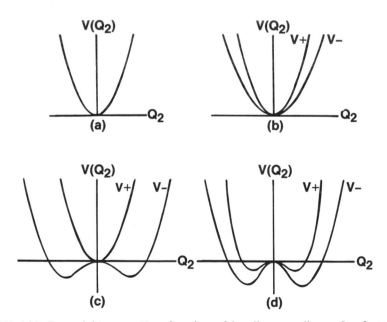

FIGURE 6.30. Potential energy V as function of bending coordinate Q_2 of a triatomic molecule: (a) in a nondegenerate electronic state and in a degenerate electronic state split by the Renner–Teller effect in which the molecule is linear in the equilibrium configuration; (b) for both components of state; (c) for only the upper component; and (d) for neither component [after Hollas (1982)].

PROBLEMS

213
*Molecular Orbital
Methods and
Polyatomic
Molecules*

6.1. The total molecular energy W of the water molecule has been studied [Bingel (1959)] as a function of the HOH angle 2α and as a function of the O—H distance R, which tends to zero near the united atom limit. The result may be written in terms of the ground-state energy of the united atom Ne, plus terms depending on R, which are shown below:

$$W(\alpha, R) = E(\text{Ne } {}^1S_g) + R^{-1}[16 + (2 \sin \alpha)^{-1}]$$
$$+ (4\pi/15)\,\rho(0)R^2(5 - \cos^2 \alpha) + \tfrac{1}{3}\pi(\partial\rho/\partial r)_{r=0}R^3$$
$$\times [(8/125) \cos^3 \alpha + 2(1 - (9/25) \cos^2 \alpha)^{3/2}] + O(R^4) \qquad \text{(P6.1)}$$

As already noted, the first term on the right-hand side is the energy of the united atom Ne, while the second term is the nuclear–nuclear potential energy. The third and fourth terms represent the variation in the electronic energy with R and α in terms of the electron density $\rho(0)$ at the nucleus of the united atom and its derivative $(\partial\rho/\partial r)_{r=0}$ at the nucleus. **(a)** Using the above result for $W(\alpha, R)$, show that the expression for W has its minimum for $2\alpha = \pi$, as R tends to zero; that is, for small O—H separation, the water molecule becomes linear. **(b)** Discuss the significance of the distance R determined by

$$R = [15/16\pi\rho(0)]^{1/3} \qquad \text{(P6.2)}$$

[N.B. You may need to appeal to Kato's theorem (see the following problem) which yields, for the Ne united atom, $(\partial\rho/\partial r)_{r=0} = -20\rho(0)/a_0$.]

6.2. Kato's theorem states that in a nonrelativistic multielectron atom of atomic number Z, the derivative $\partial\rho(r)/\partial r$ of the electron density, evaluated at the nuclear position $r = 0$, is equal to $-(2Z/a_0)$ times the density $\rho(r = 0)$. By first normalizing the ground-state wave function (A1.1.1) of the H atom, demonstrate directly the validity of Kato's theorem for this simplest case.

6.3. The theory of directed valence, approximately valid near the equilibrium configuration of the water molecule, leads to the potential energy of the water molecule as a function of bond lengths and bond angle, namely the form

$$W = f(r_1) + f(r_2) + g(r_{12}) + N \sin^2(\tfrac{1}{2}\theta - \tfrac{1}{4}\pi) \qquad \text{(P6.3)}$$

where r_1 and r_2 are the bond lengths and r_{12} is the H—H distance. The angle between the bonds is denoted by θ. For details of this form, the reader may refer to Van Vleck and Cross (1933), who obtained the minimum of this equation from the expressions for f and g given by the theory of directed valence, at $r_1^0 = r_2^0 = 1.00$ Å and $\theta = 100°$. The measured bond angle is, in fact, $104.5°$. Show, by making plausible assumptions about the expansion of f and g about their equilibrium values, that $W(r_1, r_2, \theta)$ can be expressed in the form

$$W = W(r_1^0, r_2^0, \theta) + \tfrac{1}{2}[k_1(\Delta r_1)^2 + k_2(\Delta r_2)^2 + 2k_3(r_1^0)^2(\Delta\theta)^2]$$
$$+ k_{12}\Delta r_1\Delta r_2 + 2^{1/2}k_{13}r_1^0(\Delta r_1\Delta\theta + \Delta r_2\Delta\theta) + \cdots \qquad \text{(P6.4)}$$

where k_1, k_2, k_3, k_{12}, and k_{13} are constants related to the behavior of f and g.

6.4. The secular determinant of butadiene is given in equation (6.42). **(a)** Work out this determinant to obtain the fourth-order equation

$$x^4 - 3x^2 + 1 = 0 \qquad \text{(P6.5)}$$

(b) By putting $y = x^2$, show first that the roots of the resulting quadratic, say, y_1 and y_2, can be written in the form

$$y_1 = [(1 + 5^{1/2})/2]^2 \qquad \text{(P6.6)}$$

and

$$y_2 = [(1 - 5^{1/2})/2]^2 \qquad \text{(P6.7)}$$

(c) Calling the four values of x thus obtained, x_i, with $i = 1$–4, show that the four energy levels resulting from the secular determinant have the form

$$E_i = \alpha - \beta x_i \tag{P6.8}$$

Find the molecular orbitals explicitly.

6.5. Electron density theory can be used (see Appendix A4.7) to establish that the exact electron density $\rho(\mathbf{r})$ in a molecule can be calculated from the one-electron wave functions $\psi_i(\mathbf{r})$ generated by inserting a potential energy $V(\mathbf{r})$ into the Schrödinger equation through the "elementary" relation

$$\rho(\mathbf{r}) = \sum_{\substack{\text{occupied states}}} \psi_i^*(\mathbf{r})\psi_i(\mathbf{r}) \tag{P6.9}$$

This potential energy can be written as a sum of three parts:

$$V(\mathbf{r}) = V_N(\mathbf{r}) + e^2 \int \frac{\rho(\mathbf{r}')}{|\mathbf{r} - \mathbf{r}'|} d\mathbf{r}' + V_{xc}(\mathbf{r}) \tag{P6.10}$$

where $V_N(\mathbf{r})$ is the potential energy of the bare nuclear framework, the second term clearly arises from the electrostatic potential of the electron cloud of density $\rho(\mathbf{r})$, while the third term arises from the exchange (x) and correlation (c) interactions between electrons. (a) Accepting that $V(\mathbf{r})$ in a molecule like benzene has the symmetry of the molecule, give the argument why the mere existence of the one-body potential energy $V(\mathbf{r})$ allows the total electron density in equation (P6.9) to be decomposed into two parts

$$\rho(\mathbf{r}) = \rho_\sigma(\mathbf{r}) + \rho_\pi(\mathbf{r}) \tag{P6.11}$$

where the two contributions are classified as usual solely by the σ- and π-symmetries of the one-electron wave functions generated by $V(\mathbf{r})$. (b) While the decomposition (P6.11) is formally exact, one can only translate it into energy by taking some approximate theory. Specifically considering benzene, one might employ an electron-gas model, and due to the flatness of the ring, the electron gas might be thought of as approximately two-dimensional. Using the phase-space argument of Chapter 4 and Appendix A4.1 in two dimensions, show that the kinetic energy density is now proportional to the density squared (in contrast to the density to the five-thirds power in three dimensions). (c) Using this result in conjunction with equation (P6.11), explain why the energy takes the form

$$E_{\text{total}} = E_\sigma + E_\pi + E_{\sigma-\pi \text{ interaction}} \tag{P6.12}$$

6.6. For the cyclopentadienyl system shown in the figure below, with LCAO MOs given by

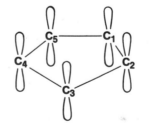

$$\psi = a_1 p_1 + a_2 p_2 + a_3 p_3 + a_4 p_4 + a_5 p_5 \tag{P6.13}$$

the secular determinant is

$$\begin{vmatrix} x & 1 & 0 & 0 & 1 \\ 1 & x & 1 & 0 & 0 \\ 0 & 1 & x & 1 & 0 \\ 0 & 0 & 1 & x & 1 \\ 1 & 0 & 0 & 1 & x \end{vmatrix} = 0 \qquad \text{(P6.14)}$$

which on evaluation gives

$$x^5 - 5x^3 + 5x + 2 = 0 \qquad \text{(P6.15)}$$

Given that the roots are $x = -2.00$, $x = -0.618$, $x = -0.618$, $x = 1.62$, and $x = 1.62$, derive the following levels:

$$\left. \begin{array}{ll} x = -2.00 & E_1 = \alpha + 2.00\beta \\[2mm] x = -0.618 & E_2 = \alpha + 0.618\beta \\[2mm] x = -0.618 & E_3 = \alpha + 0.618\beta \\[2mm] x = 1.62 & E_4 = \alpha - 1.62\beta \\[2mm] x = 1.62 & E_5 = \alpha - 1.62\beta \end{array} \right\} \qquad \text{(P6.16)}$$

Hence for positive ion, neutral, and anion species verify that

$$\left. \begin{array}{l} E_{C_5H_5}(+) = 4\alpha + 5.24\beta \\[2mm] E_{C_5H_5}(\cdot) = 5\alpha + 5.85\beta \\[2mm] E_{C_5H_5}(-) = 6\alpha + 6.47\beta \end{array} \right\} \qquad \text{(P6.17)}$$

6.7. For the π-electrons in hexatriene the LCAO MOs are represented as follows:

$$\psi = a_1 p_1 + a_2 p_2 + a_3 p_3 + a_4 p_4 + a_5 p_5 + a_6 p_6 \qquad \text{(P6.18)}$$

(a) Show that the secular determinant can be written in the form

$$\begin{vmatrix} x & 1 & 0 & 0 & 0 & 0 \\ 1 & x & 1 & 0 & 0 & 0 \\ 0 & 1 & x & 1 & 0 & 0 \\ 0 & 0 & 1 & x & 1 & 0 \\ 0 & 0 & 0 & 1 & x & 1 \\ 0 & 0 & 0 & 0 & 1 & x \end{vmatrix} = 0 \qquad \text{(P6.19)}$$

and prove that this expands to yield

$$x^6 - 5x^4 + 6x^2 - 1 = 0 \qquad \text{(P6.20)}$$

(b) Given that the roots are

$$x = -1.80, \ x = -1.25, \ x = -0.445, \ x = 0.445, \ x = 1.25, \text{ and } x = 1.80$$

show that the energy levels are

$$
\left.\begin{array}{ll}
x = -1.80 & E_1 = \alpha + 1.80\beta \\[10pt]
x = -1.25 & E_2 = \alpha + 1.25\beta \\[10pt]
x = -0.445 & E_3 = \alpha + 0.445\beta \\[10pt]
x = 0.445 & E_4 = \alpha - 0.445\beta \\[10pt]
x = 1.25 & E_5 = \alpha - 1.25\beta \\[10pt]
x = 1.80 & E_6 = \alpha - 1.80\beta
\end{array}\right\} \tag{P6.21}
$$

(c) Finally derive the π-electron ground-state energy as

$$
E = 6\alpha + 6.99\beta \tag{P6.22}
$$

6.8. The interested reader will want to look up the elegant geometrical construction for classifying the π-electron levels in organic molecules, due to Longuet–Higgins, and conveniently summarized in the book by Salem (1966). Use this construction to verify the energies for several of the π-level schemes of molecules treated in the main text.

6.9. Explain in general terms why one will expect that the π-electrons in, say, benzene, will lead to ring currents, and hence to distinctive magnetic properties of this and related molecules.

REFERENCES

M. I. Al-Joburg and D. W. Turner, Jr., *J. Chem. Soc.* **43**, 616 (1965).

N. C. Baird and M. J. S. Dewar, *J. Chem. Phys.* **50**, 1262 (1969).

L. C. Balbás, A. Rubio, J. A. Alonso, N. H. March, and G. Borstel, *J. Phys. Chem. Solids* **49**, 1013 (1988).

W. Bingel, *J. Chem. Phys.* **30**, 1250, 1254 (1959).

R. C. Bingham, M. J. S. Dewar, and D. H. Lo, *J. Am. Chem. Soc.* **97**, 1285 (1975).

R. Botter and H. M. Rosenstock, *J. Res. Nat. Bur. Stand.* **73a**, 313 (1969).

I. D. Clark and D. C. Frost, *J. Am. Chem. Soc.* **89**, 244 (1967).

W. Cochran, *Acta Crys.* **9**, 924 (1956).

C. A. Coulson, *Valence*, 2nd Ed., Oxford University Press, New York, Oxford (1961).

M. A. Coplan, J. H. Moore, and J. A. Tossell, *J. Chem. Phys.* **68**, 329 (1978).

C. W. N. Cumper, *Wave Mechanics for Chemists*, Academic, New York (1966).

M. J. S. Dewar and E. Haselbach, *J. Am. Chem. Soc.* **92**, 590 (1970).

M. J. S. Dewar and G. Klopman, *J. Am. Chem. Soc.* **89**, 3089 (1967).

R. N. Dixon, G. Duxbury, M. Horani, and J. Rostas, *Mol. Phys.* **22**, 977 (1971).

H. H. Greenwood, *Computing Methods in Quantum Organic Chemistry*, Wiley-Interscience, New York (1972).

G. Herzberg and E. Teller, *Z. Phys. Chem.* **B21**, 410 (1933).

P. J. Hiett, F. Flores, P. J. Grout, N. H. March, A. Martin Rodero, and G. Senatore, *Surf. Sci.* **140**, 400 (1984).

R. Hoffmann, *J. Chem. Phys.* **39**, 1397 (1963).

J. M. Hollas, *High Resolution Spectroscopy*, Butterworths, London (1982).

R. O. Jones and O. Gunnarsson, *Rev. Mod. Phys.* **61**, 689 (1989).

L. Karlsson, L. Mattsson, R. Jadry, T. Bergmark, and K. Siegbahn, *Phys. Scrip.* **14**, 230 (1976).

W. Kohn and L. J. Sham, *Phys. Rev.* **140**, A1133 (1965).

J. P. Lowe, *Quantum Chemistry*, Academic, New York (1978).

N. H. March, *Acta Crys.* **5**, 187 (1952).

N. H. March, Specialist Periodical Reports, *Theoretical Chemistry, Vol. A*, Royal Society of Chemistry, London (1981).

I. E. McCarthy and E. Weigold, *Contemporary Physics* **24**, 163 (1983).

A. D. McLean and M. Yoshimine, *Tables of Linear Molecule Wave Functions*, IBM, Hopewell Junction, New York (1967).

R. McWeeny, *Coulson's Valence*, 3rd Ed., Oxford University Press, Oxford (1979).

A. J. Merer and R. S. Mulliken, *Chem. Rev.* **69**, 639 (1969).

J. N. Murrell and A. J. Harget, *Semi-Empirical Self-Consistent Field Theory*, Wiley-Interscience, New York (1972).

W. von Niessen, L. S. Cederbaum, and W. P. Kraemer, *J. Chem. Phys.* **65**, 1378 (1976).

R. Pariser, *J. Chem. Phys.* **21**, 568 (1953).

R. G. Parr, *The Quantum Theory of Molecular Electronic Structure*, Benjamin, New York (1963).

J. A. Pople, *Trans. Faraday Soc.* **49**, 1375 (1953).

J. A. Pople, G. P. Santry, and G. A. Segal, *J. Chem. Phys.* **43**, 5129 (1965).

J. A. Pople and G. A. Segal, *J. Chem. Phys.* **43** (suppl) 136 (1965).

J. A. Pople and G. A. Segal, *J. Chem. Phys.* **44**, 3289 (1966).

R. Renner, *Z. Phys.* **92**, 172 (1934).

W. G. Richards. *Quantum Pharmacology, 2nd Ed.*, Butterworths, London (1983).

W. G. Richards and D. L. Cooper, *Ab Initio MO Calculations, 2nd Ed.*, Clarendon Press, Oxford (1983).

L. Salem, *Molecular Orbital Theory of Conjugated Systems*, Benjamin, New York (1966).

J. M. Schulman and J. W. Moskowitz, *J. Chem. Phys.* **47**, 3491 (1967).

J. C. Slater, *Phys. Rev.* **81**, 385 (1951).

B. Stenhouse, P. J. Grout, N. H. March, and J. Wenzel, *Phil. Mag.* **36**, 129 (1977).

A. Strietwieser, *Molecular Orbital Theory for Organic Chemists*, John Wiley and Sons, New York (1961).

D. W. Turner, C. Baker, A. D. Baker, and C. R. Brundle, *Molecular Photoelectron Spectroscopy*, John Wiley and Sons, London (1970).

J. H. Van Vleck and P. C. Cross, *J. Chem. Phys.* **1**, 357 (1933).

B. G. Williams (ed.) *Compton Scattering: The Investigation of Electron Momentum Distribution*, McGraw-Hill, New York (1977).

M. C. Zerner, *Theor. Chim. Acta* **32**, 111 (1973); **53**, 21 (1979).

Chapter Seven

CHEMICAL REACTIONS, DYNAMICS, AND LASER SPECTROSCOPY

7.1. INTRODUCTION AND BACKGROUND

Thermodynamics can often predict that a reaction should proceed almost to completion, but that reaction is not seen in the laboratory. For example, H_2 and O_2 can be kept in contact with one another without forming noticeable quantities of water, even though this reaction is accompanied by a free-energy decrease. The message here is that the rate of the reaction governs whether the formation of the products will or will not be observed.

Chemical kinetics, or equivalently chemical dynamics, is concerned with such rates. It is therefore useful to note here, in introducing the present chapter, some of the factors that influence the speed with which chemical changes occur:

1. Nature of the reactants and products.
2. Concentration of reacting species.
3. Temperature.
4. Influence of external agents (catalysts).

By a study of the factors that influence reaction rates, one can gain insight into the sequences of steps, or, equivalently, the reaction mechanism, along the path from reactants to products.

7.1.1. Rates of Chemical Reactions: Reaction Rates and Rate Laws*

Chemical kinetics involves the determination (and mathematical expressions) of the rates of chemical reactions. By way of review, we recall that the reaction rate is the increase in the concentration of products or decrease in concentration of reactants in a constant-volume system per unit time. Consequently, in a general way the rate of a reaction is proportional to the concentration of the reactants, i.e., the larger the number of reacting molecules present per unit volume, the more the reaction will occur in a given time.

---

*Following Section 7.1 topics have been chosen to illustrate the relevance of MO theory, the electron density description, and also laser spectroscopy to the study of chemical reactions.

It is helpful to express the rate in terms of reactant consumption, e.g., in the reaction

$$A \rightarrow B \tag{7.1}$$

1 mole of A is consumed per mole of B formed; hence the rate of consumption of A is numerically equal to $d[B]/dt$ but will have a negative sign as $[A]$ decreases with time. Therefore, we can write

$$\text{Rate} = -d[A]/dt \tag{7.2}$$

We now turn to the concepts of rate constant and order of chemical reactions.

(a) Rate Constant k and Order of Reaction

Consider the general reaction

$$mA + nB \rightarrow \text{Products} \tag{7.3}$$

with the rate found experimentally proportional to the concentrations as follows:

$$\text{Rate} \propto [A]^a[B]^b \tag{7.4}$$

or

$$\text{Rate} = k[A]^a[B]^b \tag{7.5}$$

where k, the proportionality factor, is the rate constant. The rate constant will have a specific value for a given process at a given temperature and pressure. The power to which the concentrations are raised, a and b, may or may not be the same as the coefficients m and n in reaction (7.3) which express the stoichiometry of the process. Note the fundamental distinction that must be made here; namely, reaction equation (7.3) indicates the number of moles of A and B that will combine whereas equation (7.5) represents the actual experimentally determined dependence of rate on concentration. The reason for this distinction is the difference between the overall reaction and the individual steps in the mechanism. The stoichiometric reaction equation usually represents the net result of a number of intermediate processes. Since each of these will have a rate which is dependent on the concentrations of the species involved (perhaps even on reactive intermediates that do not appear among the final products) the overall rate may have a different dependence on the concentrations of A and B than is expressed in the stoichiometric equation. Consistent with this point of view is the fact that in a single-step process $a = m$ and $b = n$. Hence, if the reaction, indeed, proceeds by m molecules of A coming together with n molecules of B in one event to yield the experimentally observed product(s), the rate is given by $k[A]^m[B]^n$.

The sum of a and b represents an experimentally observed number, which is a basic characteristic of the process and is known as the reaction order, i.e., the reaction order is equal to the sum of a and b, where we again stress that the reaction order is obtained experimentally. The sum of a and b can have a number of values (including fractions), but 0, 1, and 2 are the most commonly observed (i.e., if $a + b = 1$ we have a first-order reaction; if $a + b = 2$ the reaction is second order, etc.).

It is also common practice to state the order of a reaction with respect to a particular reactant, i.e., "second order in Y," where the term $[Y]^2$ appears in the rate equation.

The experimentally determined rate law for a specific reaction, allows one to identify the rate-determining step, which is all-important in determining a chemical mechanism for the reaction. Let us now consider the mathematical expressions for various ordered reactions:

(b) Zero-Order Reactions

A reaction is zero-order if the rate of the reaction is independent of the reactant concentrations. For example, say we have the reaction

$$A \rightarrow \text{Products} \tag{7.6}$$

which yields experimentally that the

$$\text{Rate} = -d[A]/dt = k \tag{7.7}$$

where k is the rate constant with the dimensions moles/s. Rearranging (7.7), we have

$$d[A] = -k \, dt \tag{7.8}$$

Integration between $t = 0$ and $t = t$ gives

$$\int_{[A]_0}^{[A]} d[A] = [A] - [A]_0 = -k \int_0^t dt = -kt \tag{7.9}$$

or

$$[A] = [A]_0 - kt \tag{7.10}$$

where $[A]_0$ and $[A]$ are the concentrations at $t = 0$ and $t - t$. A plot of $[A]_0 - [A]$ vs. t gives a straight line whose slope would determine k.

(c) First-Order Reactions

Let us say that the reaction

$$A \rightarrow \text{Products} \tag{7.11}$$

has a rate expressed by

$$\text{Rate} = -d[A]/dt = k[A] \tag{7.12}$$

and therefore

$$\int_{[A]_0}^{[A]} d[A]/[A] = -\int_0^t k \, dt \tag{7.13}$$

and

$$\ln[A]/[A]_0 = -kt \tag{7.14}$$

or

$$[A] = [A]_0 \, e^{-kt} \tag{7.15}$$

A plot of $\ln [A]/[A]_0$ *vs.* t will yield a straight line whose slope (negative) is given by $-k$.

We can determine the half-life $(t_{1/2})$ of a reaction, i.e., the time it takes for the concentration of the reactant to decrease by one-half of its original value. Setting $[A] = [A]_0/2$ in equation (7.14), we have

$$\ln \frac{[A]_0/2}{[A]_0} = -kt_{1/2} \tag{7.16}$$

and

$$t_{1/2} = \ln 2/k \tag{7.17}$$

We note that the half-life of a first-order reaction is independent of the initial concentration.

(d) Second-Order Reactions

We shall consider the two types of second-order reactions: The first type is represented by

$$2A \rightarrow \text{Reactants} \tag{7.18}$$

where

$$\text{Rate} = -d[A]/dt = k[A]^2 \tag{7.19}$$

and k is the second-order rate constant with the dimensions $\text{moles}^{-1} \text{ s}^{-1}$. Rearranging and integrating equation (7.19) we have

$$-\int_{[A]_0}^{[A]} d[A]/[A]^2 = \int_0^t k \, dt \tag{7.20}$$

and

223

*Chemical
Reactions,
Dynamics,
and Laser
Spectroscopy*

$$1/[A] - 1/[A]_0 = kt \tag{7.21}$$

We can determine the half-life by using equation (7.21) as follows

$$\frac{1}{[A]_0/2} - 1/[A]_0 = kt_{1/2} \tag{7.22}$$

and

$$t_{1/2} = 1/k[A]_0 \tag{7.23}$$

The second type of second-order reactions can be represented by

$$A + B \rightarrow \text{Products} \tag{7.24}$$

and the rate is given by

$$-d[A]/dt = -d[B]/dt = k[A][B] \tag{7.25}$$

If we let

$$[A] = [A]_0 - x, \qquad [B] = [B]_0 - x \tag{7.26}$$

where x is the number of moles of A and B used in time t, we can write

$$-d[A]/dt = -d\{[A]_0 - x\}/dt = dx/dt = k[A][B] \tag{7.27}$$

and

$$dx/dt = k\{[A]_0 - x\}\{[B]_0 - x\} \tag{7.28}$$

Rearranging, we have

$$\frac{dx}{\{[A]_0 - x\}\{[B]_0 - x\}} = k\,dt \tag{7.29}$$

Integrating by the method of partial fractions yields

$$\frac{1}{[B]_0 - [A]_0} \ln\left\{ \frac{\{[B]_0 - x\}[A]_0}{\{[A]_0 - x\}[B]_0} \right\} = kt \tag{7.30}$$

and

$$\frac{1}{[B]_0 - [A]_0} \ln\left\{\frac{[B][A]_0}{[A][B]_0}\right\} = kt \tag{7.31}$$

One could similarly determine the expression for a third-order reaction.

Table 7.1 summarizes the rate equations for the first- and second-order reactions just considered. In Appendix A7.2 we discuss theoretical considerations of chemical reactions, including: (a) the Arrhenius empirical relationship for the rate constant k; (b) the collision theory; and (c) the absolute rate theory also known as the transition state theory (TST). Appendix A7.2 should prove useful as a review of these, all of which are important as a basis for the material found in the remainder of this chapter.

7.1.2. Largely Qualitative Considerations

For a chemical change to occur during a collision between two molecules, some bonds must be broken and some new ones formed. This evidently requires, in the electron density description of molecules emphasized in Chapter 4, that the electronic charge clouds of the reacting molecules interpenetrate substantially so that the necessary electronic redistribution, appropriate when bonding changes, can occur.

While collision theory, discussed briefly in Chapter 3 (see also Appendix A.7.2), focuses attention primarily on the relationship between the reaction rate and the number of collisions per second between reactant molecules, a more detailed picture must be concerned with the energy and geometry of the reactants when they collide to form the products. This is the province of transition state theory.

7.1.3. The Transition State: Reaction Coordinate and Potential Energy Diagrams

Returning for a moment to collisions, we must stress that molecular collisions differ very significantly from collisions between billiard balls. Though the shape and size of a molecule remain qualitatively useful concepts, electronic clouds do, of course, have infinite spatial extension, even as they decay exponentially with

TABLE 7.1
Summary of Rate Equations

Order	Differential form	Integrated form	Half-life
0	$-d[A]/dt = k$	$[A]_0 - [A] = kt$	$[A]_0/2k$
1	$-d[A]/dt = k[A]$	$[A] = [A]_0\, e^{-kt}$	$\ln 2/k$
2	$-d[A]/dt = k[A]^2$	$1/[A] - 1/[A]_0 = kt$	$1/[A]_0 k$
2 [For A + B → Products]	$-\dfrac{d[A]}{dt} = k[A][B]$	$\dfrac{1}{[B]_0 - [A]_0} \ln \dfrac{[B][A]_0}{[A][B]_0} = kt$	—

225

*Chemical
Reactions,
Dynamics,
and Laser
Spectroscopy*

distance far from all nuclei at a rate governed by the appropriate ionization potential. So, as discussed in Chapter 3, as molecules approach each other they experience a gradual increase in their mutual repulsion as their initially separate charge clouds begin to overlap significantly. This repulsion causes the molecules to slow down, eventually to stop, and then to fly apart again. One important point to note already is the difference between molecules moving slowly and those moving rapidly. For slow molecules, this reversal of motion would in fact occur even without enough interpenetration of electron densities to allow bonding changes, i.e., electronic redistribution, to take place. As a consequence, slow-moving molecules simply bounce off one another without reacting. In contrast, in very rapidly moving molecules the electronic charge distributions interpenetrate sufficiently so that the necessary adjustment of electronic spatial distribution in the reacting molecules takes place to cause bond breaking and reforming in producing the products of the reaction. These high-speed molecules obviously have large kinetic energies, and can, in turn, yield large increases in potential energy during collisions. When the products fly apart, this potential energy decreases as the product molecules gain velocity, and hence kinetic energy. Thus, under these circumstances, only fast-moving molecules are able to react. In fact, there must be some minimum kinetic energy possessed jointly by the two molecules that can be transferred into potential energy. The minimum energy that must be available in a collision to cause reaction is termed the activation energy E_a.

The change in potential energy that takes place during the course of a reaction is shown in Figure 7.1. The horizontal coordinate is termed the "reaction" coordinate, and positions along this axis represent the extent to which the reaction has progressed toward completion. On the left of this potential energy diagram, one has two molecules of the species AB. As they approach, the potential energy increases to a maximum. As one continues toward the right along the reaction coordinate, the potential energy of the system decreases as the products, A_2 and B_2, move apart. When the A_2 and B_2 molecules are finally separated from each other, the total potential energy drops to what is essentially a constant value.

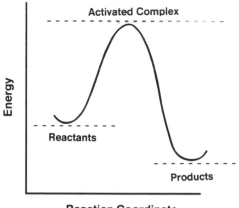

FIGURE 7.1. Potential energy *vs.* reaction coordinate for an exothermic reaction.

Suppose AB molecules decompose. Then the activation energy for the decomposition of AB corresponds to the difference between the energy of the reactants and the maximum of the potential energy curve. Slow-moving molecules, as noted above, do not possess sufficient energy to overcome this potential energy barrier, while fast-moving ones do.

In Figure 7.1 we have drawn the potential energy of the products as lower than that of the reactants. The difference between them corresponds to the heat of the reaction. In this example, since the products are at a lower energy than the reactants, the reaction is exothermic. The energy released appears as an increase in the kinetic energy of the products; therefore the temperature of the system rises as the reaction progresses.

In the reaction mixture there are also collisions between A_2 and B_2 molecules. Such collisions, if energetic enough, can lead to the reformation of AB molecules. In Figure 7.1, the activation energy for the reaction

$$A_2 + B_2 \rightarrow 2AB \tag{7.32}$$

is the difference in energy between the products and the top of the potential energy barrier. Since the forward reaction is exothermic, the reverse reaction is endothermic.

Figure 7.2 depicts the energy changes for a reaction that is endothermic in the forward direction. In this case, the products are at a higher potential energy than the reactants. The net absorption of energy that takes place as the products are formed occurs at the expense of the kinetic energy. Consequently there is a net overall decrease in the average kinetic energy as the reaction proceeds and the reaction mixture cools.

The species that exists at the top of the potential barrier (Figure 7.1) during an effective collision corresponds to neither the reactants nor the product but, instead, to some highly unstable combination of atoms that one terms the "activated complex." This complex is said to exist in a transition state along the reaction coordinate; hence the nomenclature "transition state theory" (see Appendix A.7.2).

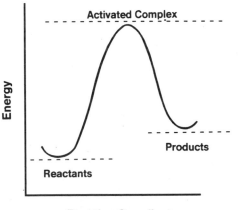

FIGURE 7.2. Potential energy *vs.* reaction coordinate for an endothermic reaction.

227
*Chemical
Reactions,
Dynamics,
and Laser
Spectroscopy*

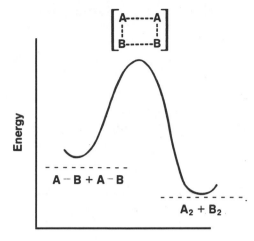

FIGURE 7.3. Potential energy *vs.* reaction coordinate showing a transition state.

As already noted, transition state theory views chemical kinetics in terms of the energy and geometry of the activated complex which, once it has formed, can either come apart to product the reactants again or can go on to yield the products. For example, let us examine again the decomposition of AB molecules to produce A_2 and B_2. The change that takes place along the reaction coordinate can be represented as

$$
\begin{matrix} A \\ | \\ B \end{matrix} + \begin{matrix} A \\ | \\ B \end{matrix} \rightleftharpoons \begin{bmatrix} A \cdots A \\ \vdots \quad \vdots \\ B \cdots B \end{bmatrix} \rightarrow \begin{matrix} A-A \\ + \\ B-B \end{matrix} \qquad (7.33)
$$

where solid dashes denote normal covalent bonds while dots indicate the partially broken and partially formed bonds in the transition state. Figure 7.3 illustrates this on the potential energy diagram for the reaction.

If the potential energy of the transition state is very high, then a good deal of energy must be available in a collision to form the activated complex. If it were possible, by some means, to produce an activated complex whose energy was closer to that of the reactants, the decreased activation energy would clearly yield a faster reaction rate. Catalysts fulfill this role by reducing the magnitude of the activation energy.

7.2. RATES OF CHEMICAL REACTIONS: ABSOLUTE RATE THEORY

It would be very useful if one could calculate the rate of a chemical reaction from some simple observables: (i) bond length, (ii) bond dissociation energy, and (iii) vibrational frequency, plus knowledge of the geometry of the atoms involved when they have reached the top of the potential energy barrier separating initial and final configurations.

The work of Pelzer and Wigner (1932) led the way to the development of the absolute rate theory of Eyring (1935) (see Appendix A7.2). In this rate theory, the species at the top of the potential energy barrier, referred to as the activated complex, is viewed (in most respects) as an ordinary chemical species in equilibrium with the reactants. The problem then separates into two parts:

1. Calculation of the equilibrium constant for the reaction forming the activated complex.
2. Calculation of the rate at which the activated complex decomposes to form the products of the reaction.

A simple bimolecular reaction will then be represented in the following manner:

$$A + B \underset{}{\overset{K^{\ddagger}}{\rightleftharpoons}} C^{\ddagger} \overset{k^{\ddagger}}{\longrightarrow} \text{Products} \tag{7.34}$$

where the activated complex is represented by $C^{\ddagger}$. The velocity of the reaction, U, say, can be expressed as

$$U = \kappa k^{\ddagger}[C^{\ddagger}] \tag{7.35}$$

where κ is called the transmission coefficient, which is the fraction of activated complexes that yields products: the remaining complexes revert to reactions. The concentration of the activated complex can be written in terms of the concentrations of the reactants and the equilibrium constant as

$$U = \kappa K^{\ddagger} k^{\ddagger}[A][B] \tag{7.36}$$

Therefore it follows that the ordinary rate constant is given by

$$k_{\text{obs}} = \kappa K^{\ddagger} k^{\ddagger} \tag{7.37}$$

7.2.1. The Methane–Chlorine-Atom Reaction

Let us illustrate the calculation of the rate of a reaction by treating

$$CH_4 + Cl^{\cdot} \rightarrow CH_3^{\cdot} + HCl \tag{7.38}$$

the necessary experimental data being available for comparison with the predictions of absolute rate theory.

Stage 1 above, namely the calculation of the equilibrium constant for the reaction leading to the activated complex will first be dealt with. Stage 2 will then be concerned with the rate of decomposition of the complex into products.

To facilitate the procedure, let us assume that methane can be considered as a diatomic molecule CH_3—H. Thus, it will be assumed that the energy of the nonreacting C—H bonds will not be changed as methane is converted to the methyl radical. Though the assumption is not in fact correct, the resulting quantitative error turns out to be quite small, and other approximations one is forced to make turn out to be much more serious.

Figure 7.4 shows the potential energy curve of a diatomic system as a function of internuclear separation (cf. the discussion of the H_2 molecule in Chapter 2). As usual, the zero of energy is defined by the atoms at infinite separation.

One way of representing this dissociation curve is by the Morse (1929) potential

$$\phi = D_0\{\{(1 - \exp[-\alpha(r - r_e)]\}^2 - 1\} \tag{7.39}$$

where D_0 is the observed dissociation energy plus the zero-point energy, r is the distance at which ϕ is calculated, and α is given by

$$\alpha = 1.218 \times 10^7 \omega_0 (\mu/D_0)^{1/2} \tag{7.40}$$

Here ω_0 is the equilibrium vibrational frequency of the bond, which may be approximated by the observed frequency, and μ is the reduced mass

$$\mu = M_1 M_2/(M_1 + M_2) \tag{7.41}$$

The quantities ω_0 and D_0 are given in cm^{-1} (1 cal/mole = 0.350 cm^{-1}) and the unit of α which is obtained is cm^{-1}. Wiberg (1964) collects the various values of the constants for the methane–halogen-atom reactions.

As the chlorine atom approaches the methane molecule, an assembly is formed which may be thought of as composed of three particles, Cl˙(X), H˙(Y), and CH_3(Z) (see Figure 7.5). We wish to represent the energy of this system of three particles as a function of the XY etc. distances $r_1 \rightarrow r_3$ shown in the figure.

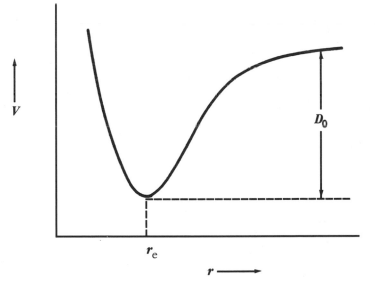

FIGURE 7.4. Potential energy *vs.* internuclear separation for a diatomic molecule.

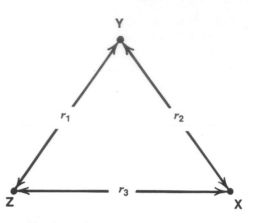

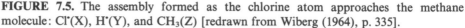

FIGURE 7.5. The assembly formed as the chlorine atom approaches the methane molecule: Cl$^{\cdot}$(X), H$^{\cdot}$(Y), and CH$_3$(Z) [redrawn from Wiberg (1964), p. 335].

We can do this approximately by means of the London equation

$$E = A + B + C - \{\tfrac{1}{2}[(\alpha - \beta)^2 + (\alpha - \gamma)^2 + (\gamma - \beta)^2]\}^{1/2} \qquad (7.42)$$

which is based on the valence bond method of Chapter 2. In this equation, A, B, and C are the Coulombic parts of the energies of the molecules YZ, XY, and XZ, all evaluated at the appropriate internuclear distances, while α, β, and γ are the corresponding exchange terms.

The total energy of each of these molecules at the desired distance can be obtained from the Morse equation, and if there were some natural way to divide the total energy into Coulombic and exchange parts, then the energy of the assembly of three particles could be calculated as a function of the three internuclear distances.

Eyring and Polanyi (1931) have merely taken 10 to 20% of the total energy as the Coulombic part and the rest as the exchange part. This proves to be a fair assumption at the larger internuclear distances but is not entirely satisfactory at the small internuclear distances at which activated complexes are often found.

Sato (1955) has made use of the Heitler–London treatment of diatomic molecules of Chapter 2 to write

$$\phi_{\text{bond}} = (Q + X)/(1 + S^2) \qquad (7.43)$$

$$\phi_{\text{anti}} = (Q - X)/(1 - S^2) \qquad (7.44)$$

where ϕ is the Coulombic contribution, X is the exchange part, and S is the overlap integral. The potential energy of the bonding state can be represented by the Morse equation, while the potential energy of the antibonding states have been well represented by Sato as

$$\phi_{\text{anti}} = (D_0/2)\{\{1 + \exp[-\alpha(r - r_e)]\}^2 - 1\} \qquad (7.45)$$

The Coulombic and exchange terms are then given by

$$Q = \frac{\phi_{bond} + \phi_{anti} + S^2(\phi_{bond} - \phi_{anti})}{2} \qquad (7.46)$$

and

$$X = \frac{\phi_{bond} - \phi_{anti} + S^2(\phi_{bond} + \phi_{anti})}{2} \qquad (7.47)$$

The value of S^2 which appears to be satisfactory in the region of the activated complex is 0.18.

To illustrate the use of this method, Table 7.2 gives the values of the bonding and antibonding potentials for methane, hydrogen chloride, and methyl chloride for different values of the internuclear distance. From this one can obtain the Coulombic and exchange terms.

Thus, one is in a position to calculate the energy of the assembly of three particles as a function of the geometry. The London treatment (Section 2.4) will be employed, but since the overlap integral was included in calculating the Coulombic and exchange terms, this equation needs to be modified to read

$$E = \frac{1}{1 + S^2} \{A + B + C - \{\tfrac{1}{2}[(\alpha - \beta)^2 + (\alpha - \gamma)^2 + (\beta - \gamma)^2]\}^{1/2}\} \qquad (7.48)$$

It should be emphasized that the energy thus calculated is at a minimum if one of the exchange terms is small compared to the others, indicating that in the approximation the activated complexes of lowest energy will be linear. If one makes this assumption for the arrangement of the activated complex, one of

TABLE 7.2
Bonding and Antibonding Potentials for Methane, Hydrogen Chloride, and Methyl Chloride[a]

	Methane			Hydrogen chloride			Methyl chloride	
r	V_{bond}	V_{anti}	r	V_{bond}	V_{anti}	r	V_{bond}	V_{anti}
1.00	−105.10	202.84	1.20	−104.14	192.62	2.20	−59.57	48.88
1.05	−107.89	179.35	1.25	−106.26	169.95	2.30	−52.69	39.89
1.10	−108.47	158.88	1.30	−106.28	150.23	2.40	−46.17	32.68
1.15	−107.35	140.99	1.35	−104.71	133.05	2.50	−40.15	26.88
1.20	−104.96	125.34	1.40	−101.96	118.04	2.60	−34.71	22.18
1.25	−101.64	111.62	1.45	−98.35	104.91	2.70	−29.87	18.35
1.30	−97.65	99.56	1.50	−94.15	93.40	2.80	−25.60	15.23
1.35	−93.21	88.94	1.55	−89.57	83.28	2.90	−21.88	12.66
1.40	−88.94	79.57	1.60	−84.75	74.37	3.00	−18.65	10.54
1.45	−83.63	71.29	1.65	−79.84	66.51	3.10	−15.87	8.80
1.50	−78.71	63.96	1.70	−74.93	59.57	3.20	−13.47	7.35
1.55	−73.85	57.45	1.75	−70.09	53.41	3.30	−11.43	6.15
1.60	−69.09	51.67	1.80	−65.38	47.96	3.40	−9.68	5.15

[a]Taken from Wiberg (1964).

the internuclear distances (r_3) will be the sum of the other two, leading to only two independent variables (r_1 and r_2). The energies thus obtained are listed in Table 7.3.

To complete representation of these data it would be useful to use a three-dimensional surface with r_1, r_2, and E as variables. However, for convenience it is often represented by a contour diagram, the contours being at constant E and functions of r_1 and r_2. Such a plot for the methane–chlorine-atom reaction is shown in Figure 7.6. It might be further noted that these plots are often constructed using an acute angle between the two axes, so that the cross terms (i.e., those involving r_1r_2) in the internal kinetic energy expression will vanish.

The energy of methane, using the model of a diatomic molecule discussed above, with a chlorine atom at infinity is -108.5 kcal/mole. The reaction could proceed by the dissociation of methane followed by the reaction of the chlorine atom with a hydrogen atom

$$CH_3 - H \ \rightarrow CH_3^{\cdot} + H^{\cdot}$$

$$H^{\cdot} + Cl^{\cdot} \ \rightarrow HCl$$

(7.49)

The classical activation energy for this reaction would be 108.5 kcal. If one examines Figure 7.6, it becomes clear that there is a much lower energy path for the reaction. The maximum energy contour over which the system must pass is -99.9 kcal. The classical activation energy would then be $108.5 - 99.9 = 8.6$ kcal.

7.2.2. A Reaction Coordinate Representation

The point made above is readily seen in a plot of the reaction coordinate, defined as the path of minimum potential energy *vs.* energy, shown in Figure 7.7.

TABLE 7.3

Calculated Energies for the Methane–Chlorine Atom Activated Complex
(energy in kcal/mole)[a]

r_{H-Cl}	r_{CH_3-H}									
	1.05	1.10	1.15	1.20	1.25	1.30	1.35	1.40	1.45	1.50
1.20	-63.77	-71.13	-76.93	-81.53	-85.23	-88.24	-90.70	-92.74	-94.44	-95.87
1.25	-72.30	-78.88	-83.94	-87.87	-90.96	-93.42	-95.40	-97.03	-98.38	-99.51
1.30	-78.90	-84.73	-89.08	-92.33	-94.80	-96.70	-98.18	-99.37	-100.34	-101.15
1.35	-84.05	-89.19	-92.85	-95.44	-97.28	-98.61	-99.57	-100.31	-100.88	-101.34
1.40	-88.13	-92.61	-95.63	-97.59	-98.82	-99.57	-100.01	-100.26	-100.41	-100.50
1.45	-91.38	-95.28	-97.70	-99.07	-99.71	-99.90	-99.80	-99.57	-99.28	-98.98
1.50	-94.01	-97.40	-99.28	-100.10	-100.20	-99.84	-99.23	-98.50	-97.76	-97.05
1.55	-96.16	-99.10	-100.51	-100.84	-100.44	-99.57	-98.46	-97.25	-96.05	-94.93
1.60	-97.94	-100.49	-101.50	-101.39	-100.54	-99.21	-97.62	-95.95	-94.31	-92.77
1.65	-99.43	-101.65	-102.30	-101.82	-100.57	-98.82	-96.81	-94.71	-92.64	-90.69
1.70	-100.67	-102.61	-102.97	-102.17	-100.57	-98.47	-96.07	-93.58	-91.11	-88.76
1.75	-101.72	-103.43	-103.53	-102.46	-100.58	-98.15	-95.43	-92.58	-89.75	-87.03
1.80	-102.61	-104.12	-104.02	-102.72	-100.59	-97.90	-94.88	-91.73	-88.57	-85.52

[a]Taken from Wiberg (1964).

233

*Chemical
Reactions,
Dynamics,
and Laser
Spectroscopy*

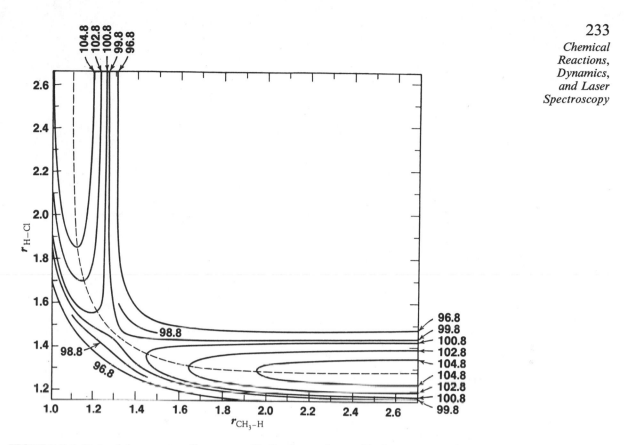

FIGURE 7.6. Potential energy *vs.* distance profile for the methane–chlorine atom reaction [redrawn from Wiberg (1964), p. 339].

In this figure, the lower parts of the dissociation energy curves for methane and hydrogen chloride are given to establish the starting and final points for the system, and the distance corresponding to 0.1 Å is also shown.

Similar calculations may be made for the reaction of methane with bromine and iodine atoms and the reader is referred to Wiberg (1964), whose account we have followed closely above, for further details of these and other reactions.

7.2.3. Determination of the Equilibrium Constant

To find the equilibrium constant $K^{\ddagger}$, we must at this point consider the absolute rate theory more closely. In the activated complex, there are $3n - 6$ vibrational degrees of freedom. However, one of these vibrations will correspond to motion along the reaction coordinate leading to products. For example, if we treat the methyl hydrogen as a pseudoatom that is attached to the carbon by a relatively high stretching force constant, vibrational modes for the activated complex will be as in Figure 7.8.

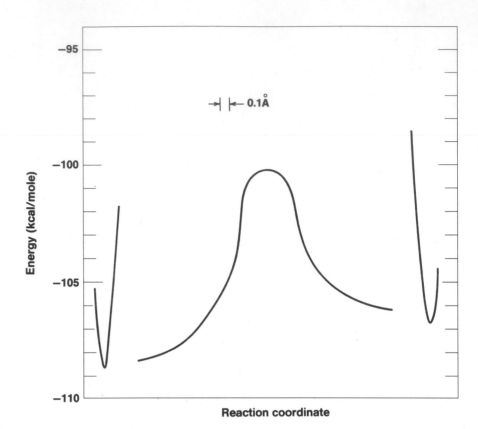

FIGURE 7.7. Potential energy as a function of distance along the reaction coordinate for the methane–chlorine reaction [redrawn from Wiberg (1964), p. 341].

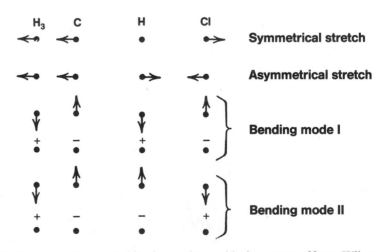

FIGURE 7.8. Stretching modes in the methane–chlorine system [from Wiberg (1964), p. 341].

235

*Chemical
Reactions,
Dynamics,
and Laser
Spectroscopy*

From this figure, it can be seen that in the asymmetric stretching mode the H and Cl move together, and this will result in product formation. Thus, this mode will be separated from the rest of the vibrational frequencies and be treated as a translation along the reaction coordinate.

If we designate the distance characterizing the activated complex as δ (see Figure 7.9), then the partition function for translation will be

$$\mathrm{pf_{trans}} = \frac{(2\pi M k_B T)^{1/2}}{h} \delta \tag{7.50}$$

This corresponds to the motion of a particle of mass M in a one-dimensional box of length δ; see the flat portion at top of barrier on Figure 7.9. The equilibrium constant in terms of the partition functions will then be given by

$$K^{\ddagger} = \frac{(2\pi M k_B T)^{1/2}}{h} \delta \frac{\mathrm{pf}^{\ddagger}}{\mathrm{pf_A pf_B}} \exp\left(\frac{-\Delta E_0}{RT}\right) \tag{7.51}$$

where A now refers to methane, B to a chlorine atom, ΔE_0 to the difference in energy between the reactants and the activated complex at 0 K and $\mathrm{pf}^{\ddagger}$ does not include the vibrational mode discussed above that leads to product formation.

To express this in terms of the classical activation energy calculated above, the zero-point energy terms may be separated out from ΔE_0, giving

$$K^{\ddagger} = \frac{(2\pi M k_B T)^{1/2}}{h} \delta \frac{\mathrm{pf}^{\ddagger}}{\mathrm{pf_A pf_B}} \prod_i^{3n^{\ddagger}-7} \exp(-\tfrac{1}{2}u_i) \prod_i^{3n-6} \exp(\tfrac{1}{2}u_i) \exp(-\Delta E_{\mathrm{class}}/RT) \tag{7.52}$$

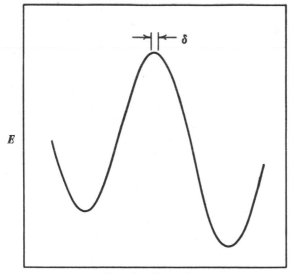

Reaction Coordinate

FIGURE 7.9. Potential energy-*vs.*-reaction coordinate relationship for a hypothetical reaction. δ characterizes the activated complex [redrawn from Wiberg (1964), p. 344].

where the products are over the $3n - 7$ remaining vibrations in the activated complex and the $3n - 6$ vibrations of each of the reactants, while $u_i = h\nu_i/k_BT$.

The rate $k^{\ddagger}$ at which the activated complex decomposes to products must now be obtained, having found $K^{\ddagger}$. Using kinetic arguments, one finds

$$k^{\ddagger} = (k_BT/2\pi M)^{1/2}/\delta \tag{7.53}$$

and hence the observed rate constant is determined as

$$k_{\text{obs}} = \kappa k^{\ddagger}K^{\ddagger} = \kappa\left(\frac{k_BT}{h}\right)\frac{\text{pf}^{\ddagger}}{\text{pf}_A\text{pf}_B}\prod_i^{3n^{\ddagger}-7}\exp(-\tfrac{1}{2}u_i)\prod_i^{3n-6}\exp(\tfrac{1}{2}u_i)\exp(-\Delta E_{\text{class}}/RT) \tag{7.54}$$

In terms of enthalpy H and entropy S this may also be expressed as

$$k_{\text{obs}} = \kappa(k_BT/h)\exp(-\Delta H^{\ddagger}/RT)\exp(\Delta S^{\ddagger}/R) \tag{7.55}$$

The problem, to this stage, has now been reduced to obtaining the thermodynamic functions. The masses, moments of inertia, and vibrational frequencies of the reactants are known, as is the mass of the activated complex. The moments of inertia may be obtained for the calculated dimensions and if a set of vibrational frequencies for the activated complex can be deduced, the thermodynamic functions can be found and the rate of reaction calculated. The way the force constants and vibrational frequencies may be obtained is set out in some detail elsewhere [see Wiberg (1964)]. For background material for this section the reader is referred to Laidler (1987), McQuarrie (1976), and Atkins (1986).

Using these results, we summarize the calculated and observed activation parameters in Table 7.4. The calculated values are in excellent agreement with the observations.

7.3. THE WOODWARD–HOFFMANN RULES

7.3.1. Introduction

Following the above discussion of an example of absolute rate theory, we turn to a different aspect of chemical reactions. Here, the question we shall raise, and which will be answered in the affirmative is: Can we use the qualitative features of MO theory to help our understanding of chemical reactions? It has been found that there are many reactions where certain symmetry characteristics of MOs control the overall course of the chemical reaction. These reactions are known as pericyclic reactions as they take place through cyclic transition states. Let us examine two basic types of reactions: electrocyclic reactions and cycloaddition reactions.

237

*Chemical
Reactions,
Dynamics,
and Laser
Spectroscopy*

TABLE 7.4

Comparison of the Calculated and Observed Activation Parameters for the
Chlorination, Bromination, and Iodination of Methane[a]

Parameter	Chlorination	Bromination	Iodination
ΔE_{class} (kcal/mole)	8.6	20.5	34.9
Reactants:			
ΔH_{tr}	2.96	2.96	2.96
S_{tr}	70.86	73.32	74.70
ΔH_r	1.78	1.78	1.78
S_r	12.65	12.65	12.65
ΔH_v	0.03	0.03	0.03
S_v	0.11	0.11	0.11
Zero-point energy for reactants	27.25	27.25	27.25
Zero-point energy for activated complex	23.37	23.23	22.58
Activated Complex:			
ΔH_{tr}	1.48	1.48	1.48
S_{tr}	37.61	39.60	40.79
ΔH_r	0.89	0.89	0.89
S_r	21.76	22.69	23.62
ΔH_v	0.64	0.97	1.61
S_v	3.10	4.98	10.42
$\Delta H_{calc}^{\ddagger}$	2.8	15.0	29.9
$\Delta S_{calc}^{\ddagger}$	−21.2	−18.8	−12.6
$\Delta H_{obs}^{\ddagger}$	2.7	17.0	34.0
$\Delta S_{obs}^{\ddagger}$	−17.8		

[a]From Wiberg (1964).

(a) Electrocyclic Reactions

In a number of reactions conjugated polyenes are transformed into cyclic
compounds. As an example,

$$\text{1,3-butadiene} \qquad \text{cyclobutene} \tag{7.56}$$

In other reactions the ring of a cyclic compound is opened and a conjugated
polyene forms:

$$\text{cyclobutene} \qquad \text{1,3-butadiene} \tag{7.57}$$

Reactions (7.56) and (7.57) are known as electrocyclic reactions, which means
that σ- and π-bonds are interconverted. For example, in (7.56) one π-bond of
1,3-butadiene becomes a σ-bond in cyclobutene; and in (7.57) the reverse is true
and a σ-bond of cyclobutene becomes a π-bond in butadiene.

(b) Cycloaddition Reactions

Several reactions of alkenes and polyenes take place in which two molecules
react to form one cyclic product. These are known as cycloaddition reactions, e.g.,

alkene + alkene   cyclobutane

(7.58)

and

diene + alkene   cyclohexene

(7.59)

Cycloaddition reactions are classified on the basis* of the number of π-electrons involved in each component. For example, reaction (7.58) is a (2 + 2) cycloaddition, whereas reaction (7.59) is (4 + 2) cycloaddition. We will consider reactions of this type later in this section.

The Woodward–Hoffmann rules proposed in (1965) [see Woodward and Hoffmann (1970)] are formulated for concerted reactions only. Recall that concerted reactions are those in which bonds are broken and formed simultaneously, and thus no intermediates are involved. The Woodward–Hoffmann rules are based on the hypothesis that in concerted reactions MOs of the reactant are continuously converted into MOs of the product. The conversion involved is not a random one, and, since MOs have symmetry, there are characteristic restrictions on which MOs of the reactant may be transformed into particular MOs of the product. We, therefore, have certain reaction paths that are symmetry-allowed while others are symmetry-forbidden. If a reaction is classified symmetry-forbidden it does not necessarily mean that it will not take place; however, it does mean that the concerted reaction would have a much higher energy of activation. The reaction could take place, but will in all probability do so through a path that is symmetry-allowed or through a nonconcerted process.

With this brief introduction to electrocyclic and cycloaddition reactions, let us turn to the details of these types of reactions and the Woodward–Hoffmann consideration.

For ease of presentation, we start with the application to unimolecular cyclization of an open conjugated molecule (e.g., *cis*-1,3-butadiene, closing to cyclobutene). This type of reaction as discussed earlier is termed electrocyclic and the details of the electrocyclic closure of *cis*-1,3-butadiene are shown in Figure 7.10.

If one can find a way of keeping track of the terminal hydrogens in butadiene (by, say, substituting H by D, as shown in the figure) then it is possible to distinguish between two products. One of them is produced if the two terminal methylene groups have rotated in the same sense, either both clockwise or both counterclockwise, to put the two inside atoms of the reactant (D atoms) on opposite sides of the plane of the four carbon atoms in the product. This is called a conrotatory closure. The other mode rotates the methylenes in opposite directions (disrotatory) to give a product wherein the inside atoms appear on the same side of the C_4 plane.

*In Section 7.3.2 we give an equivalent basis for classification of cycloaddition reactions.

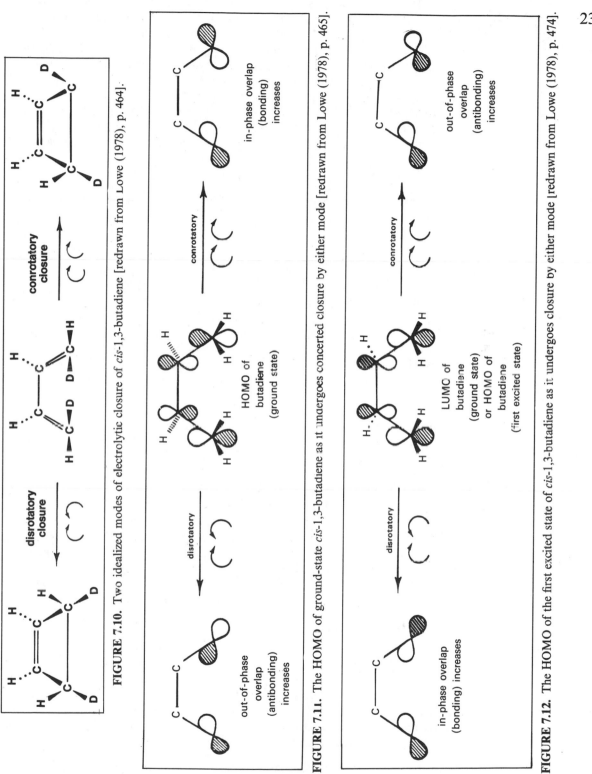

FIGURE 7.10. Two idealized modes of electrolytic closure of *cis*-1,3-butadiene [redrawn from Lowe (1978), p. 464].

FIGURE 7.11. The HOMO of ground-state *cis*-1,3-butadiene as it undergoes concerted closure by either mode [redrawn from Lowe (1978), p. 465].

FIGURE 7.12. The HOMO of the first excited state of *cis*-1,3-butadiene as it undergoes closure by either mode [redrawn from Lowe (1978), p. 474].

239

One does not know, *a priori*, whether the reaction follows either of these two paths. The figure shows processes where both the methylene groups rotate by equal amounts as the reaction proceeds. This is an example of a somewhat extreme kind of what is termed a concerted process, the two processes occurring together. The opposite extreme is a nonconcerted, or stepwise process, wherein one methylene group would rotate all the way (by 90°) and the other group would begin to rotate only after the first process is completed. If we neglect the difference between D and H, this process would lead to an intermediate having a plane of symmetry, which means that the second methylene group would be equally likely to rotate either way, giving the 50–50 mixture of the two products depicted in the figure.

One can argue in favor of some substantial degree of concertedness, which means that the second methylene group has been partly rotated before the rotation of the first group is complete. The reaction involves destruction of a four-center conjugated π-system and formation of an isolated π-bond and a new C—C σ-bond. Energy will be lost in the destruction of the old bonds, and gained in formation of the new ones. Therefore, one can anticipate that the lowest energy path between reactants and products will correspond to a reaction coordinate wherein the new bonds start to form before the old ones are completely broken. However, the new σ-bond cannot form to any significant extent until both methylene groups have undergone some rotation. Thus, concertedness in breaking old bonds and forming new bonds is assisted by some concertedness in methylene group rotations. Such concertedness does not necessarily imply the absence of an intermediate. If the reaction surface had a local minimum at a point at which both methylenes were rotated by 45°, it would not alter the argument.

It was because they knew that many electrocyclic reactions are observed to give nearly 100% of one product or the other in a reaction like that shown in Figure 7.10 that Woodward and Hoffmann looked for an explanation in terms of qualitative MO arguments. They employed frontier orbitals, i.e., the highest occupied molecular orbital (HOMO) and the lowest unoccupied molecular orbital (LUMO), and argued that their energies would change with a con- or disrotatory motion owing to changes in overlap. For butadiene in its ground state, the higher occupied MO is the π-MO shown in the middle of Figure 7.11. The figure indicates that the interaction between p–π-AOs on terminal carbons is favorable for bonding in the region of the incipient σ-bond only in the conrotatory case. It is clear then that the prediction is that for concerted electrocyclic closure, butadiene in the ground state should prefer to go by a conrotatory path. When the reaction is carried out by heating butadiene (thermal reaction), which means that the reactant is virtually all in the electronic ground state, the product is indeed that expected from conrotatory closure.

But one can also carry out electrocyclic reactions by photochemical means. The excited butadiene now has an electron in a π-MO which was unoccupied in the ground state. Indeeed, this MO was the LUMO of butadiene in its ground state, which is shown in Figure 7.12. From this, it can be seen that the change to the next-higher MO of butadiene has introduced an additional node. This reverses the phase relation between terminal π-AOs and thereby changes the predicted path from con- to disrotatory. This is in agreement with the photochemically induced reaction* when the product does correspond to purely disrotatory closure.

---

*It is not always immediately obvious though which empty MO becomes occupied in a given photochemical experiment.

241
*Chemical
Reactions,
Dynamics,
and Laser
Spectroscopy*

The above method, which focuses first on those frontier orbitals which appear to be most likely to dominate the energy change, and then secondly on those changes in these orbitals that will be different in the two different paths, is readily extended to longer systems.

As an example (cf. Lowe, 1978), hexatriene closes to cyclohexadiene in the manner predicted by these orbitals and their changes. The significant change from butadiene to hexatriene is from 4 to 6 π-electrons. This implies that the HOMO for hexatriene has more nodes than that for butadiene.* The effect is that the predictions for hexatriene are just the reverse of those for butadiene, which means that hexatriene exhibits disrotatory closure by thermal means and shows conrotatory closure photochemically.

Thus we generalize to: "The thermal electrocyclic reactions of a k π-electron system will be disrotatory for $k = 4q + 2$ and conrotatory for $k = 4q$ ($q = 0, 1, 2 \ldots$); in the first excited state these relationships are reversed." This generalization is called a Woodward–Hoffmann rule.

An alternative line of reasoning by which one can treat electrocyclic reactions was first worked out by Longuet-Higgins and Abrahamson (1965) and is summarized in Appendix A7.1.

7.3.2. Cycloaddition Reactions

As we discussed in Section 7.3.1, cycloaddition reactions are formally closely related to the electrocyclic reaction, and one example is the Diels–Alder reaction between ethylene and butadiene to give cyclohexene. Such reactions are classified in terms of the number of centers between the points of connection.† The Diels–Alder reaction is, as we have seen in Section 7.3.1, a [4 + 2] cycloaddition reaction. Several distinct geometric possibilities are conceivable for a concerted mechanism for such a reaction. The two new σ-bonds can be envisaged as being formed on the same face (suprafacial) or opposite faces (antarafacial) of each of the two reactants, the different possibilities being depicted in Figure 7.13. The qualitative arguments of MO theory are then used to judge which of the processes will be energetically the most favorable.

We shall employ the frontier orbital arguments below, though alternative arguments can be used and lead to the same conclusion. In the course of this reaction, electrons become shared between the π-systems of butadiene and ethylene. In fairly rough approximation, this situation comes about by interaction of the HOMO of butadiene and the LUMO of ethylene, and also between the LUMO of butadiene and the HOMO of ethylene. We shall consider the first of these interactions below.

The MOs are depicted in Figure 7.14 and the overlapping regions are shown for the four geometric possibilities. If one examines these diagrams, one sees that the two MOs have positive overlap in the regions of both incipient σ-bonds only for the $(4s + 2s)$ and $(4a + 2a)$ modes. The prediction is that these modes proceed with less activation energy and are therefore favored. Turning to the other pair

---

*Or the HOMO for a $2n$ π-electron system is like the LUMO for a $2n - 2$ π-electron molecule, insofar as end-to-end phase relationships are concerned.

†Recall from Section 7.3.1 an equivalent basis for classification.

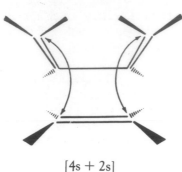

[4s + 2s]

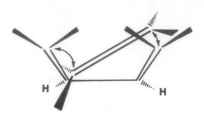

[4a + 2a]

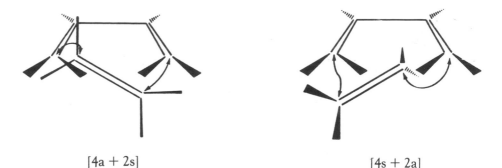

[4a + 2s]

[4s + 2a]

FIGURE 7.13. Four suprafacial–antarafacial combinations possible for the Diels–Alder [2 + 4] cycloaddition reaction [redrawn from Lowe (1978), p. 474].

of MOs, namely the LUMO of butadiene and the HOMO of ethylene, one must note that, for each MO, the end-to-end phase relationship is reversed from what it was before. Two symmetry reversals leave one with no net change in the inter-molecular phase relations. It can be concluded, therefore, that these MOs also favor the (s, s) and (a, a) modes. In cycloaddition reactions of this sort, one need analyze only one HOMO–LUMO pair in order to arrive at a prediction. Extension to longer molecules or to photochemical cycloadditions proceeds by the same kinds of arguments presented for the electrocyclic reactions.

7.3.3. Other Types of Chemical Reactions

Qualitative arguments based on MO theory have allowed the rationalization of many types of chemical reactions. For example, the association of S_N2 reactions with Walden inversion* is rationalized by arguing that an approaching nucleophile will donate electrons into the LUMO of the substrate. The LUMO for CH_3Cl is shown in Figure 7.15. A successful encounter between CH_3Cl and a base results in a bond between the base and the C atom, so that the highest MO of the base needs to overlap the p-AO of C in the LUMO of Figure 7.15a. Attack at the

---

*That is the adding group attacks the opposite side of an atom from the leaving group.

243

*Chemical
Reactions,
Dynamics,
and Laser
Spectroscopy*

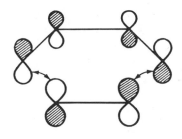

[4s + 2s]

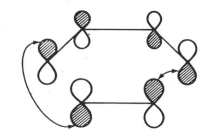

[4a + 2a]

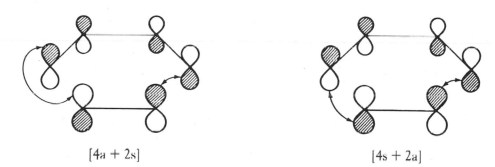

[4a + 2s]   [4s + 2a]

FIGURE 7.14. Overlap between HOMO of butadiene and LUMO of ethylene resulting from four interactive modes shown in Figure 7.13. The geometries for the four modes would all be different [redrawn from Lowe (1978), p. 475].

position labeled 1 in the figure is unfavorable because any base MO would be near a nodal surface, yielding poor overlap with the LUMO. Therefore, attack at site 2 is favored. As the previously empty LUMO of CH_3Cl becomes partially occupied, we expect a loss of bonding between C and Cl. Also negative overlap between the forming C-base bond and the three Hs on the back side should encourage the latter to migrate away from the attacked side, as indicated in Figure 7.15b.

The tendency of a high-energy, occupied MO of the base to couple strongly with the LUMO of a molecule such as CH_3Cl depends partly on the energy separation of these orbitals. If they are close in energy, they will evidently mix more readily and will give a bonded combination of much lower energy. Molecules in which the HOMO is high in energy tend to be polarizable bases. A high-energy HOMO means, of course, that the electrons are relatively weakly bound and will be appreciably affected by perturbations. Such bases react readily with molecules whose LUMO has a low energy. When the energy difference between the HOMOs and the LUMOs is substantial, orbital overlap becomes less important as a controlling mechanism and simple electrostatic interactions may become dominant.

Liotta (1975) has used qualitative MO arguments, but in a situation in which the frontier orbital is allowed to distort as the reaction proceeds. One assumes

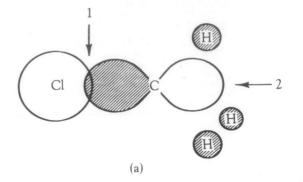

(a)

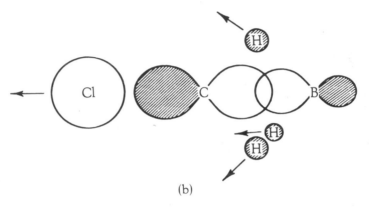

(b)

FIGURE 7.15. (a) The LUMO of CH_3—Cl. (b) Positive overlap between HOMO of the base B and LUMO of CH_3—Cl increases antibonding between C and Cl and also repels H atoms from their original positions [redrawn from Lowe (1978), p. 477].

that this distortion will occur in such a way as to minimize out-of-phase overlaps and to maximize in-phase overlaps. Let us take an example to make the point more specifically. Consider the displacement of some leaving group X from the molecule. If attack by a nucleophile Y^- occurs at C_1, we have an S_N2 reaction. Attack by a nucleophile at C_3 can also displace X as indicated in Figure 7.16. Such a displacement is denoted by S_N2'. The question arises as to whether Y attacks C_3 suprafacially or antarafacially with respect to the leaving group X. One

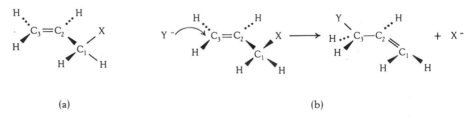

(a)   (b)

FIGURE 7.16. (a) Displacement of leaving group X; if attack by a nucleophile Y^- occurs at C_1, the reaction is S_N2. (b) The attack by a nucleophile at C_3 in the displacement of X is S_N2' [redrawn from Lowe (1978), p. 478].

245

*Chemical
Reactions,
Dynamics,
and Laser
Spectroscopy*

first draws the LUMO for the substrate (Figure 7.17a). This is antibonding between C_2 and C_3, bonding between C_1 and C_2, and antibonding between C_1 and X.* Now one imagines σ–π mixing to occur in this MO as it becomes occupied, so as to minimize the antibonding between C_1 and X, as shown in Figure 7.17b. Next, one carries the σ–π mixings on down the molecule, always keeping the in-phase overlaps maximized and the out-of-phase minimized, but at the same time maintaining the basic nodal structure of the original LUMO (Figure 7.17c). Since the MO at C_3 bulges out on the same side of the molecular plane as X,

---

*The X AO is shown to be of *s*-type. It would be some mixture of *s* and *p*, depending on which atom or group of atoms X is. One can obtain the LUMO in a molecule of this sort from a semiempirical calculation such as extended Hückel or complete neglect of differential overlap (see Chapter 6).

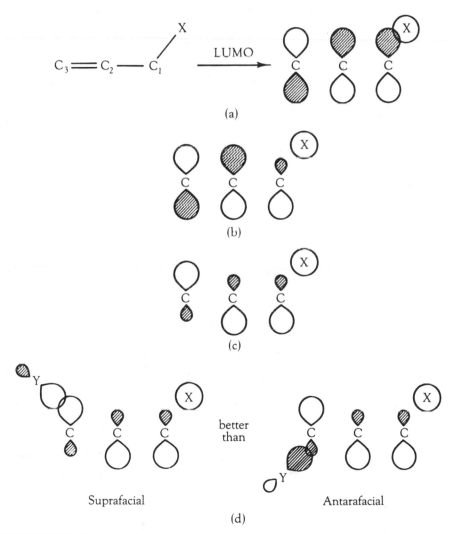

FIGURE 7.17. Suprafacial and antarafacial attacks [redrawn from Lowe (1978), p. 479].

attack is favored in a suprafacial mode (Figure 7.17d). Liotta cites experimental results to support predictions made in this way. It should be noted that the method also predicts that ordinary S_N2 reactions should proceed with attack from the rear side of C_1. Basically, Liotta's argument is to find which of two possible modes allows for the existence of the lower-energy HOMO of the intermediate complex.

Naturally, we have singled out but a few examples of the way in which the qualitative features of MO theory can be used to gain insight into chemical reactions. Various specialized accounts of this area exist in the literature; and the interested reader is referred to those for further details but especially to Lowe (1978), whose treatment we have followed closely in this section.

7.4. LOCALIZATION ENERGY AND THE RATE OF REACTION FOR AROMATIC HYDROCARBONS

A topic of long-standing interest in chemical reactivity studies is that involving localization energy and hydrocarbons. In the transition state discussed in Section 7.1, it has been suggested that the π-system is interrupted at the site of the attack, as, for example, in an addition reaction on hydrocarbons. The localization energy is defined as the π-energy lost in the process that interrupts the π-system. The smaller the resultant loss of π-electron energy, the faster the reaction.

It is worthwhile in this context to consider the case of benzene. It should be recalled from Chapter 6 that the energy of the π-electron system in Hückel theory is $(6\alpha + 8\beta)$. If we isolate one C atom, five are left conjugated, as for example in the pentadienyl system. In this latter case the orbital energies in Hückel theory are $(\alpha + 1.732\beta)$, $(\alpha + \beta)$, α, $(\alpha - \beta)$, and $(\alpha - 1.732\beta)$. If a nucleophilic attack occurs, two electrons are required to bind the reagent molecule to the benzene ring. This leaves four π-electrons to be distributed in these orbitals, the corresponding energy being

$$2(\alpha + 1.732\beta) + 2(\alpha + \beta) = (4\alpha + 5.464\beta) \tag{7.60}$$

Consequently the localization energy L is

$$L = (6\alpha + 8\beta) - (4\alpha + 5.464\beta) = 2\alpha + 2.536\beta \tag{7.61}$$

In the case of a radical attack, five π-electrons are left to the pentadienyl system and its localization energy is easily calculated along similar lines. The same method can also be employed to calculate the localization energy for an electrophilic attack (where the electrophilic reagent supplies both electrons that are required to bond to the benzene ring). In general, therefore, the localization energy L can be easily estimated, and has the form

$$L = x\alpha + y\beta \tag{7.62}$$

where x is dependent on the class of reagent while y depends on the aromatic C atom that is under attack.

247

*Chemical
Reactions,
Dynamics,
and Laser
Spectroscopy*

As an example of experimental results involving localization energy, we note that the work of Kooyman and Farenhorst (1953) has yielded a linear plot for the relative reactivities of the trichloromethyl radical $CCl_3^{\cdot}$ toward aromatic hydrocarbons and the radical localization energies, denoted $L_r^{\cdot}$ in Figure 7.18, at the most reactive position.

7.5. OXIDATION–REDUCTION AND ORBITAL ENERGIES OF HYDROCARBONS

It is of interest to turn now to another study of considerable significance for quantum chemistry. Using standard electrochemical techniques, several conjugated hydrocarbons can be oxidized or reduced in solution. By definition, oxidation involves the removal of an electron from the HOMO. As we saw in Chapter 6, these are π-electrons for hydrocarbons. We would expect molecules having lower-energy HOMOs to have higher oxidation potentials. In like manner, we would anticipate reduction (gain of electrons) to be easier for molecules where the LUMO is lower in energy.

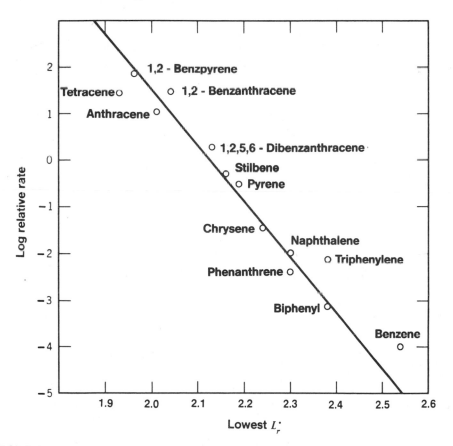

FIGURE 7.18. Relative reactivities of the trichloromethyl radical $CCl_3^{\cdot}$ toward aromatic hydrocarbons *vs.* radical localization energies $L_r^{\cdot}$ [redrawn from Streitwieser (1961), p. 400].

Evidently, then, it is to be expected that oxidation–reduction potentials correlate with the HOMO and LUMO energies. Figure 7.19 is a plot of the experimental oxidation potential *vs.* the Hückel HOMO energy (in units of the Hückel β) for several conjugated hydrocarbons. In Figure 7.20, the reduction potential is depicted *vs.* the Hückel LUMO energy (also in units of β) for numerous hydrocarbons. The correlation is quite good considering the relatively crude character of Hückel calculations, solvent effects, and the nature of the molecules involved.

7.6. LASER SPECTROSCOPY AND CHEMICAL DYNAMICS*

One important advance afforded by tunable lasers is to facilitate the finding of relative probabilities for forming product molecules in specific quantum states from reagents in selected states. This represents the attainment of one major objective of chemical dynamics.

7.6.1. Rate Coefficient Analysis

Related to the above, it is true that kineticists had to be content for a long period with observing the overall rates of chemical reactions as a function of temperature T.

In the simple bimolecular gas-phase reaction

$$A + BC \rightarrow AB + C \tag{7.63}$$

gaseous atoms A come into contact with BC molecules. One then monitors the rate of removal of a reactant (A or BC) or the rate of appearance of a product (AB or C). The rate R is expressed in terms of the rate of change of the various number densities n as a function of time t:

$$R = - dn_A/dt = - dn_{BC}/dt = dn_{AB}/dt = dn_C/dt \tag{7.64}$$

The initial rate of reaction depends on the product of the reagent concentrations (see also equation (7.25)):

$$R = kn_A n_{BC} \tag{7.65}$$

where $k = k(T)$ is the rate coefficient. It is to be noted that $k(T)$ may be represented as

$$k(T) = \sum_i F_i(T) k_i(T) \tag{7.66}$$

---

*Though some more basic material on photons and chemical reactions is presented in Section 7.8.2, it is helpful to present this application at this particular point.

249
*Chemical
Reactions,
Dynamics,
and Laser
Spectroscopy*

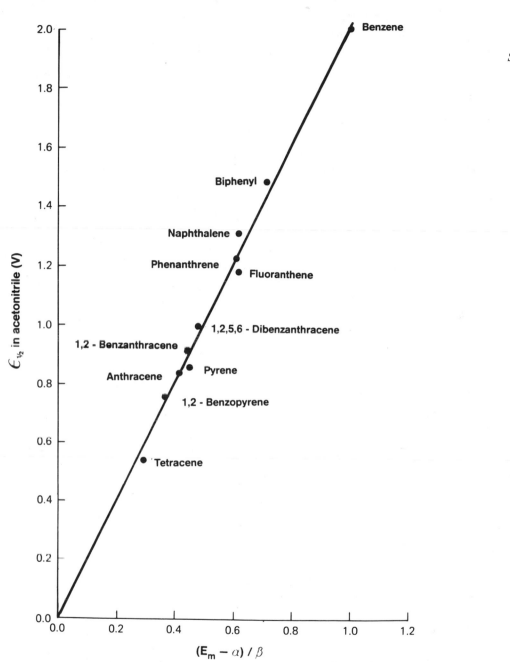

FIGURE 7.19. Polarographic oxidation potential *vs.* Hückel HOMO energy (in β units) [redrawn from Lowe (1978), p. 236].

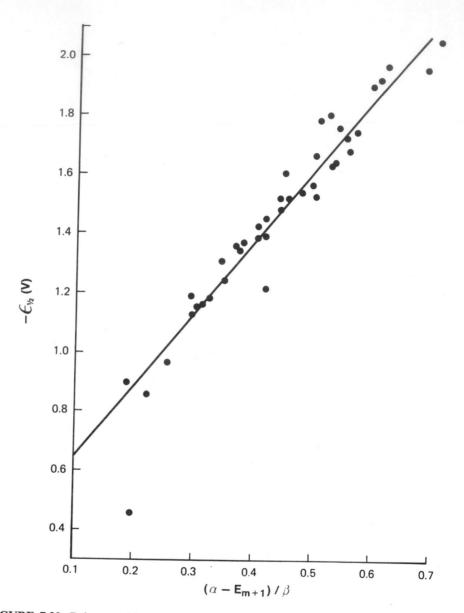

FIGURE 7.20. Polarographic reduction potential *vs.* Hückel LUMO energy (also in β units) [redrawn from Lowe (1978), p. 237].

where $F_i(T)$ is the fraction of reagents in some initial set of internal states i, while $k_i(T)$ is the total rate constant for all processes out of the set i into the final set of states f of the products [see Zare and Bernstein (1980)]:

$$k_i(T) = \sum_f k_{fi}(T) \tag{7.67}$$

One can expect to gain further insight into chemical dynamics by exploring k_i and k_{fi} in addition to the well studied overall reaction rate k.

7.6.2. State-to-State Chemical Reaction Dynamics: Hydrogen Exchange Reaction

251

*Chemical
Reactions,
Dynamics,
and Laser
Spectroscopy*

To illustrate the use of lasers in investigating state-to-state chemical reaction dynamics, we now consider briefly the hydrogen exchange reaction. This is the simplest example of neutral chemistry, namely

$$H + H_2 \rightarrow H_2 + H \tag{7.68}$$

and the analogues with isotopes. As shown in Figure 7.21, this "atom exchange" reaction has an energy barrier, the threshold for reaction being 32 kJ/mole, and it is to be noted that a single quantum of vibrational excitation in H_2 (50 kJ/mole) easily exceeds this threshold energy. Thus this reaction allows an assessment of the effect of vibrational excitation on the reaction dynamics. Direct experimental investigation was not possible for a long time, owing to the difficulty of preparing known concentrations of H_2 ($v = 1$). Since H_2 is a nonpolar, infrared-inactive species, direct laser photoexcitation is not possible.

However, Kneba *et al.* (1979) were able to measure the rate of the D + H_2 ($v = 1$) exchange reaction in a so-called "discharge flow" system. A low-pressure flow of HF and H_2 in a He carrier gas was exposed to short pulses (10 μs) from an HF laser tuned to the fundamental transition, HF ($v = 0 \rightarrow 1$). Rapid population of H_2 ($v = 1$) takes place via near-resonant vibrational ($V \rightarrow V$) transfer from HF ($v = 1$). D atoms which were generated by dissociation of D_2 in a microwave discharge were introduced upstream. The absolute concentrations of vibrationally excited HF molecules were found from time-resolved infrared emission, and the concentration of H atoms formed in the exchange reaction was determined from time-resolved absorption using Lyman-alpha resonance radiation. Owing to the fact that H_2 ($v - 0$) molecules react relatively slowly with D

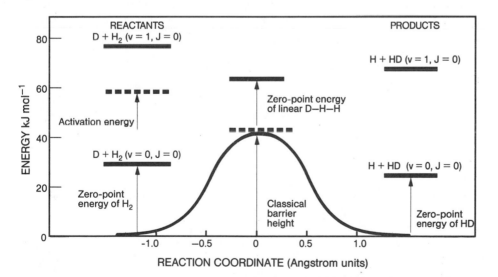

FIGURE 7.21. Energy *vs.* reaction coordinate for the reaction D + H_2 ($v = 0, 1$). The approach is in collinear orientation [redrawn from Zare and Bernstein (1980), p. 47].

atoms at room temperature, the concentrations of H_2 and D could be adjusted so that the formation of H atoms from the $D + H_2$ ($v = 0$) exchange reaction gave no strong background absorption of the Lyman-alpha radiation without laser excitation. Furthermore, the concentrations of H_2 and HF were chosen such that only a small fraction of the H_2 molecules was excited to $v = 1$, thus ensuring that the concentration of H_2 ($v > 1$) is in practice negligible.

This study yielded a bimolecular rate coefficient k for HD production at 298 K of some 10^{-11} cm^3/s. This value is about 4×10^4 times faster than the thermal H_2 ($v = 0$) rate constant. To our knowledge there are, at the time of writing, no accurate quantum-mechanical calculations of this rate. Quasi-classical trajectory calculations using an accurate H_3 potential energy surface yield a slower rate than that observed by much more than an order of magnitude. The interested reader is referred to a review of bimolecular reactions of vibrationally excited molecules found in *Physics Today* (Special Issue—Laser Chemistry), American Institute of Physics, New York (November 1980).

7.7. ELECTRON DENSITY THEORY AND CHEMICAL REACTIVITY

We now return to the electron density theory set out in Chapter 4. Though, as emphasized there, this theory is basically about the ground state, the chemical potential of the theory, defined "thermodynamically" in equation (4.42), does allow the chemical concepts of hardness and softness to be put into this density framework. These concepts, as emphasized in the pioneering work of Pearson (1966), are relevant to some aspects of chemical reactivity, as we outline below.

7.7.1. Hardness and Softness

The chemical potential is "thermodynamically" the derivative of the ground-state energy E with respect to the number of electrons N, according to equation (4.43). This already presents some problems, unless one thinks, following Mulliken (1935), of a smooth curve joining, say, for atomic species with nuclear charge Ze, the positive ion energy, the neutral atom, and the negative ion. Then μ is the slope of this curve of E *vs.* N at the appropriate point ($N = Z$ for the neutral atom) (see also Figure 4.5).

Parr and Pearson (1983) have assumed that not only the first derivative of such an E-*vs.*-N curve can be specified, but also the second derivative. They then have defined the "absolute hardness" of a species, η, say, as

$$\eta = \tfrac{1}{2}\partial^2 E/\partial N^2 = \tfrac{1}{2}\partial\mu/\partial N \tag{7.69}$$

the latter form following from equation (4.43). The inverse of hardness is "softness" S [cf. Parr and Yang (1989)]

$$S = 1/2\eta = \partial N/\partial\mu \tag{7.70}$$

They next note that the finite difference approximation for hardness, the analogue of the Mulliken form $(I + A)/2$ for electronegativity, is

$$\eta = (I - A)/2 \tag{7.71}$$

253

*Chemical
Reactions,
Dynamics,
and Laser
Spectroscopy*

TABLE 7.5

Classification of Lewis Acids and Bases as Hard, Borderline, or Soft[a]

	Hard	Borderline	Soft
Acids	H^+, Li^+, Na^+, K^+, Be^{2+}, Mg^{2+}, Ca^{2+}, Cr^{2+}	Fe^{2+}, Co^{2+}, Ni^{2+}, Cu^{2+}, Zn^{2+}, Pb^{2+}	Cu^+, Ag^+, Au^+, Tl^+, Hg^+, Pd^{2+}, Cd^{2+}, Pt^{2+}, Hg^{2+}
Bases	F^-, OH^-, H_2O, NH_3, CO_3^{2-}, NO_3^-, O^{2-}, SO_4^{2-}, PO_4^{3-}, ClO_4^-	NO_2^-, SO_3^{2-}, Br^-, N_3^-, N_2	H^-, R^-, CN^-, CO, I^-, SCN^-, R_3P, C_6H_6, R_2S

[a]From Atkins (1991).

Table 7.5 lists some examples of hardness for various species. This concept is relevant to acid–base reactions, as discussed below.

7.7.2. Acids and Bases

The concept of acids and bases is widely useful in chemistry. In fact, nearly all chemical reactions can be broadly classified as either reactions between acids and bases, or as reactions involving oxidation and reduction.[*]

Several properties are characteristic of acids and bases in general. These include:

1. Neutralization: Acids and bases react with one another so as to cancel, or neutralize, their acidic and basic characters.
2. Reaction with indicators: Certain organic dyes, called indicators, give different colors depending on whether they are in an acidic or basic medium.
3. Catalysis: Many chemical reactions are catalyzed by the presence of acids or bases.

As Parr and Yang (1989) particularly emphasize, the terms "hardness" and "softness" have been in the vocabulary of chemistry since the early 1950s [see, e.g., Mulliken (1952)], but they cite the definitive work introducing the concepts as that of Pearson (1963). Whereas there was a long period when there was not a consensus as to the precise meaning of "hard" and "soft," there has long been agreement as to the general characteristics of hard and soft species [Pearson (1966), Klopman (1968)]. These can be summarized as follows:

1. *Soft Base*: The donor atom has high polarizability and low electronegativity, is easily oxidized, and is associated with empty low-lying orbitals.
2. *Hard Base*: In complete contrast this has low polarizability, large electronegativity, is hard to oxidize, and is associated with empty orbitals having high energy.
3. *Soft Acid*: The acceptor atom has small positive charge, is large in extent, and has easily excited outer electrons.

---

[*]Often there is an overlap between them, so that for some reactions it is sometimes convenient to view them as acid–base reactions while at other times it may be best to describe them in terms of oxidation–reduction.

4. *Hard Acid*: In contrast the atom has a high positive charge, small size, and no readily excited outer electrons.

As to Pearson's motivation for introducing these concepts, he found that it was not possible to understand the fundamentals of acid–base reactions with one parameter (electronegativity) per species. He therefore employed hardness or softness as a second parameter, and enunciated the principle that, "both in their thermodynamic and kinetic properties, hard acids prefer hard bases and soft acids prefer soft bases."

Some reference can also be made to the important work of Fukui (1952) on the frontier-electron theory of reactivity by defining the derivative of the electron density $\rho(\mathbf{r})$ itself with respect to N. This derivative has been termed the Fukui index and its significance is discussed at some length in the book by Parr and Yang (1989).

7.8. OTHER STUDIES IN CHEMICAL KINETICS

To conclude, we shall briefly consider, following the treatment of Laidler (1987) some further areas of interest for chemical kinetics: (i) electron impact studies, (ii) photons and chemical reactions, and (iii) rate studies for reactions of the hydrated electron.

7.8.1. Electron Impact Studies

When a high-energy electron encounters a molecule, positive ions are frequently produced:

$$e^- + M \rightarrow M^+ + 2e^- \tag{7.72}$$

Of course, there is then an increase in the number of electrons in the reaction (7.72), which means that an electron can bring about the formation of a number of ions. As the electron is slowed down, it will no longer be able to cause ejection of another electron from the molecule M. When this situation is reached, a number of processes may take place as follows:

(i) A slow electron may attach to M to form a negative ion:

$$\underset{\text{slow}}{e^- + M \rightarrow M^-} \tag{7.73}$$

(ii) Electron capture may take place, bringing about the rupture of bonds and the dissociation of a molecule, say R_2:

$$e^- + R_2 \rightarrow R^- + R \tag{7.74}$$

(iii) An electron may neutralize a positive ion:

$$e^- + M^+ \rightarrow M \tag{7.75}$$

(iv) Following neutralization, dissociation may occur:

$$e^- + R_2^+ \rightarrow R + R \qquad (7.76)$$

255

*Chemical
Reactions,
Dynamics,
and Laser
Spectroscopy*

It is of interest to note that some of these possible reactions (7.73) to (7.76) have been analyzed in terms of potential energy curves.

Figure 7.22 depicts such curves for H_2 and H_2^+, indicating how positive ions are formed by electron impact. This figure shows that electrons with an energy of 16 eV bring about excitation to the $^2\Sigma_g^+$ state (the lower state of the H_2^+ ion). Electrons having an energy of 26 eV can yield molecules in the $^2\Sigma_u^+$ state of H_2^+, with the result that H and H^+ can be produced with high kinetic energies.

7.8.2. Photons and Chemical Reactions

Three types of processes that are brought about when photons are absorbed by molecules will now be considered:

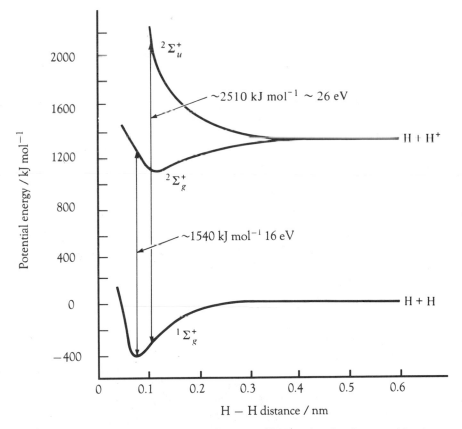

FIGURE 7.22. Potential energy curves for H_2 and H_2^+, showing how positive ions are formed by electron impact [redrawn from Laidler (1987), p. 368].

(i) Photon bombardment leads to an excited molecule, which in turn may dissociate into atoms and radicals:

$$hv + M \rightarrow M^* \rightarrow 2R \tag{7.77}$$

(ii) Photons (high-energy) may cause ionization and the following process is possible:

$$hv + M \rightarrow M^+ + e^- \text{ (with energies } \leq 5 \times 10^5 \text{ eV)} \tag{7.78}$$

(iii) The following reaction can occur:

$$hv + M \rightarrow M^+ + e^- + hv' \tag{7.79}$$

which is a type of Compton scattering of the photon. As opposed to (ii), here the photon is not annihilated.

Whereas with material particles one encounters a stepwise loss of energy, photons are usually absorbed and annihilated in a single step. The use of photons to bring about reactions is important in the study of chemical kinetics, and, as we noted in Chapter 1, multiphoton processes can lead to various chemical and physical applications involving absorption, dissociation, and ionization. The reader is referred to the books by Hollas (1987) and Steinfeld (1985) for fuller details of multiphoton processes.

7.8.3. Rate Studies for Reactions of the Hydrated Electron

The radiolysis of liquid water is brought about using various types of ionizing radiation:

$$\text{Energy} + H_2O \rightarrow H_2O^+ + e^- \tag{7.80}$$

If the e^- in equation (7.80) remains free, one usually has the following reactions:

$$e^- + H_2O \rightarrow H + OH^- \tag{7.81}$$

and

$$e^- + H_2O^+ \rightarrow H_2O^* \rightarrow H + OH \tag{7.82}$$

A hydrated electron e_{aq}^- was predicted a long time ago by Platzman (1953), and evidence for its existence was provided by the studies of Hart and Boag [Hart and Boag (1962), Boag and Hart (1963)]. These researchers observed the transient spectrum of the species that was produced by high-energy pulses to bring about the formation of the hydrated electron. Various studies since then have yielded rate constants k as listed in Table 7.6.

257

*Chemical
Reactions,
Dynamics,
and Laser
Spectroscopy*

TABLE 7.6
Rates Constants for Reactions Involving
Hydrated Electrons[a]

Reaction	$k/dm^3/mole\ s$
$e_{aq}^- + e_{aq}^- \rightarrow H_2 + 2OH^-$	5×10^9
$e_{aq}^- + H_2O \rightarrow H + OH^-$	16
$e_{aq}^- + H_3O^+ \rightarrow H + H_2O$	2.06×10^{10}

[a]Taken from Laidler (1987).

7.8.4. Topics for Further Study Involving Chemical Reactivity

In a single chapter, it is only possible to be very selective in highlighting some areas of importance in the now vast subject of chemical reactivity. The student will find it fruitful to make further detailed study of some of the topics listed below:

1. Frontier orbital theory.
2. Perturbation MO theory.
3. Free molecule theory of reaction rates.
4. Aromatic reactivity and Hückel's rule.
5. Localization energy involving nucleophilic, electrophilic, and addition reactions.
6. Substituent effects.
7. Aliphatic nucleophilic substitution.
8. Addition reactions.
9. Elimination reactions.
10. Chemiluminescence.
11. Gas-phase combustion.
12. *Ab initio* MO calculations and the reaction pathway.
13. Reactions on surfaces.
14. Molecular beams and chemical reactions.

Some useful references are given at the end of this chapter.

PROBLEMS

7.1. The gaseous dimerization reaction $2A \rightarrow A_2$ follows second-order kinetics at 400°C. From the following data determine the rate constant:

t (s)	0	100	200	300	400	500
p (mm Hg)	600	500	450	420	400	385

7.2. The rate law for the reaction $A \rightarrow$ Products may be represented by $dx/dt = k_n(a_0 - x)^n$, where a_0 is the initial concentration, x is the decrease in concentration after time t, k_n is the rate constant, and n is the order of the reaction. If $n \neq 1$, **(a)** integrate the expression to determine k_n and derive the relationship between the half-life and the rate constant. **(b)** Indicate how the latter expression could be used to determine n.

7.3. Distinguish between molecularity and the order of a chemical reaction.

7.4. The following rate constants were found for the decomposition of N_2O_5 at various temperatures.

t (°C)	25	45	65
k (s^{-1})	3.46×10^{-5}	4.98×10^{-4}	4.85×10^{-3}

Determine graphically the activation energy and the Arrhenius pre-exponential factor A.

7.5. A substance A may decompose by either first- or second-order kinetics. If both reactions have the same half-life for the same initial concentration of A, calculate the ratio of the first-order rate to the second-order rate as follows: **(a)** initially, **(b)** at the time of one half-life.

7.6. The rate equation for the general reaction $A + B + C \rightarrow$ Products is

$$-d[A]/dt = -d[B]/dt = -d[C]/dt = k_3[A][B][C]$$

since $[A] = [A]_0 - x$, $[B] = [B]_0 - x$, and $[C] = [C]_0 - x$ we can write that

$$dx/dt = k_3([A]_0 - x)([B]_0 - x)([C]_0 - x)$$

In the special case when the three initial concentrations are equal we can write $dx/dt = k_3([A_0] - x)^3$. Show that

$$2k_3t = 1/\{[A]_0 - x\}^2 - 1/[A]_0^2$$

7.7. **(a)** State the Arrhenius equation and indicate the significance of the various factors. **(b)** Outline the collision theory of reaction rates and indicate its limitations. **(c)** Outline the absolute rate theory and indicate how and why it is superior to the collision theory.

7.8. The Woodward–Hoffmann rules for electrocyclic reactions can be summarized as follows:

Number of π-electrons	Reaction	Motion
$4n$ (n = any integer)	thermal	conrotatory
$4n$	photochemical	disrotatory
$4n + 2$	thermal	disrotatory
$4n + 2$	photochemical	conrotatory

Explain the above rules.

7.9. The Woodward–Hoffmann rules for $[i + j]$ *cycloadditions* can be summarized as follows:

$[i + j]$	Thermal	Photochemical
$4n$	supra–antara antara–supra	supra–antara antara–antara
$4n + 2$	supra–supra antara–antara	supra–antara antara–supra

Explain the above rules.

7.10. Starting from first principles derive expressions for the translational, rotational, and vibrational partition functions for molecules.

REFERENCES

259

*Chemical
Reactions,
Dynamics,
and Laser
Spectroscopy*

P. W. Atkins, *Physical Chemistry, 3rd Ed.*, W. H. Freeman, New York (1986).

P. W. Atkins, *Quanta, 2nd Ed.*, Oxford University Press, Oxford (1991).

J. W. Boag and E. J. Hart, *Nature* **197**, 45 (1963).

H. Eyring, *Chem. Revs.* **17**, 651 (1935).

H. Eyring and M. Polanyi, *Z. Physik Chem.* **B12**, 279 (1931).

W. H. Flygare, *Molecular Structure and Dynamics*, Prentice-Hall, Englewood Cliffs, NJ (1978).

K. Fukui, *Theory of Orientation and Stereoselection*, Springer-Verlag, Berlin (1975).

K. Fukui, *Science* **218**, 747 (1987).

K. Fukui, T. Yonezawa, C. Nagata, and H. Shingu, *J. Chem. Phys.* **22**, 1433 (1954).

K. Kukui, T. Yonezawa, and H. Shingu, *J. Chem. Phys.* **20**, 722 (1952).

E. J. Hart and J. W. Boag, *J. Am. Chem. Soc.* **84**, 4080 (1962).

J. M. Hollas, *Modern Spectroscopy*, John Wiley and Sons, New York (1987).

G. Klopman, *J. Am. Chem. Soc.* **90**, 223 (1968).

M. Kneba, U. Wellhausen, and J. Wolfrum, *Ber. Bunsenges Phys. Chem.* **83**, 940 (1979).

E. C. Kooyman and E. Farenhorst, *Trans. Faraday Soc.* **49**, 58 (1953).

K. J. Laidler, *Chemical Kinetics, 3rd Ed.*, Harper and Row, New York (1987).

C. L. Liotta, *Tetrahedron Lett.* **8**, 519, 523 (1975).

F. London, *Z. Elektrochem.* **35**, 552 (1929).

H. C. Longuet-Higgins and E. W. Abrahamson, *J. Am. Chem. Soc.* **87**, 2045 (1965).

J. P. Lowe, *Quantum Chemistry*, Academic Press, New York (1978).

D. A. McQuarrie, *Statistical Mechanics*, Harper and Row, New York (1976).

P. Morse, *Phys. Rev.* **34**, 57 (1929).

R. S. Mulliken, *J. Chem. Phys.* **3**, 573 (1935).

R. S. Mulliken, *J. Am. Chem. Soc.* **64**, 811 (1952).

R. G. Parr and R. G. Pearson, *J. Am. Chem. Soc.* **105**, 7512 (1983).

R. G. Parr and W. Yang, *Density-Functional Theory of Atoms and Molecules*, Oxford University Press, New York (1989).

R. G. Pearson, *J. Amer. Chem. Soc.* **85**, 3533 (1963).

R. G. Pearson, *Science* **151**, 172 (1966).

II. Pelzer and E. Wigner, *Z. Physik Chem.* **B15**, 445 (1932).

R. I. Platzman, *Basic Mechanisms in Radiology*, National Research Council Publication 305, Washington, D.C. (1953).

S. Sato, *J. Chem. Phys.* **23**, 592, 2465 (1955).

A. Sreitwieser, Jr., *Molecular Orbital Theory*, John Wiley and Sons, New York (1961).

J. I. Steinfeld, *Molecules and Radiation: An Introduction to Modern Spectroscopy, 2nd Ed.*, MIT Press, Cambridge, MA (1985).

K. B. Wiberg, *Physical Organic Chemistry*, John Wiley and Sons, New York (1964).

R. B. Woodward and R. Hoffmann, *The Conservation of Orbital Symmetry*, Academic Press, New York (1970).

R. N. Zare and R. B. Bernstein, *Physics Today* **33**, 43 (November 1980).

Further Reading

M. J. S. Dewar, *The Molecular Orbital Theory of Organic Chemistry*, McGraw-Hill, New York (1969).

H. F. Hameka, *Quantum Theory of the Chemical Bond*, Hafner Press–MacMillan, New York (1975).

K. Higasi, II. Baba, and A. Rembaum, *Quantum Organic Chemistry*, Interscience Publishers, New York (1965).

R. E. Lehr and A. P. Marchand, *Orbital Symmetry*, Academic Press, New York (1972).

A. Liberles, *Introduction to Theoretical Organic Chemistry*, MacMillan, New York (1968).

R. D. Levine and R. B. Bernstein, *Molecular Reaction Dynamics and Chemical Reactivity*, Oxford University Press, Oxford (1987).

R. McWeeny, *Coulson's Valence, 3rd Ed.*, Oxford University Press, Oxford (1979).

J. N. Murrell, S. F. A. Kettle, and J. M. Tedder, *The Chemical Bond, 2nd Ed.*, John Wiley and Sons, New York (1985).

W. G. Richards, *Quantum Pharmacology, 2nd Ed.*, Butterworths, London (1983).

APPENDIX

A1.1. WAVE FUNCTIONS FOR THE HYDROGEN ATOM

In this appendix we summarize the form of the wave functions for one electron moving in the bare Coulomb field of a nucleus of charge Ze. For $Z = 1$, this is simply, of course, the hydrogen atom.

When the potential energy $V(\mathbf{r})$ in the Schrödinger equation is dependent only on the radial distance r from the nucleus and not on the spherical polar angles θ and ϕ, then the wave equation separates into the form

$$\psi(r, \theta, \phi) = R(r)S(\theta, \phi) \tag{A1.1.1}$$

Writing the Schrödinger equation

$$\nabla^2 \psi + \frac{2m}{\hbar^2}[E - V(r)]\psi = 0 \tag{A1.1.2}$$

in spherical polar coordinates, shown in Figure A.1, and related to Cartesian coordinates x, y, and z by

$$x = r \sin \theta \cos \phi$$

$$y = r \sin \theta \sin \phi \tag{A1.1.3}$$

$$z = r \cos \theta$$

with $0 < r < \infty, 0 \leq \theta \leq \pi, 0 \leq \phi \leq 2\pi$, the operator ∇^2 in equation (A1.1.2) takes the form [see, e.g., Rutherford (1940)]:

$$\nabla^2 = \frac{1}{r^2}\frac{\partial}{\partial r}\left(r^2 \frac{\partial}{\partial r}\right) + \frac{1}{r^2 \sin \theta}\frac{\partial}{\partial \theta}\left(\sin \theta \frac{\partial}{\partial \theta}\right) + \frac{1}{r^2 \sin^2 \theta}\frac{\partial^2}{\partial \phi^2} \tag{A1.1.4}$$

As indicated in equation (A1.1.1), even when the potential energy $V(r)$ depends solely on r, and not on θ and ϕ, this is not true of the wave function $\psi(\mathbf{r})$. The wave functions can evidently have a lower symmetry than the potential energy V in the Schrödinger equation.

261

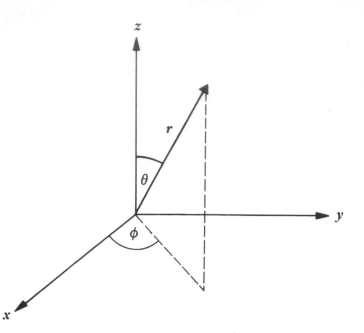

FIGURE A.1. Spherical polar coordinates.

Substituting the form (A1.1.1) into the wave equation (A1.1.2), and dividing both sides of the resulting Schrödinger equation by (RS/r^2), it follows that

$$\frac{1}{R}\frac{\partial}{\partial r}\left(r^2\frac{\partial R}{\partial r}\right) + \frac{1}{S\sin\theta}\frac{\partial}{\partial\theta}\left(\sin\theta\frac{\partial S}{\partial\theta}\right) + \frac{1}{S\sin^2\theta}\frac{\partial^2 S}{\partial\phi^2}$$

$$+ \frac{2m}{\hbar^2}[E - V(r)]r^2 = 0 \qquad \textbf{(A1.1.5)}$$

The first and fourth terms on the left-hand side of equation (A1.1.5) are evidently functions solely of the radial distance r, while the other two terms are functions of the angular variables θ and ϕ alone. Thus, if equation (A1.1.5) is to hold for all r, θ, and ϕ, we must have separately that

$$\frac{1}{R}\frac{\partial}{\partial r}\left(r^2\frac{\partial R}{\partial r}\right) + \frac{2m}{\hbar^2}[E - V(r)]r^2 = \text{constant} = \lambda, \quad \text{say} \qquad \textbf{(A1.1.6)}$$

and

$$\frac{1}{\sin\theta}\frac{\partial}{\partial\theta}\left(\sin\theta\frac{\partial S}{\partial\theta}\right) + \frac{1}{\sin^2\theta}\frac{\partial^2 S}{\partial\phi^2} = -\lambda S \qquad \textbf{(A1.1.7)}$$

It is now quite clear that while equation (A1.1.7) can be solved generally, and that the shapes of atomic wave functions for one electron moving in a central field can be obtained once for all, the radial wave functions $R(r)$ can only be determined for each explicit choice of $V(r)$. Nevertheless, before studying the solutions $S(\theta, \phi)$ of equation (A1.1.7), it is of interest to examine equation (A1.1.6) a little further. To do this, let us rewrite it in a way which makes clear that it is, in essence, a one-dimensional Schrödinger equation, but with the potential energy $V(r)$ modified in a particular manner. Thus, after a little manipulation, one obtains the form

$$\frac{\partial^2}{\partial r^2}(rR) + \frac{2m}{\hbar^2}\left[E - V(r) - \frac{\hbar^2}{2m}\frac{\lambda}{r^2}\right](rR) = 0 \qquad \text{(A1.1.8)}$$

which is indeed a Schrödinger equation for the product rR, with an "effective" potential energy

$$V_{\text{eff}}(r) = V(r) + (\hbar^2/2m)(\lambda/r^2) \qquad \text{(A1.1.9)}$$

To understand the origin of the term modifying $V(r)$ in equation (A1.1.9), let us treat the elementary classical-mechanical problem of an electron moving in a circular orbit of radius r around a fixed proton, say. If the tangential velocity of the electron is v, then the Newtonian equation of motion yields for the radial force F:

$$F = mv^2/r = (mvr)^2/mr^3 = L^2/mr^3 \qquad \text{(A1.1.10)}$$

where we have used the fact that the magnitude L of the orbital angular momentum vector $\mathbf{L}$, defined generally in terms of position vector $\mathbf{r}$ and momentum vector $\mathbf{p}$ by

$$\mathbf{L} = \mathbf{r} \times \mathbf{p} \qquad \text{(A1.1.11)}$$

is simply in this example given by $L = mvr$. Since potential energy and force are related by

$$F = -\partial V/\partial r \qquad \text{(A1.1.12)}$$

for a central field, we see that corresponding to equation (A1.1.10) there is a potential energy term of the form $L^2/2mr^2$. Thus, the interpretation of the separation constant λ in equations (A1.1.6) and (A1.1.7) appears from this argument as

$$\lambda = L^2/\hbar^2 \qquad \text{(A1.1.13)}$$

We shall see below from a fully quantum-mechanical argument that this is indeed correct and, furthermore, L^2 is quantized, with allowed values $l(l+1)\hbar^2$, where $l = 0, 1, 2$, etc. The basic unit in which angular momentum is measured in quantum mechanics is $\hbar$.

The effective potential energy $V_{\text{eff}}(r)$ is indicated schematically in Figure A.2 for the case when $V(r) = -e^2/r$, appropriate to the hydrogen atom. It can be seen that except for $l = 0$, in which case $V_{\text{eff}} = V(r)$, the modification due to the "centrifugal potential energy" $l(l+1)\hbar^2/2mr^2$ is major at sufficiently small r. We must therefore expect that the behavior of atomic wave functions near the nucleus will be crucially dependent on the angular momentum of the state in question.

Dependence of Wave Functions on ϕ and Magnetic Quantum Number m

In order to examine equation (A1.1.7) in more detail, let us take the further step of writing

$$S(\theta, \phi) = \Theta(\theta)\Phi(\phi) \tag{A1.1.14}$$

Then we immediately find that equation (A1.1.7) separates, the new equations for Θ and Φ being, after dividing through by $[\Theta\Phi/\sin^2\theta]$,

$$\frac{1}{\Phi}\frac{\partial^2\Phi}{\partial\phi^2} = -m^2 \tag{A1.1.15}$$

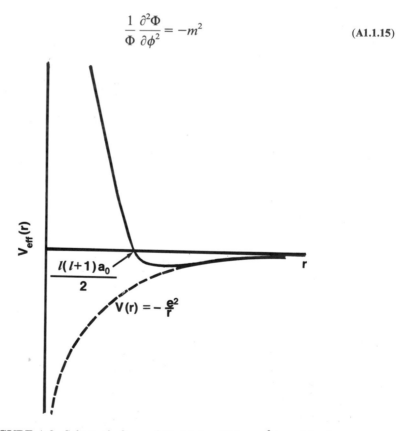

FIGURE A.2. Schematic form of $V_{\text{eff}}(r)$ for $V(r) = -e^2/r$ and $l \neq 0$.

$$\frac{\sin\theta}{\Theta}\frac{\partial}{\partial\theta}\left(\sin\theta\,\frac{\partial\Theta}{\partial\theta}\right)+\lambda\sin^2\theta=m^2 \tag{A1.1.16}$$

The form of the separation constant adopted in equation (A1.1.16), namely $-m^2$, where m is taken to be real, has anticipated the fact that we want periodic solutions for $\Phi(\phi)$, simply because the addition of multiples of 2π to ϕ (see Figure A.1) could not change the physical properties of the system. Therefore the solution of equation (A1.1.15) that we adopt is

$$\Phi(\phi) = \exp(im\phi) \tag{A1.1.17}$$

and the requirement that $\Phi(\phi) = \Phi(\phi + 2\pi s)$, where s is an integer, yields

$$m = 0, \pm 1, \pm 2, \text{ etc.} \tag{A1.1.18}$$

where m is the so-called magnetic quantum number, determining the way the energy levels of an atom split in an applied magnetic field. For many purposes in atomic theory, it is important to relate m to the properties of the angular momentum operator L. Evidently, from equation (A1.1.11), we can write the z-component of angular momentum L_z as

$$L_z = xp_y - yp_x \tag{A1.1.19}$$

and, since $p_x \equiv (\hbar/i)\,\partial/\partial x$ etc., the quantum-mechanical operator has the form

$$L_z = \frac{\hbar}{i}\left(x\frac{\partial}{\partial y} - y\frac{\partial}{\partial x}\right) \tag{A1.1.20}$$

with i as usual denoting the $\sqrt{-1}$.

Transforming equation (A1.1.20) to spherical polar coordinates, using the relations (A1.1.3), one finds after some calculation that

$$L_z = \frac{\hbar}{i}\frac{\partial}{\partial\phi} \tag{A1.1.21}$$

Hence from equation (A1.1.17) it follows that

$$L_z\Phi = m\hbar\Phi \tag{A1.1.22}$$

and, therefore, from the definition of eigenfunction and eigenvalue, it follows immediately that $\Phi(\phi)$ is an eigenfunction of the operator L_z, with eigenvalue $m\hbar$. Thus, since m is an integer, L_z is quantized in units of $\hbar$. This completes the discussion of the ϕ-dependence of the central-field wave functions and we turn next to the θ-dependence.

To solve equation (A1.1.16) it is convenient to make the substitutions

$$x = \cos \theta, \qquad \Theta(\theta) = P(x) \tag{A1.1.23}$$

Then one readily finds that the differential equation satisfied by $P(x)$ is

$$\frac{d}{dx}\left[(1 - x^2)\frac{dP}{dx}\right] + \left[\lambda - \frac{m^2}{1 - x^2}\right]P = 0 \tag{A1.1.24}$$

and x is such that $-1 \le x \le +1$, corresponding to $0 \le \theta \le \pi$.

Investigation of equation (A1.1.24) shows that, for a general value of λ, the solutions become infinite at $x = \pm 1$, as might be anticipated from the presence of the term $m^2/(1 - x^2)$. As these solutions do not lead to well-behaved wave functions they are not of chemical or physical relevance. Only in the special case when $\lambda = l(l + 1)$, $l = 0$, 1, 2, etc., $|m| \le l$, does the equation for P possess one solution which is everywhere finite. These solutions are, in fact, the associated Legendre functions of order l [see, e.g., Sneddon (1951)]. They can be defined in the following manner:

$$P_l^m(x) = (1 - x^2)^{|m|/2} \frac{d^{|m|}}{dx^{|m|}} P_l(x) \tag{A1.1.25}$$

$$|m| < l, \qquad l = 0, 1, 2, \text{ etc.}$$

where the Legendre polynomials $P_l(x)$ are given by

$$P_l(x) = \frac{1}{2^l l!} \frac{d^l}{dx^l}[(x^2 - 1)^l], \qquad l = 0, 1, 2, \text{ etc.} \tag{A1.1.26}$$

The Legendre polynomials therefore satisfy equation (A1.1.24) with $\lambda = l(l + 1)$ and $m = 0$, whereas the associated Legendre functions, satisfying equation (A1.1.24) for general m, have the form $(1 - x^2)^{|m|/2} \times$ (a polynomial of degree $(l - |m|)$. The first four Legendre polynomials are readily found from equation (A1.1.26) to be:

$$\left. \begin{aligned} P_0(x) &= 1 \\[4pt] P_1(x) &= x \\[4pt] P_2(x) &= \tfrac{1}{2}(3x^2 - 1) \\[4pt] P_3(x) &= \tfrac{1}{2}(5x^3 - 3x) \end{aligned} \right\} \tag{A1.1.27}$$

and in general the polynomials $P_{2s}(x)$ are even in x, while the polynomials $P_{2s+1}(x)$ are themselves odd in x.

The first few associated Legendre functions are:

$$P_1^{\pm 1}(x) = (1 - x^2)^{1/2}$$

$$P_2^{\pm 1}(x) = (1 - x^2)^{1/2}3x$$

$$P_3^{\pm 1}(x) = (1 - x^2)^{1/2}\tfrac{3}{2}[5x^2 - 1]$$

$$P_2^{\pm 2}(x) = 3(1 - x^2)$$

(A1.1.28)

In conformity with a general property of the solutions of the Schrödinger equation discussed in Appendix A1.4, the associated Legendre functions are orthogonal over the interval $-1 \le x \le 1$, i.e.,

$$\int_{-1}^{1} P_l^m(x)P_{l'}^m(x) = 0, \qquad l' \ne l$$

(A1.1.29)

For $l = l'$ we have

$$\int_{-1}^{1} [P_l^m(x)]^2 \, dx = \left[\frac{2}{2l + 1}\frac{(l + |m|)!}{(l - |m|)!}\right]$$

(A1.1.30)

which allows us to define normalized associated Legendre functions, apart from an arbitrary phase factor.

Eigenfunctions of L^2 and L_z

The complete solutions $S(\theta, \phi)$ for the cases $\lambda = l(l + 1), l = 0, 1, 2$, etc. and $|m| \le l$ are, in fact, the so-called spherical harmonics $Y_{lm}(\theta, \phi)$. Hence, one can write

$$Y_{lm}(\theta, \phi) = N_{lm}P_l^m(\cos \theta)\Phi_m(\phi)$$

(A1.1.31)

There is no universally adopted definition of the normalization factor N_{lm}, but if one requires that

$$\int_0^{2\pi} d\phi \int_0^{\pi} d\theta \sin \theta \; Y_{lm}^*(\theta, \phi)Y_{l'm'}(\theta, \phi) = 0$$

(A1.1.32)

unless $l = l'$ and $m = m'$ and is unity in this case, then N_{lm} is determined by

$$\frac{1}{|N_{lm}|^2} = \left[\frac{4\pi(l + |m|)!}{(2l + 1)(l - |m|)!}\right]$$

(A1.1.33)

As two examples, we have

$$Y_{10} = (3/4\pi)^{1/2} \cos \theta$$

and

$$Y_{11} = (3/8\pi)^{1/2} \sin \theta \exp(i\phi) \tag{A1.1.34}$$

We now return to the discussion of the angular momentum operator L^2. In spherical polar coordinates we have, corresponding to equation (A1.1.21) for L_z, the results

$$L_x = i\hbar\left(\sin \phi \frac{\partial}{\partial \theta} + \cot \theta \cos \phi \frac{\partial}{\partial \phi}\right) \tag{A1.1.35}$$

and

$$L_y = i\hbar\left(-\cos \phi \frac{\partial}{\partial \theta} + \cot \theta \sin \phi \frac{\partial}{\partial \phi}\right) \tag{A1.1.36}$$

Forming $L^2 = L_x^2 + L_y^2 + L_z^2$, we find from equations (A1.1.35), (A1.1.36), and (A1.1.21),

$$L^2 = -\hbar^2\left[\frac{1}{\sin \theta} \frac{\partial}{\partial \theta}\left(\sin \theta \frac{\partial}{\partial \theta}\right) + \frac{1}{\sin^2 \theta} \frac{\partial^2}{\partial \phi^2}\right] \tag{A1.1.37}$$

Returning to equation (A1.1.7) satisfied by $S = Y_{lm}(\theta, \phi)$ when $\lambda = l(l + 1)$, we see that the operator on the left-hand side acting on S is in fact $-L^2/\hbar^2$ from equation (A1.1.37). Thus, we are led to the result

$$L^2S \equiv L^2 Y_{lm}(\theta, \phi) = l(l + 1)\hbar^2 Y_{lm}(\theta, \phi) \tag{A1.1.38}$$

However we already know that $\Phi(\phi) = \exp(im\phi)$ is an eigenfunction of L_z with eigenvalue $m\hbar$, and hence we conclude that the spherical harmonic $Y_{lm}(\theta, \phi)$ is a simultaneous eigenfunction of L^2 and L_z with eigenvalues $l(l + 1)\hbar^2$ and $m\hbar$, respectively. The former result confirms the earlier deduction in equation (A1.1.13) made on purely classical grounds.

Thus the central field wave functions correspond to sharp values of L^2 and L_z, and this is the quantal analogue of the classical result that the angular momentum is constant for a central field of force. All three components of **L** can be precisely specified at once in classical theory. However, in general only L^2 and L_z are specified in the quantum-mechanical central field problem, since $Y_{lm}(\theta, \phi)$ is not an eigenfunction of L_x and L_y except when $l = 0$. It should be added, however, that the choice of the direction of the polar axis that distinguishes L_z from L_x and L_y is arbitrary.

Though our main interest here is in the atomic hydrogen problem, we have dealt with the problem of the shape of atomic wave functions at length because these results are applicable to every central field problem. The shapes of these orbitals are of considerable importance in the theory of molecules, and also when one wishes to describe atoms bound in a crystal. We show in Figure A.3 a few examples of the shapes of particular wave functions, the spectroscopic notation s, p, d, f, g, h (historically: sharp, principal, diffuse, fundamental, while g, h ... follow alphabetically after f) corresponding to $l = 0, 1, 2, 3$, etc., respectively. The s-states are spherically symmetric, while the p-, d-, etc. states all exhibit angularity. It is difficult to overemphasize the role of angular momentum (not only orbital, but also spin) in atomic theory [for an advanced discussion, see, e.g., Condon and Shortley (1951)].

Form of Radial Wave Functions for the Hydrogen Atom

We shall conclude this appendix with a quite brief discussion of the form of the radial wave functions for the different states of the hydrogen atom, i.e., when $V(r) = -e^2/r$.

Actually, for realistic forms of the central potential energy $V(r)$ in multielectron atoms, the radial Schrödinger equation (A1.1.8) has to be solved numerically, an easy task on a modern electronic computer. However, when $V(r) = -e^2/r$, the radial wave functions can be obtained analytically. For completeness, we merely record below the general result that determines the radial wave function $R(r)$, and then discuss a few simple examples briefly.

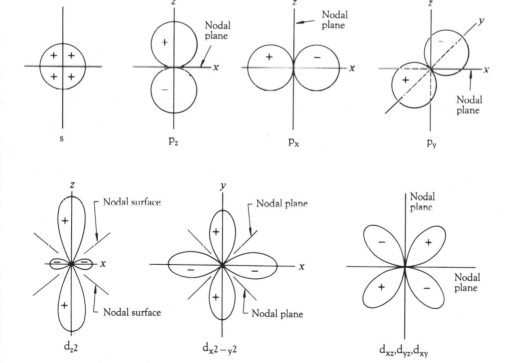

FIGURE A.3. Shapes of particular wave functions.

Putting $Q = rR$, it follows from equation (A1.1.6), with $V = -e^2/r$, that if we use the abbreviations

$$\rho = 2[(2m/\hbar^2)|E|]^{1/2}r, \qquad \tau = (me^2/\hbar^2)/[(2m/\hbar^2)|E|]^{1/2} \qquad \text{(A1.1.39)}$$

having written $E = -|E|$ since E is negative for the bound states of the electron under discussion here, then $Q(\rho)$ satisfies the differential equation

$$\frac{d^2Q}{d\rho^2} + \left[-\frac{1}{4} + \frac{\tau}{\rho} - \frac{l(l+1)}{\rho^2} \right] Q = 0 \qquad \text{(A1.1.40)}$$

Well-behaved solutions exist only if the energies E are determined by the Bohr formula

$$E_n = -(1/2n^2)(e^2/a_0), \qquad a_0 = \hbar^2/me^2 \qquad \text{(A1.1.41)}$$

where n, the usual principal quantum number takes the values 1, 2, 3, etc. Then Q is given by

$$Q_{nl} = \rho^{l+1} \exp(-\rho/2) L_{n+l}^{2l+1}(\rho) \qquad \text{(A1.1.42)}$$

where L_{n+l}^{2l+1} is the associated Laguerre function and ρ is simply $2r/na_0$ when we make use of equations (A1.1.39) and (A1.1.41). Here a_0 is the first Bohr radius for hydrogen, defined in equation (A1.1.41).

To conclude, we note first that radial wave functions do have some common properties, independent of the detailed potential energy $V(r)$. In particular, for $l = 0$, $m = 0$ the radial wave functions $R_{n,l=0}(r)$, classified by the principal quantum number n, have $n - 1$ radial nodes. The forms for $n = 1$, 2, and 3 for the s-states ($l = 0$) are shown in Figure A.4, while two p-state ($l = 1$) functions and one d function are also plotted.

From the properties of the Laguerre polynomials [see, e.g., Sneddon (1951) or Pauling and Wilson (1935)], the ground-state radial wave function for the hydrogen atom has the (un-normalized) form

$$R_{10}(r) = \exp(-r/a_0) \qquad \text{(A1.1.43)}$$

The s-wave function associated with the (degenerate) first excited state is, explicitly,

$$R_{20}(r) = (2 - r/a_0) \exp(-r/2a_0) \qquad \text{(A1.1.44)}$$

its radial node evidently occurring at twice the Bohr radius. The position of this node will, of course, move when the potential energy is changed, as in a multielectron atom, from the bare Coulomb form $-e^2/r$ appropriate to hydrogen; that is, the radial extent of the wave functions will depend on the potential energy $V(r)$. A further general property worthy of note is that $R_{nl}(r) \sim r^l$ for sufficiently small r [cf. equation (A1.1.42)], the centrifugal potential energy term in equation (A1.1.9) dominating the Coulomb term for $l \neq 0$ (see Figure A.2). This property can be seen from Figure A.4 for s- and p-states.

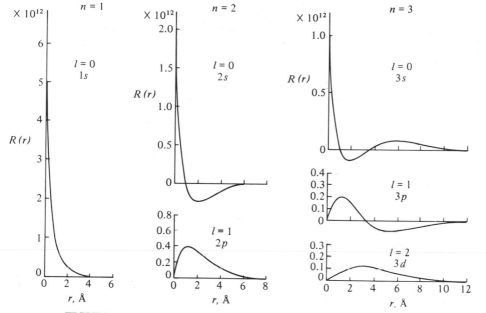

FIGURE A.4. Forms of radial wave functions for *s*, *p*, and *d* states.

A1.2. THE PERIODIC TABLE AND ATOMIC GROUND STATES

The purpose of this appendix is (a) to set out the Periodic Table of the elements and (b) to list their ground state configurations.

Periodic System (Excluding H; *see p. 272)*

O	I	II	III	IV	V	VI	VII							
He	Li	Be	B	C	N	O	F							
2	3	4	5	6	7	8	9							
Ne	Na	Mg	Al	Si	P	S	Cl							
10	11	12	13	14	15	16	17							

O	Ia	IIa	IIIa	IVa	Va	VIa	VIIa	VIII			Ib	IIb	IIIb	IVb	Vb	VIb	VIIb
Ar	K	Ca	Sc	Ti	V	Cr	Mn	Fe	Co	Ni	Cu	Zn	Ga	Ge	As	Se	Br
18	19	20	21	22	23	24	25	26	27	28	29	30	31	32	33	34	35
Kr	Rb	Sr	Y	Zr	Nb	Mo	Tc	Ru	Rh	Pd	Ag	Cd	In	Sn	Sb	Te	I
36	37	38	39	40	41	42	43	44	45	46	47	48	49	50	51	52	53
Xe	Cs	Ba	La–Lu	Hf	Ta	W	Re	Os	Ir	Pt	Au	Hg	Tl	Pb	Bi	Po	At
54	55	56	57–71	72	73	74	75	76	77	78	79	80	81	82	83	84	85
Rn	Fr	Ra	Ac	Th	Pa	U	Np										
86	87	88	89	90	91	92	93										

Atomic Ground States

In the table below, we list information for atomic configurations and atomic ground states for all atoms in the Periodic Table.

Element	Atomic number	Ground configuration	I_1 (in eV)	Ground state
H	1	$1s^1$	13.595	$^2S_{1/2}$
He	2	$1s^2$	24.481	$^1S_0$
Li	3	$K2s^1$	5.39	$^2S_{1/2}$
Be	4	$K2s^2$	9.32	$^1S_0$
B	5	$K2s^22p^1$	8.296	$^2P^0_{1/2}$
C	6	$K2s^22p^2$	11.256	$^3P_0$
N	7	$K2s^22p^3$	14.53	$^4S^0_{3/2}$
O	8	$K2s^22p^4$	13.614	$^3P_2$
F	9	$K2s^22p^5$	17.418	$^2P^0_{3/2}$
Ne	10	$K2s^22p^6$	21.559	$^1S_0$
Na	11	$KL3s^1$	5.138	$^2S_{1/2}$
Mg	12	$KL3s^2$	7.644	$^1S_0$
Al	13	$KL3s^23p$	5.984	$^2P_{1/2}$
Si	14	$KL3s^23p^2$	8.149	$^3P_0$
P	15	$KL3s^23p^3$	10.484	$^4S^0_{3/2}$
S	16	$KL3s^23p^4$	10.357	$^3P_2$
Cl	17	$KL3s^23p^5$	13.01	$^2P^0_{3/2}$
Ar	18	$KL3s^23p^6$	15.755	$^1S_0$
K	19	$KL3s^23p^64s^1$	4.339	$^2S_{1/2}$
Ca	20	$KL3s^23p^64s^2$	6.111	$^1S_0$
Sc	21	$KL3s^23p^63d^14s^2$	6.54	$^2D_{3/2}$
Ti	22	$KL3s^23p^63d^24s^2$	6.82	$^3F_2$
V	23	$KL3s^23p^63d^34s^2$	6.74	$^4F_{3/2}$
Cr	24	$KL3s^23p^63d^54s^2$	6.764	$^7S_3$
Mn	25	$KL3s^23p^63d^54s^2$	7.432	$^6S_{5/2}$
Fe	26	$KL3s^23p^63d^64s^2$	7.87	$^5D_4$
Co	27	$KL3s^23p^63d^74s^2$	7.86	$^4F_{9/2}$
Ni	28	$KL3s^23p^63d^84s^2$	7.633	$^3F_4$
Cu	29	$KLM4s^1$	7.724	$^2S_{1/2}$
Zn	30	$KLM4s^2$	9.391	$^1S_0$
Ga	31	$KLM4s^24p^1$	6.0	$^2P^0_{1/2}$
Ge	32	$KLM4s^24p^2$	7.88	$^3P_0$
As	33	$KLM4s^24p^3$	9.81	$^4S^0_{3/2}$
Se	34	$KLM4s^24p^4$	9.75	$^3P_2$
Br	35	$KLM4s^24p^5$	11.84	$^2P^0_{3/2}$
Kr	36	$KLM4s^24p^6$	13.996	$^1S_0$
Rb	37	$KLM4s^24p^65s^1$	4.176	$^2S_{1/2}$
Sr	38	$KLM4s^24p^65s^2$	5.692	$^1S_0$
Y	39	$KLM4s^24p^64d^15s^2$	6.38	$^3D_{3/2}$
Zr	40	$KLM4s^24p^64d^25s^2$	6.84	$^3F_2$
Nb	41	$KLM4s^24p^64d^45s^1$	6.88	$^6D_{1/2}$
Mo	42	$KLM4s^24p^64d^55s^1$	7.10	$^7S_3$
Tc	43	$KLM4s^24p^64d^55s^2$	7.28	$^6S_{5/2}$
Ru	44	$KLM4s^24p^64d^75s^1$	7.364	$^5F_5$
Rh	45	$KLM4s^24p^64d^85s^1$	7.46	$^4F_{9/2}$
Pd	46	$KLM4s^24p^64d^{10}$	8.33	$^1S_0$
Ag	47	$KLM4s^24p^64d^{10}5s^1$	7.574	$^2S_{1/2}$
Cd	48	$KLM4s^24p^64d^{10}5s^2$	8.991	$^1S_0$
In	49	$KLM4s^24p^64d^{10}5s^25p^1$	5.785	$^2P^0_{1/2}$
Sn	50	$KLM4s^24p^64d^{10}5s^25p^2$	7.342	$^3P_0$

(con't)

Element	Atomic number	Ground configuration	I_1 (in eV)	Ground state
Sb	51	$KLM4s^24p^64d^{10}5s^25p^3$	8.639	$^4S^0_{3/2}$
Te	52	$KLM4s^24p^64d^{10}5s^25p^4$	9.01	$^3P_2$
I	53	$KLM4s^24p^64d^{10}5s^25p^5$	10.454	$^2P^0_{3/2}$
Xe	54	$KLM4s^24p^64d^{10}5s^25p^6$	12.127	$^1S_0$
Cs	55	$KLM4s^24p^64d^{10}5s^25p^66s^1$	3.893	$^2S_{1/2}$
Ba	56	$KLM4s^24p^64d^{10}5s^25p^66s^2$	5.21	$^1S_0$
La	57	$KLM4s^24p^64d^{10}5s^25p^65d^16s^2$	5.61	$^2D_{3/2}$
Ce	58	$KLM4s^24p^64d^{10}4f^15s^25p^65d^16s^2$	5.6	$^1G_4$
Pr	59	$KLM4s^24p^64d^{10}4f^35s^25p^66s^2$	5.46	$^4I^0_{9/2}$
Nd	60	$KLM4s^24p^64d^{10}4f^45s^25p^66s^2$	5.51	$^5I_4$
Pm	61	$KLM4s^24p^64d^{10}4f^55s^25p^66s^2$	5.55	$^6H_{5/2}$
Sm	62	$KLM4s^24p^64d^{10}4f^65s^25p^66s^2$	5.6	$^7F_0$
Eu	63	$KLM4s^24p^64d^{10}4f^75s^25p^66s^2$	5.67	$^8S^0_{7/2}$
Gd	64	$KLM4s^24p^64d^{10}4f^75s^25p^65d^16s^2$	6.16	$^9D^0_2$
Tb	65	$KLM4s^24p^64d^{10}4f^95s^25p^66s^2$	5.98	$^6H^0_{15/2}$
Dy	66	$KLM4s^24p^64d^{10}4f^{10}5s^25p^66s^2$	6.8	$^5I_8$
Ho	67	$KLM4s^24p^64d^{10}4f^{11}5s^25p^66s^2$	6.02	$^4I^0_{15/2}$
Er	68	$KLM4s^24p^64d^{10}4f^{12}5s^25p^66s^2$	6.08	$^3H_6$
Tm	69	$KLM4s^24p^64d^{10}4f^{13}5s^25p^66s^2$	5.81	$^2F^0_{7/2}$
Yb	70	$KLMN5s^25p^66s^2$	6.2	$^1S_0$
Lu	71	$KLMN5s^25p^65d^16s^2$	5.43	$^2D_{3/2}$
Hf	72	$KLMN5s^25p^65d^26s^2$	7.0	$^3F_2$
Ta	73	$KLMN5s^25p^65d^36s^2$	7.88	$^4F_{3/2}$
W	74	$KLMN5s^25p^65d^46s^2$	7.98	$^5D_0$
Re	75	$KLMN5s^25p^65d^56s^2$	7.87	$^6S_{5/2}$
Os	76	$KLMN5s^25p^65d^66s^2$	8.5	$^5D_4$
Ir	77	$KLMN5s^25p^65d^76s^2$	9.0	$^4F_{9/2}$
Pt	78	$KLMN5s^25p^65d^96s^1$	9.0	$^3D_3$
Au	79	$KLMN5s^25p^65d^{10}6s^1$	9.22	$^2S_{1/2}$
Hg	80	$KLMN5s^25p^65d^{10}6s^2$	10.43	$^1S_0$
Tl	81	$KLMN5s^25p^65d^{10}6s^26p^1$	6.106	$^2P^0_{1/2}$
Pb	82	$KLMN5s^25p^65d^{10}6s^26p^2$	7.415	$^3P_0$
Bi	83	$KLMN5s^25p^65d^{10}6s^26p^3$	7.287	$^4S^0_{3/2}$
Po	84	$KLMN5s^25p^65d^{10}6s^26p^4$	8.43	$^3P_2$
At	85	$KLMN5s^25p^65d^{10}6s^26p^5$	9.5	$^2P^0_{3/2}$
Rn	86	$KLMN5s^25p^65d^{10}6s^26p^6$	10.746	$^1S_0$
Fr	87	$KLMN5s^25p^65d^{10}6s^26p^67s^1$	4.0	$^2S_{1/2}$
Ru	88	$KLMN5s^25p^65d^{10}6s^26p^67s^2$	7.364	$^1S_0$
Ac	89	$KLMN5s^25p^65d^{10}6s^26p^66d^17s^2$	6.9	$^2D_{3/2}$
Th	90	$KLMN5s^25p^65d^{10}6s^26p^66d^27s^2$	6.95	$^3F_2$
Pa	91	$KLMN5s^25p^65d^{10}5f^26s^26p^66d^17s^2$	5.9	$^4K_{11/2}$
U	92	$KLMN5s^25p^65d^{10}5f^36s^26p^66d^17s^2$	6.08	$^5L^0_6$
Np	93	$KLMN5s^25p^65d^{10}5f^46s^26p^66d^17s^2$	6.2	$^6L_{11/2}$

A1.3. TOTAL ENERGIES OF HEAVY ATOMIC IONS—COULOMB FIELD MODEL

We start from the total energy of an atomic ion with nuclear charge Ze and N electrons. In the ground state under consideration we will denote this by $E(Z, N)$. Reference to the solution of the hydrogen atom problem presented in Appendix

A1.1 will make it clear that for the case of a bare Coulomb field, corresponding to potential energy $V(r) = -Ze^2/r$ in the Schrödinger equation (A1.1.2), the wave functions and energy levels can be obtained for any arbitrary value of Z, not simply for Z integral. Thus one can regard $E(Z, N)$ introduced above as an ordinary, continuous function of Z.

This situation is exploited in a method introduced by Hylleraas in the early days of quantum theory and developed especially by Layzer (1959). In this method $E(Z, N)$ is, essentially, expanded in a power series in $1/Z$ by writing

$$E(Z, N) = Z^2 \left(\varepsilon_0(N) + \frac{1}{Z} \varepsilon_1(N) + \frac{1}{Z^2} \varepsilon_2(N) + \cdots \right) \tag{A1.3.1}$$

There are two points to note here. As shown in equation (A1.3.1), the coefficients $\varepsilon_n(N)$ are functions only of the number of electrons N, and, in fact, as we shall demonstrate below, the first term in the expansion (A1.3.1) can be calculated once for all from the bare Coulomb field problem solved in Appendix A1.1. Secondly, we simply note here that Kato (1957) has shown that, for fixed N, the convergence of the expansion (A1.3.1) is assured for sufficiently large Z.

Let us now calculate the leading term in equation (A1.3.1) from the bare Coulomb field solution. The way to proceed is from the Bohr formula (A1.1.41) for the energy levels, generalized to apply to one electron moving in the field of a nuclear charge Ze, which simply multiplies the hydrogenic results for $Z = 1$ by Z^2. Assuming that there are $\mathscr{N}$ closed shells occupied, and using the fact that there are $2n^2$ electrons in a closed shell of principal quantum number n, it follows that the total energy per shell is $-(Z^2/2n^2)2n^2(e^2/a_0) = -(Z^2e^2/a_0)$. Thus for the bare Coulomb field, each closed shell contributes the same amount to the energy, independent of n. The total energy of such a noninteracting model system is therefore the energy per shell times the number of closed shells $\mathscr{N}$ and hence the total energy is $-(Z^2e^2\mathscr{N}/a_0)$. The number of electrons N filling the $\mathscr{N}$ closed shells is given by $N = \sum_1^{\mathscr{N}} 2n^2$, which can be summed to yield

$$N = \mathscr{N}(\mathscr{N} + 1)(2\mathscr{N} + 1)/3 \tag{A1.3.2}$$

For large $\mathscr{N}$ we find the result that $N \sim 2\mathscr{N}^3/3$ and hence

$$E_{\text{Coulomb}} \sim -(\tfrac{3}{2})^{1/3}(Z^2e^2/a_0)N^{1/3} \tag{A1.3.3}$$

From equation (A1.3.1) we can therefore write

$$\varepsilon_0(N) \sim -(\tfrac{3}{2})^{1/3}N^{1/3}e^2/a_0, \quad \text{for large } N \tag{A1.3.4}$$

If equation (A1.3.4) is inserted into equation (A1.3.1), and higher terms neglected for the moment (see Appendix A4.5), the neutral atom energy $E(Z, Z)$ is readily seen in this Coulomb field model to vary with atomic number as $Z^{7/3}$, an important result which proves to have more general validity than this model calculation might imply.

A1.4. ORTHOGONALITY OF SOLUTIONS OF THE SCHRÖDINGER EQUATION

We shall now prove the orthogonality of solutions of the Schrödinger equation in one dimension. In the three-dimensional case, solutions of the wave equation corresponding to the same degenerate level need not be orthogonal. For wave functions corresponding to different energy levels, however, the theorem is equally true in three dimensions.

The wave equation for the nth level takes the form

$$\frac{d^2}{dx^2}\psi_n + \frac{2m}{\hbar^2}[E_n - V(x)]\psi_n = 0 \tag{A1.4.1}$$

Since $V(x)$ is real, the following identities can be written immediately from equation (A1.4.1):

$$\psi_m^*\frac{d^2\psi_n}{dx^2} + \frac{2m}{\hbar^2}\psi_m^*[E_n - V(x)]\psi_n = 0 \tag{A1.4.2}$$

and

$$\psi_n\frac{d^2\psi_m^*}{dx^2} + \frac{2m}{\hbar^2}\psi_n[E_m - V(x)]\psi_m^* = 0 \tag{A1.4.3}$$

By subtraction of equation (A1.4.3) from equation (A1.4.2), and integration over all space, the potential energy $V(x)$ can be eliminated to yield

$$\int\left[\psi_m^*\frac{d^2\psi_n}{dx^2} - \psi_n\frac{d^2\psi_m^*}{dx^2}\right]dx + \frac{2m}{\hbar^2}(E_n - E_m)\int\psi_m^*\psi_n\,dx = 0 \tag{A1.4.4}$$

Use may now be made of the identity

$$\frac{d}{dx}\left[\psi_m^*\frac{d\psi_n}{dx} - \psi_n\frac{d\psi_m^*}{dx}\right] = \psi_m^*\frac{d^2\psi_n}{dx^2} - \psi_n\frac{d^2\psi_m^*}{dx^2} \tag{A1.4.5}$$

This can be integrated over x, and provided that $\psi_m^*(d\psi_n/dx) - \psi_n(d\psi_m^*/dx)$ vanishes at the limits of integration, as is usually the case, we can write

$$(E_n - E_m)\int\psi_m^*\psi_n\,dx = 0 \tag{A1.4.6}$$

Thus for different energy levels, that is, $E_n \neq E_m$, it follows that

$$\int\psi_m^*\psi_n\,dx = 0 \tag{A1.4.7}$$

which is the orthogonality property we wished to establish. In three dimensions, in place of the x integration in equation (A1.4.7), the integration must be performed over the entire volume of the three-dimensional system, i.e., in Cartesian coordinates also over y and z.

A2.1. VARIATION PRINCIPLE

Suppose we have a system described by a Hamiltonian H. The wave equation may be written

$$H\psi = E\psi \tag{A2.1.1}$$

and, in particular, if ψ_0 is the wave function and E_0 the energy of the lowest state, then

$$H\psi_0 = E_0\psi_0 \tag{A2.1.2}$$

In quantum mechanics the energy of the ground state is often of particular interest and the variation method enables us to approximate it.

Consider any function ϕ which is well behaved and we will assume also normalized. Now let us form the quantity $\mathscr{E}$ defined by

$$\mathscr{E} = \int \phi^* H\phi \, d\tau \tag{A2.1.3}$$

We show below that

$$\mathscr{E} \geq E_0 \tag{A2.1.4}$$

which is the basis of the variation principle.

Proof: If ϕ is not equal to ψ_0 we can expand ϕ in terms of the complete set of normalized orthogonal functions $\psi_0, \psi_1, \ldots, \psi_n$, which are the solutions of (A2.1.1), obtaining for a normalized ϕ,

$$\phi = \sum_n a_n\psi_n \tag{A2.1.5}$$

with $\sum_n a_n^* a_n = 1$. From equations (A2.1.3) and (A2.1.5)

$$\mathscr{E} = \int \sum_{n'} a_{n'}^* \psi_{n'}^* H \sum_n a_n\psi_n \, d\tau \tag{A2.1.6}$$

However from equation (A2.1.1),

$$H\psi_n = E_n\psi_n \tag{A2.1.7}$$

Thus it follows that

$$\mathscr{E} = \int \sum_{n'} a_{n'}^* \psi_{n'}^* \sum_n E_n a_n\psi_n \, d\tau \tag{A2.1.8}$$

but again (cf. Appendix A1.4)

$$\int \psi_{n'}^* \psi_n \, d\tau = \begin{cases} 0, & n' \neq n \\ 1, & n' = n \end{cases} \qquad \text{(A2.1.9)}$$

Therefore equation (A2.1.8) immediately reduces to

$$\mathcal{E} = \sum_n a_n^* a_n E_n \qquad \text{(A2.1.10)}$$

Using equation (A2.1.5) we can write this in the form

$$\mathcal{E} - E_0 = \sum_n a_n^* a_n (E_n - E_0) \qquad \text{(A2.1.11)}$$

However, E_0 is the lowest energy level by definition and hence

$$E_n - E_0 \geq 0 \qquad \text{(A2.1.12)}$$

Clearly

$$a_n^* a_n \geq 0 \qquad \text{(A2.1.13)}$$

and, thus, from equation (A2.1.11) we obtain the desired result (A2.1.4).

A2.2. INTEGRALS INVOLVED IN LCAO TREATMENT OF THE H_2^+ GROUND STATE

The explicit forms of the integrals needed to plot the energies of the states of H_2^+ associated with the LCAO wave functions ψ_g and ψ_u set up in Section 2.3 will be collected below.

H_{aa}, in units in which the Hamiltonian without the nuclear–nuclear potential energy reads

$$\mathcal{H} = -\tfrac{1}{2}\nabla^2 - 1/r_a \quad 1/r_b \qquad \text{(A2.2.1)}$$

is explicitly

$$H_{aa} = \int 1s_a(-\tfrac{1}{2}\nabla^2 - 1/r_a - 1/r_b)1s_a \, d\tau \qquad \text{(A2.2.2)}$$

This can evidently be separated into two parts:

$$H_{aa} = \int 1s_a(-\tfrac{1}{2}\nabla^2 - 1/r_a)1s_a \, d\tau + \int 1s_a(-1/r_b)1s_a \, d\tau \qquad \text{(A2.2.3)}$$

The first term on the right-hand side of equation (A2.2.3) is readily identified as the energy of the hydrogen atom, since $1s_a$ is taken as the exact hydrogenic $1s$ function centered on nucleus a. Thus one can write

$$H_{aa} = E_H + \int 1s_a(-1/r_b)1s_a \, d\tau \qquad \text{(A2.2.4)}$$

where E_H is the hydrogenic energy. Including the internuclear repulsion $1/R$, and using the explicit form of the $1s$ orbital given in Appendix A1.1 to evaluate the integral, one can show that

$$H_{aa}^{\text{total}} = -\tfrac{1}{2} - 1/R + \exp(-2R)(1 + 1/R) \qquad \text{(A2.2.5)}$$

By a closely related procedure, one first writes H_{ab} in the form

$$H_{ab} = E_H S + \int 1s_a(-1/r_a)1s_b \, d\tau \qquad \text{(A2.2.6)}$$

Again including the $1/R$ term, and making use of the explicit forms of $1s_a$ and $1s_b$, one can show that

$$H_{ab}^{\text{total}} = -S/2 - \exp(-R)(1 + R) \qquad \text{(A2.2.7)}$$

Finally, the overlap integral S is required in the evaluation of H_{ab}. After some calculation with the explicit $1s$ orbitals one arrives at

$$S = \exp(-R)[1 + R + (R^2/3)] \qquad \text{(A2.2.8)}$$

Using these results, one can readily construct analytic forms for E_g and E_u from equations (A2.2.7) and (A2.2.8). From these forms, it is a straightforward matter to plot the energy of each of these two states ψ_g and ψ_u as functions of the internuclear distance R for comparison with the exact wave-mechanical solutions.

A2.3. BORN–OPPENHEIMER APPROXIMATION

In the approximation in which all nonelectrostatic interactions are ignored, the Hamiltonian of any molecule can be written

$$H_{en} = T_n + T_e + V_{en} + V_{ee} + V_{nn} \qquad \text{(A2.3.1)}$$

Here the individual terms on the right-hand side correspond, in order of appearance to (i) nuclear kinetic energy (ii) electronic kinetic energy, (iii) electron–nuclear electrostatic interaction, (iv) electron–electron Coulomb repulsion, and (v) bare nucleus–nucleus Coulomb repulsion.

A major simplification of this Hamiltonian (A2.3.1) now invokes the small ratio of electron to nuclear mass; at its largest (for proton mass) this ratio is 1/2000. In the lowest-order treatment it is thus natural enough to take the limit when the nuclear masses are thought to tend to infinity so that the nuclear kinetic energy drops out of equation (A2.3.1). Of course, this assumption must clearly be relaxed at a later stage in order to discuss nuclear vibrations.

With this assumption of stationary nuclei, one is led immediately to seek the stationary-state solutions of the electronic Schrödinger equation

$$(H_e - E_i(\mathbf{R}))\Psi_i(\mathbf{R}, \mathbf{r}) = 0 \qquad \text{(A2.3.2)}$$

where the nuclei, now fixed in space, enter only through the set of nuclear positions denoted in totality in equation (A2.3.2) by $\mathbf{R}$. Evidently, these are no longer "variables" in the partial differential equation (A2.3.2), but appear "merely" as parameters. The Hamiltonian H_e in equation (A2.3.2) is the same as H_{en} in equation (A2.3.1) except that the nuclear kinetic energy has been set equal to zero because of the assumption of infinite nuclear masses. The energy $E_i(\mathbf{R})$ in equation (A2.3.2) corresponds to the total energy of the molecule with nuclei fixed in configuration $\mathbf{R}$, the corresponding wave function being $\Psi_i(\mathbf{R}, \mathbf{r})$.

This simplification in which one has forced complete separation of the motion of the electrons from that of the nuclei by proceeding to the limit of infinite mass is in the spirit of the so-called Born–Oppenheimer approximation.

It is then, of course, important to ascertain the error introduced into the theory of electronic structure of molecules when using this complete separation of electronic from nuclear motions. If one returns to the full Hamiltonian (A2.3.1), one has the corresponding stationary-state Schrödinger equation, but now for both electronic and nuclear motion:

$$(H_{en} - W)\Phi(\mathbf{R}, \mathbf{r}) = 0 \qquad \text{(A2.3.3)}$$

where in contrast to equation (A2.3.2) the quantities $\mathbf{R}$ and $\mathbf{r}$ are on the same footing as variables in the wave equation.

At this stage, it will be assumed that the fixed-nucleus problem has been completely solved for the wave functions $\Psi_i(\mathbf{R}, \mathbf{r})$. Then, it is a mathematical fact that the total wave function including nuclear and electronic motion can be expanded in terms of these fixed nuclei wave functions as

$$\Phi(\mathbf{R}, \mathbf{r}) = \sum_i C_i(\mathbf{R})\Psi_i(\mathbf{R}, \mathbf{r}) \qquad \text{(A2.3.4)}$$

The "coefficients" $C_i(\mathbf{R})$ in this expansion are clearly functions of the nuclear coordinates $\mathbf{R}$.

Substitution of the exact expansion (A2.3.4) into the full Schrödinger equation will clearly lead to a (complicated) system of differential equations to determine these expansion coefficients $C_i(\mathbf{R})$. With the use of judicious approximation, these equations simplify to a form of the Schrödinger equation, but nowsolely for the nuclear motion. To a useful degree of approximation, this Schrödinger equation can be written for the coefficient $C_j(\mathbf{R})$ as

$$[T_n + E_j(\mathbf{R})]C_j(\mathbf{R}) = WC_j(\mathbf{R}) \tag{A2.3.5}$$

The outcome of this should be clear. The electronic energy $E_j(\mathbf{R})$, calculated for fixed nuclei at positions $\mathbf{R}$, is playing the role of the "potential energy" in the Schrödinger equation for the nuclear motion.

Actually there are some corrections to this "simple" potential energy $E_j(R)$ which take into account the weak coupling of both motions. When this modification is introduced into equation (A2.3.5) the result is usually termed the adiabatic approximation, a term used to reflect the fact that the electrons, being light, respond almost instantaneously to the sluggish motions of the heavy nuclei.

A more detailed analysis, such as that made by Longuet-Higgins (1956), shows that when one is proceeding in the spirit of the Born–Oppenheimer approximation, one is expanding in some power of the ratio of electronic to nuclear mass. This power turns out to be 1/4, and though the mass ratio at its largest is 1/2000, this raised to the power 1/4 is about 1/7, which is not a particularly small number. Thus, corrections are often required. The reader who wants to take the matter further is referred to the treatment of Longuet-Higgins (1956).

A2.4. SLATER- AND GAUSSIAN-TYPE ORBITALS

In this appendix, we deal briefly with two types of orbitals that are valuable in LCAO self-consistent field calculations, as well as in approximations transcending one-electron theory, e.g., configuration interaction (see Section 2.5).

We start with Slater-type orbitals, which are essentially modeled on the radial wave function $R_{nl}(r)$ in hydrogen-like atoms (cf. Appendix A1.1).

Slater Orbitals*

Atomic orbitals are assumed to be of the form

$$\left. \begin{aligned} \psi_{1s} &= N_{1s}\exp(-cr), & \psi_{2p_x} &= N_{2p}x\exp(-cr/2) \\ \psi_{2s} &= N_{2s}r\exp(-cr/2), & \psi_{3p_x} &= N_{3p}xr\exp(-cr/3) \\ \psi_{3s} &= N_{3s}r^2\exp(-cr/3), & \psi_{3d_{xy}} &= N_{3d}2xy\exp(-cr/3) \\ \psi_{3d_{x^2-y^2}} &= N_{3d}(x^2-y^2)\exp(-cr/3), & \psi_{3d_{z^2}} &= N_{3d}[(3z^2-r^2)/3^{1/2}]\exp(-cr/3) \end{aligned} \right\} \tag{A2.4.1}$$

*The general unnormalized radial form is $r^{n-1}\exp(-\zeta r)$, where n is the usual principal quantum number.

where the normalization factors are given explicitly by

$$N_{1s} = (c^3/\pi)^{1/2}, \qquad\qquad N_{2s} = (c^5/96\pi)^{1/2}$$

$$N_{3s} = (2c^7/5 \cdot 3^9 \cdot \pi)^{1/2}, \qquad N_{2p} = (c^5/32\pi)^{1/2} \qquad (A2.4.2)$$

$$N_{3p} = (2c^7/5 \cdot 3^8 \cdot \pi)^{1/2}, \qquad N_{3d} = (c^7/2 \cdot 3^8 \cdot \pi)^{1/2}$$

In these equations, the essential point to recognize is that

$$c = Z - s \qquad\qquad (A2.4.3)$$

where Z is the atomic number, and s is termed the screening or shielding constant. There are specific rules for obtaining these quantities, to be found, e.g., in the book by Eyring *et al.* (1944).

*Gaussian-Type Orbitals**

Though the above Slater orbitals do not have the same form in detail, the fact that they are modeled on the radial wave functions of hydrogen-like atoms endows them with numerous favorable physical and chemical properties. For example, the electron density $\rho(r)$ calculated at small r has at least the correct form of expansion around $r = 0$, as required by Kato's theorem (see Problem 6.2). However, the main drawback is that the multicenter integrals required in molecules (with forms like those given in Section 2.4) cannot be readily evaluated with such Slater orbitals.

Boys (1950) proposed the use of Gaussian-type orbitals based on the form $\exp(-\zeta r^2)$. With such orbitals, integral evaluation is greatly simplified. For details the reader is referred to the works of Boys and Handy (1969) and to later papers by Handy *et al.* (1972). In spite of numerous advantages, the point made above about the behavior of orbitals and electron density near an atomic nucleus remains a serious defect in using Gaussians. Notwithstanding this, these orbitals are much favored now in programing accurate wave function and energy calculations in molecules.

A2.5. THE COULSON–FISCHER WAVE FUNCTION FOR THE GROUND STATE OF THE H_2 MOLECULE

In order to understand the range of validity of the LCAO–MO method for the ground state of the H_2 molecule as the internuclear distance R is varied beyond its equilibrium value R_e, Coulson and Fischer (1949) proposed setting up a spatial ground-state wave function $\Psi(1, 2)$ using asymmetric orbitals. Thus, they envisaged an orbital of the form $1s_a + \lambda 1s_b$, and if λ is, say, much less than unity then this orbital is predominantly localized around nucleus a. Likewise one can construct a similar orbital, centered mainly on nucleus b, namely $1s_b + \lambda 1s_a$.

---

*Numerous types of Gaussian orbitals exist, e.g., spherical: $r^{n-1}\exp(-\zeta r^2)$ and Cartesian: $x^a y^b z^c \exp(-\zeta r^2)$, with a, b, and c nonnegative integers.

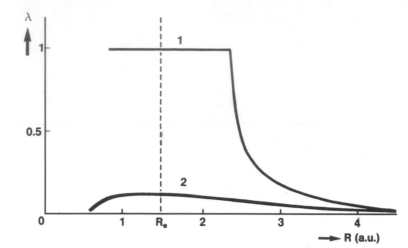

FIGURE A.5. Values of the parameter λ in asymmetrical MOs giving energy minimum [after Coulson and Fischer (1949)].

The idea now is to construct a variational trial wave function for the ground state of H_2 by putting electron 1 into the first of these two asymmetric orbitals, and electron 2 into the second. One can then write an acceptable trial wave function as

$$\Psi_{\text{trial}}(1, 2) = [1s_a(1) + \lambda 1s_b(1)][1s_b(2) + \lambda 1s_a(2)] \qquad \text{(A2.5.1)}$$

Notwithstanding the way in which equation (A2.5.1) has been set up, it should be noted here that it contains the LCAO–MO treatment, which corresponds to the choice $\lambda = 1$. The procedure of Coulson and Fischer was to use λ as the variational parameter in the trial wave function (A2.5.1) for each value of the internuclear separation R.

Their main finding, at first sight surprising, was that the LCAO–MO choice $\lambda = 1$ minimized the ground-state energy until R exceeded $1.6R_e$. However, as can be seen from Figure A.5, taken from the work of Coulson and Fischer, the value of λ for internuclear distances larger than $1.6R_e$ rapidly decreases with increasing R. Loosely, one can say that electrons tend to "go back on to their own atoms" when the internuclear distance exceeds 1.6 times the equilibrium internuclear separation. Curve 1 represents this assertion quantitatively.

A2.6. VIRIAL AND HELLMANN–FEYNMAN THEOREMS

Two theorems will be established in this appendix which, though valid in quantum mechanics, exhibit close similarities to classical ideas. They are: (i) the virial theorem, which in classical mechanics goes back to Clausius; and (ii) the Hellmann–Feynman theorem.

As we shall demonstrate, (i) relates the kinetic energy T to the force $\mathbf{F}$, via the scalar product of position $\mathbf{r}$ and force $\mathbf{F}$, denoted $\mathbf{r} \cdot \mathbf{F}$, which was termed by Clausius the virial of the forces.

As to (ii), its essential content can be stated as follows: We can calculate the force on a nucleus (or the absence of it at equilibrium), by using classical electrostatics to calculate the electric field at this nucleus due to (a) the nuclear framework of the molecule and (b) the electronic charge cloud in the molecule. The force on the nucleus is then calculated by multiplying the electric field at the nucleus by the charge on the nucleus.

We shall prove the virial theorem for the case of a single particle, acted on by a force X, say, and moving along the x-axis. The result can be readily generalized to three dimensions, and the generalization to many-particles is straightforward as well.

For a particle of mass M, Newton's equation of motion reads then

$$M\ddot{x} = X \qquad\qquad (A2.6.1)$$

After multiplying both sides of (A2.6.1) by x, $Mx\ddot{x}$ is eliminated by using the identity

$$M\frac{d}{dt}(x\dot{x}) = Mx\ddot{x} + M\dot{x}^2 \qquad\qquad (A2.6.2)$$

when one obtains

$$-2T + M\frac{d}{dt}(x\dot{x}) = xX \qquad\qquad (A2.6.3)$$

T being the kinetic energy $M\dot{x}^2/2$.

Following Clausius, we now take a time average of equation (A2.6.3) where, for example, the time average of T is defined in the usual way by

$$\bar{T} = \frac{\int_0^\tau T\, dt}{\int_0^\tau dt} \qquad\qquad (A2.6.4)$$

with similar definitions of time average for xX, etc. Unless we are dealing with a periodic motion, in which case it would be natural to choose τ as the period, we consider τ to be very large. Then we find from equations (A2.6.3) and (A2.6.4) that

$$2\bar{T} = -\overline{xX} \qquad\qquad (A2.6.5)$$

In writing equation (A2.6.5), we note that when we average $M\, d(x\dot{x})/dt$ we obtain $M[x\dot{x}]_0^\tau/\tau$, which clearly becomes small as $1/\tau$ for large τ provided $x\dot{x}$ remains bounded, as it almost always does. Of course, in any cases of doubt about the boundary conditions, one must always go back to equation (A2.6.3) and average that.

The generalization to three dimensions is immediate, the result being

$$2\bar{T} = -\overline{\mathbf{r} \cdot \mathbf{F}} \tag{A2.6.6}$$

Quantum-Mechanical Virial Theorem

Just as we did in the classical case, let us consider, to minimize mathematical complexity, one particle moving in one dimension (along the x-axis as before). Then the Schrödinger equation for the wave function ψ reads

$$-\frac{\hbar^2}{2M}\frac{\partial^2\psi}{\partial x^2} + [V(x) - E]\psi = 0 \tag{A2.6.7}$$

where $V(x)$ is the potential energy in which the particle of mass M moves, while E is the total energy. We follow Slater (1933) by (i) differentiating with respect to x and (ii) multiplying by x times the complex conjugate ψ^* of ψ to obtain

$$-\frac{\hbar^2}{2M}x\psi^*\frac{\partial^3\psi}{\partial x^3} + x\frac{\partial V}{\partial x}\psi^*\psi + x(V - E)\psi^*\frac{\partial\psi}{\partial x} = 0 \tag{A2.6.8}$$

It will be seen that the virial $xX = -x\,\partial V/\partial x$ now appears in equation (A2.6.8) weighted by the probability function $\psi^*\psi$. Using the Schrödinger equation for $(V - E)\psi^*$ in the last term on the left-hand side of equation (A2.6.8) then yields

$$\frac{-\hbar^2}{2M}\left(x\psi^*\frac{\partial^3\psi}{\partial x^3} - x\frac{\partial^2\psi^*}{\partial x^2}\frac{\partial\psi}{\partial x}\right) + x\frac{\partial V}{\partial x}\psi^*\psi = 0 \tag{A2.6.9}$$

We now make use of the identity [cf. Slater (1933)]

$$\left(x\psi^*\frac{\partial^3\psi}{\partial x^3} - x\frac{\partial^2\psi^*}{\partial x^2}\frac{\partial\psi}{\partial x}\right) = -2\psi^*\frac{\partial^2\psi}{\partial x^2} + \frac{\partial}{\partial x}\left[\psi^{*2}\frac{\partial}{\partial x}\left(\frac{x\,\partial\psi/\partial x}{\psi}\right)\right] \tag{A2.6.10}$$

Substituting this identity into equation (A2.6.9) and integrating over the whole space of the system then yields

$$\frac{-\hbar^2}{2M}\int \psi^*\frac{\partial^2\psi}{\partial x^2}\,dx = \frac{1}{2}\int x\frac{\partial V}{\partial x}\psi^*\psi\,dx \tag{A2.6.11}$$

which is evidently equivalent to

$$2\langle T \rangle = -\langle xX \rangle, \qquad X = -\partial V/\partial x \tag{A2.6.12}$$

T denoting the operator $-(\hbar^2/2M)(\partial^2/\partial x^2)$ for kinetic energy, and the angle brackets indicating expectation values. Thus, again, we have twice the average kinetic energy related to the average virial, the averages now being, of course, quantum mechanical. The virial theorem has then precisely the form of the classical result (A2.6.5) provided only that the time averages there are replaced by quantum-mechanical averages.

The generalization to three dimensions is almost immediate, yielding

$$2\langle T \rangle = -\langle \mathbf{r} \cdot \mathbf{F} \rangle, \qquad \mathbf{F} = -\mathrm{grad}\ V \qquad \text{(A2.6.13)}$$

and as before the many-particle generalization follows straightforwardly.

We caution, as we did in the classical case, that if there is any doubt about dropping the last term

$$\frac{\partial}{\partial x}\left[(\psi^*)^2\, \frac{\partial}{\partial x}\left(\frac{x\, \partial\psi/\partial x}{\psi} \right) \right]$$

on the right-hand side of equation (A2.6.10) as a result of integration, then one must check by explicit evaluation. If ψ vanishes at the boundaries of the system then there is no contribution. While it is true that in a periodic crystal, say, care is needed, for free molecules equation (A2.6.13) is the basic result we are seeking.

Binding of Diatomic Molecule

We immediately illustrate the use of equation (A2.6.13) as applied to an electronic system with fixed nuclei (cf. Appendix A2.3) by considering the binding of a diatomic molecule; specifically, we have in mind a homonuclear case.

Anticipating an approximate form of the force law between atoms (cf. Chapter 3), let us write the change in energy $\Delta E(R)$ in bringing two such atoms from infinite separation to distance R apart in the form

$$\Delta E(R) = -a/R^n + b/R^m \qquad \text{(A2.6.14)}$$

or for the total energy

$$E(R) = E(\infty) - a/R^n + b/R^m \qquad \text{(A2.6.15)}$$

$E(\infty)$ evidently being the sum of the separated atom energies. A typical choice would be a Lennard-Jones 6-12 potential (cf. Problem 2.1), but for the present let us consider general powers of n and m.

As we have seen in Chapter 3, the form of equation (A2.6.14) is to be used as the potential energy $\phi(R)$ in which the nuclei move and the force acting on the nuclei is evidently of magnitude $-\partial\phi/\partial R \equiv -\partial E/\partial R$. This times R is the virial of the forces required to hold the nuclei fast at separation R, and hence from the virial theorem we can write

$$2\langle T \rangle = -R\, dE/dR + \text{Contribution from virial of the Coulomb forces} \quad \text{(A2.6.16)}$$

However, for the Coulomb force, $-r\,\partial V/\partial r = -Z\,e^2 r/r^2 =$ potential energy U and hence we obtain for a diatomic molecule the virial theorem as

$$2\langle T\rangle + \langle U\rangle = -R\,dE/dR \qquad \text{(A2.6.17)}$$

Only for the equilibrium distance R_e, say, at which $dE/dR = 0$, is the system in equilibrium under purely Coulomb forces, in which case we have $2\langle T\rangle + \langle U\rangle = 0$, and hence $E = -\langle T\rangle = \langle U\rangle/2$. Otherwise, as discussed above, the extra term in the virial arises from the forces required to hold the nuclei fast.

If the form (A2.6.15) is now inserted into equation (A2.6.17), a certain amount of insight into the process of molecular binding can be gained by calculating $\langle T\rangle$ and $\langle U\rangle$ as functions of R. The results are

$$\langle T(R)\rangle = -E(\infty) - (n-1)(a/R^n) + (m-1)b/R^m \qquad \text{(A2.6.18)}$$

and

$$\langle U(R)\rangle = 2E(\infty) + (n-2)(a/R^n) - (m-2)b/R^m \qquad \text{(A2.6.19)}$$

Following Slater (1933), typical curves are shown in Figure A.6, and one of the points he stressed at that time was that as two atoms are brought together, at first (i.e., at sufficiently large R) the kinetic energy is reduced, this often being an important ingredient in the process of molecular bonding (see also Ruedenberg, 1962). At this stage we go on to consider the second of the force theorems.

Hellmann–Feynman Theorem

As anticipated above, it is very satisfying that, if we could solve for the electron density $\rho[\mathbf{r}, \{\mathbf{R}\}]$ in the ground state, say, of the system where nuclei are fixed in the configuration $\{\mathbf{R}\}$, then the force acting on a particular nucleus which we choose to focus on is simply what we would calculate from classical electrostatics, as arising from all other nuclei plus the electronic density $\rho[\mathbf{r}, \{\mathbf{R}\}]$ above.

Though this theorem has to be used in practice when only approximate results for ρ are available, which is the usual situation, and this often creates severe difficulty, it is a pillar of the fundamental theory and therefore we shall prove the theorem below.

Consider ν fixed nuclei at positions $\mathbf{R}_1 \ldots \mathbf{R}_\nu$ and N electrons with coordinates $\mathbf{r}_1 \ldots \mathbf{r}_N$. Suppose that the ground-state wave function is $\Psi(\mathbf{r}_1 \ldots \mathbf{r}_N; \mathbf{R}_1 \ldots \mathbf{R}_\nu)$ and let $E(\mathbf{R}_1 \ldots \mathbf{R}_\nu)$ be the ground-state energy which, according to the separation of electronic and nuclear motions we discussed in Chapter 2 (see also Appendix A2.3), is the potential energy $\Phi(\mathbf{R}_1 \ldots)$ for the nuclear motions.

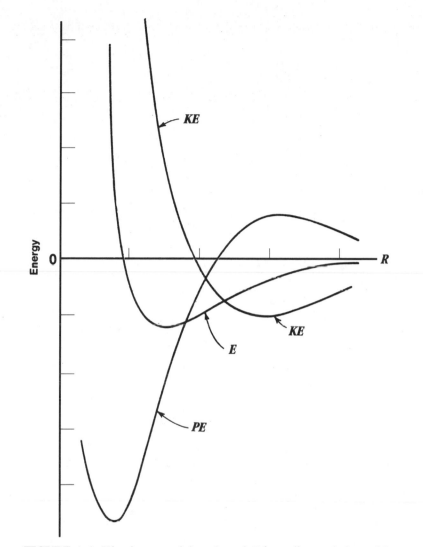

FIGURE A.6. Kinetic, potential, and total E for a diatom (schematic).

Evidently $E \equiv \Phi$ is given in terms of the wave function Ψ and the electronic Hamiltonian H_e as

$$\Phi(\mathbf{R}_1 \ldots \mathbf{R}_v) = \int \Psi^* H_e \Psi \, d\mathbf{r}_1 \ldots d\mathbf{r}_N \qquad \text{(A2.6.20)}$$

Clearly the force on the nth nucleus, $\mathbf{F}_n$, say, is given by

$$\mathbf{F}_n = -\frac{\partial \Phi}{\partial \mathbf{R}_n} \qquad \text{(A2.6.21)}$$

and in particular the component in the x-direction F_{nX} is explicitly

$$F_{nX} = -\int \Psi^* \frac{\partial H_e}{\partial X_n} \Psi \, d\mathbf{r}_1 \ldots d\mathbf{r}_N - \int \frac{\partial \Psi^*}{\partial X_n} H_e \Psi \, d\mathbf{r}_1 \ldots d\mathbf{r}_N$$
$$- \int \Psi^* H_e \frac{\partial \Psi}{\partial X_n} \, d\mathbf{r}_1 \ldots d\mathbf{r}_N \qquad \text{(A2.6.22)}$$

However, since H_e is a real Hermitian operator [see, e.g., Atkins (1983)], we can write

$$\int \Psi^* H_e \frac{\partial \Psi}{\partial X_n} \, d\mathbf{r}_1 \ldots d\mathbf{r}_N = \int \frac{\partial \Psi}{\partial X_n} H_e \Psi^* \, d\mathbf{r}_1 \ldots d\mathbf{r}_N \qquad \text{(A2.6.23)}$$

Using this result, plus the fact that Ψ^* satisfies the wave equation

$$H_e \Psi^* = \Phi \Psi^* \qquad \text{(A2.6.24)}$$

we can write the last two terms on the right-hand side of equation (A2.6.22) as

$$-\Phi \left(\int \frac{\partial \Psi^*}{\partial X_n} \Psi \, d\mathbf{r}_1 \ldots d\mathbf{r}_N + \int \Psi^* \frac{\partial \Psi}{\partial X_n} \, d\mathbf{r}_1 \ldots d\mathbf{r}_N \right)$$
$$= -\Phi \left(\frac{\partial}{\partial X_n} \int \Psi^* \Psi \, d\mathbf{r}_1 \ldots d\mathbf{r}_N \right) \qquad \text{(A2.6.25)}$$

As we are differentiating the normalization integral with respect to X_n here, the result is zero, and since H_e is the electron Hamiltonian we can write

$$F_{nX} = -\int \Psi^* \frac{\partial H_e}{\partial X_n} \Psi \, d\mathbf{r}_1 \ldots d\mathbf{r}_N$$
$$= -\int \Psi^* \frac{\partial \Phi_e}{\partial X_n} \Psi \, d\mathbf{r}_1 \ldots d\mathbf{r}_N \qquad \text{(A2.6.26)}$$

the last line following as shown in Bethe and Jackiw (1968). Here Φ_e is explicitly given by

$$\Phi_e = \frac{1}{2} \sum_{i=1}^{N} \sum_{j=1}^{N} \frac{e^2}{r_{ij}} - \sum_{i=1}^{N} \sum_{n=1}^{v} \frac{Z_n e^2}{r_{in}} + \frac{1}{2} \sum_{\alpha=1}^{v} \sum_{\beta=1}^{v} \frac{Z_\alpha Z_\beta e^2}{r_{\alpha\beta}} \qquad \text{(A2.6.27)}$$

It follows in a straightforward manner that

$$\frac{\partial \Phi_e}{\partial X_n} = \frac{\partial}{\partial X_n} \left(\sum_{\alpha' > \beta} \frac{Z_{\alpha'} Z_\beta e^2}{R_{\alpha'\beta}} \right) - \sum_i Z_n e^2 \frac{\partial}{\partial X_n} \left(\frac{1}{r_{in}} \right) \qquad \text{(A2.6.28)}$$

Let us now denote the electric field in the x-direction produced by the nth nucleus at the position of the ith electron by $\mathscr{E}_{nx}(\mathbf{r}_i)$:

$$\mathscr{E}_{nx}(\mathbf{r}_i) = Z_n e \frac{\partial}{\partial X_n}\left(\frac{1}{r_{in}}\right) \tag{A2.6.29}$$

Substituting in equation (A2.6.28), one obtains

$$F_{nX} = -\frac{\partial}{\partial X_n}\sum_{\beta}{}' \frac{Z_n Z_\beta e^2}{R_{n\beta}} + \sum_i e \int \Psi^* \Psi \mathscr{E}_{nx}(\mathbf{r}_i)\, d\mathbf{r}_1 \ldots d\mathbf{r}_N \tag{A2.6.30}$$

The electron density associated with electron 1 is given by

$$\rho_1(\mathbf{r}_1) = \int \Psi^* \Psi\, d\mathbf{r}_2 \ldots d\mathbf{r}_N \tag{A2.6.31}$$

and since electrons are indistinguishable the total density at $\mathbf{r}$ is

$$\rho(\mathbf{r}) = N\rho_1(\mathbf{r}) \tag{A2.6.32}$$

Hence we reach the formal statement of the Hellmann–Feynman theorem, using equations (A2.6.31) and (A2.6.32) in equation (A2.6.30):

$$F_{nX} = -\frac{\partial}{\partial X_n}\sum_{\beta} \frac{Z_n Z_\beta e^2}{R_{n\beta}} + \int \rho \mathscr{E}_{nX}\, d\mathbf{r} \tag{A2.6.33}$$

This is just the result anticipated: We calculate the force according to classical electrostatics from the nuclear fields plus the electronic charge cloud. It should be emphasized again that while the theorem is exact, small departures of $\rho(\mathbf{r})$ from the exact ground-state density can lead to large errors in calculating the forces from equation (A2.6.33). A great deal of caution must therefore be exercised in its application in conjunction with inevitably approximate charge densities for molecules.

A3.1. TOPICS RELEVANT TO THE TREATMENT OF INTERMOLECULAR FORCES

In this appendix we extend the introductory treatment of intermolecular forces presented in Chapter 3 by listing several pertinent topics to be studied. Some key references are listed in Appendix A3.2.

1. Derivations of expressions involving the interaction between: (a) two ions, (b) an ion and permanent dipole, and (c) two permanent dipoles.
2. Quadrupole moments.
3. The radial distribution function.

4. The potential energy and the Mie potential.
5. Further details of methods of determining intermolecular energy constants.
6. Further details on experimental molecular beam studies.
7. Further study of potential functions for (a) monatomic systems, (b) polyatomic systems, (c) open shell atoms, and (d) large polyatomic systems.
8. Further consideration of spectroscopic methods.
9. *ab initio* calculations and intermolecular forces.

A3.2. BIBLIOGRAPHY FOR FURTHER STUDY OF INTERMOLECULAR FORCES

P. W. Atkins, *Physical Chemistry*, Oxford University Press and W. H. Freeman, New York (1990).

R. S. Berry, S. A. Rice, and J. Ross, *Physical Chemistry*, John Wiley and Sons, New York (1980).

U. Buck, Elastic Scattering, *Adv. Chem. Phys.* **30**, 313 (1975).

M. S. Child, *Molecular Collision Theory*, Academic, London (1974).

M. S. Child, *Semiclassical Mechanics*, Clarendon Press, Oxford (1991).

B. Chu, *Molecular Forces*, Wiley-Interscience, New York (1967).

J. H. Dymond and E. B. Smith, *The Virial Coefficients of Gases and Mixtures: A Critical Compilation*, Clarendon Press, Oxford (1980).

M. A. D. Fluendy and K. P. Lawley, *Molecular Beams in Chemistry*, Chapman and Hall, London (1974).

J. O. Hirschfelder, C. F. Curtiss, and R. B. Bird, *The Molecular Theory of Gases and Liquids*, John Wiley and Sons, New York (1954).

K. J. Laidler, *Chemical Kinetics*, *3rd. Ed.*, Harper and Row, New York (1987).

G. C. Maitland, M. Rigby, E. B. Smith, and W. A. Wakeham, *Intermolecular Forces: Their Origin and Determination*, Clarendon Press, Oxford (1981).

E. A. Moelwyn-Hughes, *Physical Chemistry*, *2nd Ed.*, Pergamon Press, Oxford (1961).

M. Rigby, E. B. Smith, W. A. Wakeham, and G. C. Maitland, *The Forces Between Molecules*, Clarendon Press, Oxford, 1986.

J. Ross (ed.), Molecular Beams, *Adv. Chem. Phys.* **10** (1966).

A4.1. THE CORRESPONDENCE BETWEEN CELLS IN PHASE SPACE AND QUANTUM-MECHANICAL ENERGY LEVELS

The object of this appendix is to verify the result employed in establishing the electron-density–potential-energy relation (4.6): A cell of size h^3 in phase space is equivalent to one quantum-mechanical energy level and, therefore, in accordance with the Pauli exclusion principle, can hold two electrons, provided they have opposed spins. This result is valid for three dimensions, but we give the proof here of the analogous result in one-dimension, in which case a cell of "area" h corresponds to a quantum-mechanical energy level. That this is true, at least in a semiquantitative way, is clear from Heisenberg's uncertainty principle as stressed in Chapter 4. Thus, since phase space is combined position–momentum space, we can form an element of area in one dimension as the product $\Delta p_x \Delta x$, with p_x denoting the x-component of momentum. However, according to the uncertainty principle, in quantum mechanics we cannot make $\Delta p_x \Delta x$ less than of the order of Planck's constant, and hence there is a "minimum size" of a cell in phase space, corresponding to a quantum-mechanical energy level. This rough statement will be made precise by referring specifically to two "model" problems, namely the free-particle in a one-dimensional box and the simple harmonic oscillator. We treat these two cases, in turn, using the phase-space description.

Phase Space Description of a Particle in a One-Dimensional Box

For a particle moving freely along the x-axis, we can write $E = p_x^2/2m$ in classical mechanics as the energy–momentum relation. It follows immediately that the statement that a particle has a specified energy E_1, say, means classically that the x-component of momentum in this one-dimensional box problem is given by

$$p_x = \pm(2mE_1)^{1/2} \tag{A4.1.1}$$

Referring to Figure A.7, the point Q denotes the representative point in the phase space, lying on one of the lines of constant momentum in equation (A4.1.1), and, of course, bounded by the box edges at $x = 0$ and l.

Now consider a second energy, E_2, say, greater than E_1. If the particle has this energy, the representative point in phase space specifying momentum p_x and

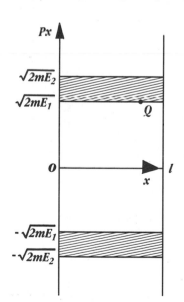

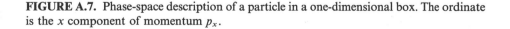

FIGURE A.7. Phase-space description of a particle in a one-dimensional box. The ordinate is the x component of momentum p_x.

position x (at each time t in the classical description) must lie on either of the lines labeled $(2mE_2)^{1/2}$, apart from the sign.

Evidently the statement that the particle has energy E such that $E_1 \leq E \leq E_2$ means that Q can lie only in one or the other of the shaded regions in Figure A.7.

The energy levels obtained from Schrödinger's equation for a particle in a one-dimensional box (see any book on quantum mechanics) can be arrived at appropriately in the present context by fitting de Broglie standing waves, of wavelength $\lambda = h/p_x$, into the box of length l by the requirement that an integral number of half-wavelengths (the quantum number n) shall just fit into the box length. Thus

$$p_x = h/\lambda, \qquad n\lambda/2 = l, \qquad E_n = p_x^2/2m = n^2h^2/8ml^2 \qquad \text{(A4.1.2)}$$

where $n = 1, 2$, etc.

Returning to Figure A.7, we now choose E_1 and E_2 to be the quantally allowed adjacent energy levels corresponding to quantum numbers $n = 1$ and $n = 2$, respectively in equation (A4.1.2), or more generally to n and $n + 1$. Treating this latter general case, the sum of the two (equal) shaded areas in Figure A.7 plainly leads to the result that the total shaded area A into which phase space is divided by adjacent quantal energy levels is given by

$$\begin{aligned}
A &= 2l[(2mE_2)^{1/2} - (2mE_1)^{1/2}] \\
&= 2l[(n+1)h/2l - nh/2l] \\
&= h
\end{aligned} \qquad \text{(A4.1.3)}$$

It should be noted that, provided the levels E_1 and E_2 are adjacent, the division of phase space corresponds to an area A independent of n. In writing the second line of equation (A4.1.3), explicit use has been made of the energy level equation (A4.1.2).

To summarize, for this admittedly elementary example, successive quantally allowed energy levels divide phase space into areas precisely equal to Planck's constant h. Here then, in this example, we have the basis for the association of an area h of phase space with one quantal energy level, making the qualitative statement $\Delta p_x \Delta x \sim h$ quantitatively precise through this example. To show that such an argument holds for a particle moving in a force field, we treat below a second example by describing a one-dimensional harmonic oscillator in phase space.

Phase Space Description of an Harmonic Oscillator

Again we start out from the classical mechanics of a simple harmonic oscillator, vibrating along the x-axis. The solution of the Newtonian equation of motion [see equation (5.20)] subject to the condition that the displacement x from the origin is zero at time $t = 0$ is

$$x = A \sin \omega t \qquad \text{(A4.1.4)}$$

where the angular frequency ω is given by equation (5.22).

Now we want to discuss the motion of the representative point $(x, p_x) \equiv Q$ in the phase space for this example, analogous to the motion depicted in Figure A.7 for the particle in a box. To do so, we construct the momentum p_x from equation (A4.1.4) as

$$p_x = m\dot{x} = m\omega A \cos \omega t \qquad \text{(A4.1.5)}$$

Evidently, we can plot a figure analogous to Figure A.7 for the free-particle case if we eliminate the time t between equations (A4.1.4) and (A4.1.5). Forming $\sin^2 \omega t$ from the first of these equations and $\cos^2 \omega t$ from the second, and using $\sin^2 + \cos^2 = 1$, we readily find

$$x^2/A^2 + p_x^2/(m\omega A)^2 = 1 \qquad \text{(A4.1.6)}$$

Equation (A4.1.6) represents an ellipse, so that Q describes such a path in the phase space, depicted, say, by the ellipse labeled by energy E_1 in Figure A.8. It should be noted that whereas we have labeled this ellipse by its energy E_1, this is, in classical mechanics, determined by the amplitude A of the motion, in this case, say, A_1. Evidently, at this amplitude, the energy of the oscillator is entirely potential energy, $V(x) = \frac{1}{2}kx^2$, with x equal to the amplitude A_1, and k being the force constant [see equations (5.20) and (5.22)]. Hence we have the relation for energy E_1 labeling the ellipse in Figure A.8 with a semimajor axis A_1,

$$E_1 = \tfrac{1}{2}kA_1^2 = \tfrac{1}{2}m\omega^2 A_1^2 \qquad \text{(A4.1.7)}$$

Since the area enclosed by an ellipse with semimajor axis a and semiminor axis b is πab, we have from Figure A.8 and the above discussion

$$\text{Area of ellipse bounded by } E_1 = \pi m\omega A_1^2 \qquad \text{(A4.1.8)}$$

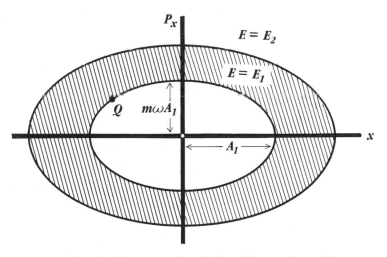

FIGURE A.8. Phase-space description of harmonic oscillator.

Analogous to the previous example, we have drawn a second ellipse in Figure A.8, labeled by energy $E_2 > E_1$. The shaded area enclosed between these two constant energy curves follows from equation (A4.1.8) and its counterpart for the ellipse labeled E_2 as

$$\text{Area enclosed between } E_2 \text{ and } E_1 = \pi m \omega (A_2^2 - A_1^2) \qquad \text{(A4.1.9)}$$

Rewriting the semimajor axes A_1 and A_2 in terms of E_1 and E_2 using equation (A4.1.7) leads from (A4.1.9) to the desired result:

$$\text{Area enclosed between } E_2 \text{ and } E_1 = (2\pi/\omega)(E_2 - E_1) \qquad \text{(A4.1.10)}$$

The final step, as before, is to pass to quantally allowed adjacent energy levels, and since for the harmonic oscillator, the spacing between adjacent energy levels $(E_2 - E_1)$ is $\hbar\omega$, with ω the classical angular frequency of the oscillator, we have again the same conclusion—that quantally allowed adjacent energy levels divide the phase space into cells of area h.

Though these two examples lead to precisely the same result, the reader should recognize that for an arbitrary force field $V(x)$, one will find the above result to hold accurately only when the correspondence between classical and quantum mechanics becomes close. According to Bohr's correspondence principle, this will occur in the limit of large quantum numbers. This should be borne in mind when applying the phase-space theorem, which states that in d dimensions, a cell of "size" h^d in phase space corresponds to one quantal energy level, which is the correct generalization of the one-dimensional result treated in this appendix.

A4.2. THE KINETIC ENERGY DENSITY OF AN INHOMOGENEOUS ELECTRON GAS

The original work of Thomas (1926) and Fermi (1928) was based on the idea of treating the inhomogeneous electronic charge cloud in an atom as though, locally, it behaved like an electron gas. Then the phase space argument set out in Section 4.1, based on cells of volume h^3 being equivalent to one energy level and able to hold two electrons with opposed spins, leads to equation (4.5). Our aim here is to extend this line of argument to the result (4.15) and thereby to evaluate the kinetic constant c_k.

To obtain the kinetic energy in terms of the electron density $\rho(\mathbf{r})$, the same phase-space argument summarized above, applied to a free Fermi gas of electrons, tells us that the number of electrons having momentum of magnitude lying between p and $p + dp$ is the corresponding phase-space volume, i.e., $4\pi p^2 dp V$, for spatial volume V, divided by the cell volume h^3 and multiplied by 2 for opposed spin electrons. Hence the total electronic kinetic energy is given by multiplying the above number by the kinetic energy of an electron with momentum p, namely $p^2/2m$, and then integrating through the sphere in momentum space out to the maximum momentum p_m [see equation (4.4)]. Then one has

$$\text{Kinetic energy per unit volume} = t = \int_0^{p_m} \frac{8\pi p^2}{h^3} \frac{p^2}{2m} \, dp \qquad \text{(A4.2.1)}$$

which is evidently proportional to p_m^5, the precise result being

$$t = (4\pi/5mh^3)\bar{p}_m^5 \qquad \text{(A4.2.2)}$$

With equation (4.5), this is readily shown to lead to equation (4.15), namely

$$\text{Kinetic energy density } t = c_k\{\rho(\mathbf{r})\}^{5/3}$$

$$\qquad \text{(A4.2.3)}$$

$$c_k = (3h^2/10m)(3/8\pi)^{2/3}$$

Hence the total kinetic energy $T = \int t \, d\tau$ of the inhomogeneous electron gas is

$$T = c_k \int [\rho(\mathbf{r})]^{5/3} \, d\tau \qquad \text{(A4.2.4)}$$

This result is correct for constant electron density. For an inhomogeneous electron gas, it is clear that one ought to correct this equation (A4.2.4) with terms involving the density gradient (cf. Problem 4.6). However, equation (A4.2.4) is already of use in calculating interatomic forces, examples being the work of Gordon and Kim (1972) on rare gas interactions [see Appendix A4.11, also LeSar (1989)]. However, this kinetic energy approximation (A4.2.4) must then be supplemented by expressions for the exchange and correlation energies (see also Section 6.9) in terms of the electron density $\rho(\mathbf{r})$ before such force laws can be calculated with useful accuracy.

A4.3. THE CHEMICAL POTENTIAL, TELLER'S THEOREM, AND SCALING OF ENERGIES OF HOMONUCLEAR DIATOMS

In Appendix A4.5, we will consider the scaling properties of the total energy $E(Z, N)$ of atomic ions. In this appendix, we summarize an approximate generalization of the $1/Z$ expansion to homonuclear diatoms. Here Z is the atomic number of each atom in the diatom, N is the total number of electrons in the molecule, while the energy evidently now depends on the internuclear distance, say, R. In fact, the series in $1/Z$ [compare equation (A1.3.1) for atomic ions] is argued by March and Parr (1980) to have the approximate sum

$$E(Z, N, R) = Z^{7/3}F_1(N/Z, N^{2/3}/ZR) + Z^2F_2(N/Z, N^{2/3}/ZR) + O(Z^{5/3})$$

$$\qquad \text{(A4.3.1)}$$

Just as for atoms, the leading term in equation (A4.3.1) is the self-consistent Thomas–Fermi (TF) answer, obtained by solving the density-potential relation (4.6) (i.e., the equation for constant chemical potential throughout the molecular electronic cloud) self-consistently for the potential V and density ρ. That this energy scales as shown in equation (A4.3.1) is known from the work of Hund (1932) plus the later work of Townsend and Handler (1962). We can expect that the major contribution to the term in Z^2 will come from correcting the TF energy for the errors in the method when applied to the K-shell electrons around each nucleus. In contrast, the term of $0(Z^{5/3})$ is expected to come predominantly from the exchange energy, which can be approximated in terms of the electron density $\rho(\mathbf{r})$ by the Dirac–Slater exchange form $\propto \rho^{4/3}$ derived by microscopic electron-gas theory in Appendix A4.7 (see also Problem 4.5).

We note here that Teller has shown that, in fact, if one requires complete internal consistency, the TF treatment does not lead to molecular binding. March and Parr (1980) consider the consequences of this in their work, but we must refer the interested reader to their paper for the details (see also Problem 4.6).

A4.4. THE SELF-CONSISTENT FIELD IN THE HELIUM ATOM

In this appendix we shall make quite explicit how a self-consistent field (SCF) can be established, by reference to the He atom. Here, the spatial wave function in the SCF approximation (cf. the exact form discussed in Appendix AVI) is a simple product (we deal with the ground state below):

$$\Psi(1, 2) = \psi(1)\psi(2) \tag{A4.4.1}$$

with both electrons in the same space orbital, with, of course, opposed spins.

As an example, in equation (A4.4.1) we could choose the hydrogen-like form for the $1s$ orbital:

$$\psi = N \exp(-Z'r/a_0) \tag{A4.4.2}$$

with N the usual normalization factor. Then Z', an "effective" nuclear charge, can be found by using the variational principle of Appendix A2.1. The Z' found in this way for a two-electron atomic ion with nuclear charge Ze is in fact $Z - 5/16$, which is $27/16$ for He with $Z = 2$. It is clear on physical grounds why Z' should lie between 1 and 2. If only one electron were present, i.e., for the He$^+$ ion, Z' would be 2. But in the neutral atom, the second electron spends some time inside the first one, that is, it screens the second electron from the full attractive power of the nucleus. Thus it is clear that in this case $1 < Z' < 2$. This already tells us that the potential energy to insert into a one-electron Schrödinger equation to determine the "best possible" ψ in equation (A4.4.1) is not $-2e^2/r$, as it would be in the He$^+$ ion, but this, weakened in its attraction by the presence of a term due to shielding by the second electron.

We will return to this point below and write a new potential energy on physical grounds, with the nuclear potential energy $V_N = -2e^2/r$ modified in this way. For the present let us merely note that we could, in fact, insert the product form (A4.4.1) as it stands into the variational principle and minimize the energy with respect to the entire one-electron atomic orbital. This minimization with respect to a function, which is, in elementary terms, equivalent to an infinite number of variation parameters, can be carried out by means of the calculus of variations [see, e.g., the book by Hartree (1928, 1957)]. But we need not do so because the result is readily understood in the language used above. In short, the orbital ψ in equation (A4.4.1) must be obtained by solving the one-electron

$$\nabla^2\psi + \frac{2m}{\hbar^2}[E - V(r)]\psi = 0$$

(A4.4.3)

$$V(r) = -\frac{2e^2}{r} + e^2 \int \frac{\psi^2(r')}{|\mathbf{r} - \mathbf{r}'|}\, d\tau'$$

This is seen to have the nuclear potential energy $-2e^2/r$ screened by a term involving the electron density ψ^2 of the orbital into which the second electron is also placed. Thus, we have corrected the nuclear potential V_N by a term which is $-e$ times the classical electrostatic potential created by the electron not under consideration. This is how the screening of the nucleus comes about.

However, it will rightly be argued that the new screened potential energy involves the wave function we are trying to find and this is indeed the case! We could try to proceed by inserting the approximate wave function (A4.4.2) with $Z' = 27/16$ into equation (A4.4.3) inside the potential energy term. Clearly, we can then solve the Schrödinger equation (numerically) for ψ. This will not be the assumed wave function, so we calculate a new potential energy based on the next approximation to ψ, and repeat the process. We must go on until the ψ inserted in the potential energy term reproduces itself to a desired specified accuracy, at which point we have established the SCF in this atom, and in fact this process was carried out by Wilson and Lindsay (1935). Although this is a pretty simple example, it should be sufficient to establish clearly the concept of the SCF, first introduced into atomic theory by Hartree [see Hartree (1957) for all the relevant references to the early work].

A4.5. THE SELF-CONSISTENT FIELD TREATMENT OF BINDING ENERGIES OF HEAVY POSITIVE ATOMIC IONS

From the density-potential relation (4.6), we can use the SCF solution (see Appendix A4.4 for the idea behind that) to calculate the total energy $E(Z, N)$ of heavy positive ions with nuclear charge Ze and N electrons. This appendix will focus mainly on the neutral-atom case, for which $N = Z$, though the conclusions for $N < Z$ will also be summarized.

To give the essence of the argument, however, let us return to the bare Coulomb potential energy $-Ze^2/r$ treated in Appendix A1.3. We know that for this case the energy levels are

$$\varepsilon_n = -(Z^2/2n^2)e^2/a_0$$

(A4.5.1)

There are $2n^2$ electrons in a closed shell of principal quantum number n and hence the energy per shell from equation (A4.5.1) is $-Z^2e^2/a_0$, independent of n.

Now suppose that we fill these levels with N electrons, chosen specifically such that the levels are doubly filled until $\mathcal{N}$ closed shells are completed. Then the total energy E in this model is the sum of these occupied energy levels, and hence

$$E = (-Z^2e^2/a_0)\mathcal{N}$$

(A4.5.2)

However, from the above results (cf. equation (A1.3.2)):

$$N = \sum_1^{\mathcal{N}} 2n^2 = \mathcal{N}(\mathcal{N} + 1)(2\mathcal{N} + 1)/3 \tag{A4.5.3}$$

In the limit appropriate to the approximate density theory based on statistical mechanical arguments in phase space, we take N and Z large, so that equation (A4.5.3), neglecting 1 compared with $\mathcal{N}$, yields $N = \frac{2}{3}\mathcal{N}^3$. Working to the next order, we can solve equation (A4.5.3) more accurately to find

$$\mathcal{N} = (\tfrac{3}{2})^{1/3}N^{1/3} - \tfrac{1}{2} + \cdots \tag{A4.5.4}$$

Inserting this into equation (A4.5.2) we find for this bare Coulomb field model,

$$E = (-Z^2 e^2/a_0)(\tfrac{3}{2})^{1/3}N^{1/3} + \tfrac{1}{2}(Z^2 e^2/a_0) + \cdots \tag{A4.5.5}$$

If we use the TF theory in equation (4.6) with the bare Coulomb potential to calculate the density, and thus the energy E, we find the result to be precisely the first term on the right-hand side of equation (A4.5.5).

When we go to the case $N = Z$, we obtain

$$E_{\text{Coul}}(Z, N) = -(\tfrac{3}{2})^{1/3}Z^{7/3}e^2/a_0 \tag{A4.5.6}$$

The $Z^{7/3}$ power law for the dependence of the energy on atomic number in equation (A4.5.6) is the important consequence of the above argument, valid for large Z.

When we solve equation (4.6) to find the self-consistent potential energy $V(r)$ and the corresponding density $\rho(r)$, which has to be done numerically [though no iteration is now required, in contrast to the case of the previous appendix; see, e.g., March (1975)], we can again calculate the energy $E(Z, N)$ and, hence, in particular the neutral atom energy $E(Z, Z)$ in this SCF framework. Only one modification occurs from equation (A4.5.6): The $Z^{7/3}$ power is found again, but obviously inclusion of the SCF reduces the binding of the outer electrons because they are shielded from the nucleus. The result is found to be

$$E_{\text{TF}}(Z, Z) = -0.7687Z^{7/3}e^2/a_0: \tag{A4.5.7}$$

To be useful in the range of the Periodic Table, i.e., for $Z \lesssim 100$, one has to add correction terms to equation (A4.5.7), as in equation (A4.5.5). As Scott (1952) was the first to demonstrate, the term $\frac{1}{2}Z^2 e^2/a_0$ in equation (A4.5.5) comes predominantly from the K-shell. Here shielding effects are negligible for heavy atoms, and one can immediately add this same term to the energy in equation (A4.5.7) appropriate to the SCF case. However, a further term in this expansion comes from the exchange energy discussed in Appendix A4.7. If this term is evaluated with the TF self-consistent density $\rho(r)$ referred to above, one finds a value $-0.221(e^2/a_0)Z^{5/3}$. Actually another small correction comes into equation (A4.5.4) also proportional to $Z^{5/3}$, and adding such a correction we find the final binding energy formula to be [March and Plaskett (1956)]*

$$E_{\text{SCF}}/(e^2/a_0) = -0.7687Z^{7/3} + \tfrac{1}{2}Z^2 - 0.26Z^{5/3} + \cdots \qquad \text{(A4.5.8)}$$

It is interesting to compare this first principle and largely analytical formula with the numerical HF results using the method of Appendix A4.6. These numerical results can be fitted by a least-squares method to a formula like equation (A4.5.8). Fixing the coefficient of the $Z^{7/3}$ term as in equation (A4.5.8), one finds that the least-squares fit yields

$$E_{\text{HF}}(e^2/a_0) = -0.7687Z^{7/3} + 0.4904Z^2 - 0.2482Z^{5/3} \qquad \text{(A4.5.9)}$$

in excellent agreement with the analytical equation (A4.5.8).

Finally we return to heavy positive ions with $N < Z$. We can evidently rewrite equation (A1.3.3) in the form

$$E_{\text{Coul}}(Z, N) = -(e^2/a_0)Z^{7/3}(N/Z)^{1/3}(\tfrac{3}{2})^{1/3} \qquad \text{(A4.5.10)}$$

In view of the preceding discussion it will occasion little surprise that the SCF TF electron density theory yields a result

$$E_{\text{SCF}}/(e^2/a_0) = Z^{7/3}f(N/Z) \qquad \text{(A4.5.11)}$$

The function f has to be obtained numerically from equation (4.6) for each value of N/Z in the range between 0 and 1 for positive ions (negative ions are not stable in this theory). The numerical results for f can be readily obtained, for example, from the work of Grout *et al.* (1983).

The final comment is that the formula (A4.5.11) exhibits a scaling property valid for heavy positive ions treated by nonrelativistic theory. The energy $E(Z, N)$, which is in general a function of both Z and N, becomes such that, for large N and Z and $N/Z \leq 1$, $E(Z, N)/Z^{7/3}$ is a function of the single variable N/Z, obviously an important simplification. Naturally the TF result (A4.5.11) needs correcting analogously to equation (A4.5.8) for $E(Z, Z)$ but we shall not pursue that further here.

A4.6. THE HARTREE–FOCK SELF-CONSISTENT FIELD METHOD

The basic principle of any method employing SCF theory is to consider a set of one-electron Schrödinger equations, one for each electron in an N-electron atom or molecule, the selected electron moving in the field generated by the nuclear framework plus the average field of the other electrons. The SCF methods derived below from quantum mechanics have a direct semiclassical analogue in the density–potential relation (4.6) of Section 4.1. In what follows, the first of these quantum-mechanical methods, introduced by Hartree (1928), will be summarized, followed by the extension due primarily to Fock and Slater (HF formalism).

*March and Wind [*Molecular Phys.* **77**, 791 (1992)] discuss the linear dependence on Z of atomic correlation energies.

We have seen in Appendix A4.4 that Hartree's method leads to a one-electron Schrödinger equation in which, as noted immediately above, electron 1, say, moves in the nuclear potential energy $-Ze^2/r$ plus the potential energy created by the charge distribution associated with the second electron. We saw in Appendix A4.4, for He and for heavy atoms in Appendix A4.5, how the SCF could be established.

Therefore, we note here that the Hartree method for two electrons given in Appendix A4.4 can be extended to an N-electron atom or molecule. In its most general form the approach leads to N coupled integrodifferential equations, namely,

$$\mathscr{H}_i^{\text{eff}} \psi_i(\mathbf{r}) = \varepsilon_i \psi_i(\mathbf{r}), \qquad i = 1, \ldots, N$$

$$\mathscr{H}_i^{\text{eff}} = -\frac{\hbar^2}{2m} \nabla^2 + V_i(\psi_1, \psi_2, \ldots, \psi_{i-1}, \ldots, \psi_{i+1}, \ldots, \psi_N)$$

(A4.6.1)

$$V_i = -\frac{Ze^2}{r} + \sum_{j \neq i} e^2 \int \frac{\psi_j(\mathbf{r}')^2}{|\mathbf{r} - \mathbf{r}'|} \, d\mathbf{r}'$$

Though equation (A4.6.1) constitutes a fairly evident intuitive generalization of equation (A4.4.3), it is instructive to consider the variational basis of the Hartree theory, characterized by one-electron wave functions $\psi_i(\mathbf{r})$ obtained from equations (A4.6.1). To proceed, the important thing is to recognize what is the total spatial wave function of the N-electron atom or molecule in the Hartree approximation. It is, in fact, a straightforward generalization of the product form (A4.4.1) for the two-electron He atom, namely

$$\Psi(\mathbf{r}_1, \ldots, \mathbf{r}_N) = \psi_1(\mathbf{r}_1) \psi_2(\mathbf{r}_2), \ldots, \psi_N(\mathbf{r}_N)$$

(A4.6.2)

The variational basis of equations (A4.6.1) should now be apparent: one inserts the total wave function (A4.6.2) into the variational integral

$$E = \frac{\int d\tau_1 \ldots d\tau_N \, \Psi^* \mathscr{H} \Psi}{\int d\tau_1 \ldots d\tau_N \, |\Psi|^2}$$

(A4.6.3)

where the total Hamiltonian $\mathscr{H}$ for, say, the case of an atomic ion of nuclear charge Z is

$$\mathscr{H} = \sum_{i=1}^{N} \left(-\tfrac{1}{2}\nabla_i^2 - \frac{Ze^2}{r_i} \right) + \sum_{j>i}^{N} \sum_{i=1} \frac{e^2}{r_{ij}}$$

(A4.6.4)

or, more generally, for any atom or molecule with N electrons

$$\mathscr{H} = \sum_{i=1}^{N} \hat{H}_i + \sum_{i<j}^{N} \sum_{i=1} \frac{e^2}{r_{ij}}$$

(A4.6.5)

The energy E in equation (A4.6.3), with Ψ taken as in equation (A4.6.2), is readily found to take the form

$$E = \sum_{i=1}^{N} H_i + \tfrac{1}{2} \sum_{i \neq j} \sum_{i=1}^{N} J_{ij} \qquad (\text{A4.6.6})$$

where H_i is the expectation value with respect to $\psi_i(r)$ of the one-electron operator $\hat{H}_i$ introduced in equation (A4.6.5) while J_{ij}, defined by

$$J_{ij} = \int \frac{|\psi_i(\mathbf{r})|^2 |\psi_j(\mathbf{r}')|^2}{|\mathbf{r} - \mathbf{r}'|} \, d\tau \, d\tau' \qquad (\text{A4.6.7})$$

evidently denotes the Coulomb integral, with interpenetrating charge clouds.

To carry out the minimization of E with respect to the one-electron wave functions $\psi_i(\mathbf{r})$ is a technical matter which can be handled by the calculus of variations. The interested reader is referred, for example, to the book by Fomin (1963). But there is one point that it is useful to emphasize here. This minimization is not unconditional because in carrying it out with respect to the one-electron wave functions $\psi_i(\mathbf{r})$ one must preserve the N normalization conditions on the one-electron functions, namely,

$$\int |\psi_i(\mathbf{r})|^2 \, d\tau = 1, \qquad i = 1, \ldots, N \qquad (\text{A4.6.8})$$

Instead of minimizing E, subject to the restrictive conditions (A4.6.8), the method of Lagrange undetermined multipliers, discussed, for example, by Rushbrooke (1949) can be used to minimize the quantity

$$E - \sum_{i=1}^{N} \varepsilon_i \int d\tau_i \, \psi_i^*(\mathbf{r}_i) \psi_i(\mathbf{r}_i) \qquad (\text{A4.6.9})$$

where the N Lagrange multipliers have been denoted by ε_i, in anticipation of their identification as eigenvalues in the Hartree equations. Given the introduction of these N independent quantities, one can forget the restrictive conditions (A4.6.8) and can go ahead with unconditional minimization of the quantity (A4.6.9) with respect to the N one-electron wave functions $\psi_i(\mathbf{r})$. The outcome, which we shall not discuss in further detail, is to lead back to equations (A4.6.1). It is, however, in view of the interest in the sum of the one-electron energies over occupied states, worthy of note that one can show additionally that

$$\sum_{i=1}^{N} \varepsilon_i = \sum_{i=1}^{N} H_i + \sum_{i \neq j} \sum_{i=1}^{N} J_{ij} \qquad (\text{A4.6.10})$$

and, hence, from equations (A4.6.6) and (A4.6.10) one has finally

$$E = \sum_{i(\text{occupied})} \varepsilon_i - \tfrac{1}{2} \sum_{i \neq j} \sum_{i=1}^{N} J_{ij} \qquad (\text{A4.6.11})$$

In summary, the Hartree method is contained, essentially, in equations (A4.6.1), which, when solved, determine the N one-electron wave functions, as well as the N Lagrange multipliers, or the one-electron energies ε_i. These one-electron energies, when suitably corrected for counting electron–electron interactions twice over, leading to the subtraction of the Coulomb integrals exhibited quite explicitly in equation (A4.6.11), yield the total energy E in the Hartree theory. The parallel with the electron density description is clear; in that case the energy depends just on the electron density $\rho(\mathbf{r})$. Maintaining its normalization, through $\int \rho(\mathbf{r}) \, d\mathbf{r} = N$, one needs only one Lagrange multiplier, which in electron density theory is the important chemical potential μ. There the total energy is again related to the sum of the one-electron energies, but in the TF approximation, the correction for double counting the electron–electron interactions in $E_s = \sum_i \varepsilon_i$ turns out to be simply related to the total energy, leading, for neutral atoms (and homonuclear molecules) in that simplest density description to the result $E = (\frac{3}{2})E_s$. In fact, when refined, the chemical potential does enter "corrections" to this relation [March and Deb (1987)].

Electron Spin and Slater Determinants

The description of the electronic ground state, say, with which we are predominantly concerned here, is, of course, incomplete without the explicit introduction of spin, through the functions α and β already met in the discussion of the total wave function of a two-electron atom or molecule.

In writing the product Hartree wave function (A4.6.2), we have not introduced spin, and have therefore not been able to impose the generalized statement of the Pauli exclusion principle that the total wave function, including both space and spin coordinates, shall be antisymmetrical (i.e., shall change sign) on interchange of the space plus spin coordinates of any two electrons.

The mathematics of determinants proves ideal for handling this requirement, essentially from the property that when two rows or two columns of a determinant are interchanged, the determinant changes sign [see, for example, Aitken (1954)]. This property was exploited to great effect by Slater in constructing antisymmetric approximate many-electron wave functions, and such use of determinantal wave functions has led to their being termed Slater determinants. In particular, we can write an antisymmetric total wave function $\Phi(1, \ldots, N)$ in terms of spin orbitals* ϕ_i which form an orthonormal set; i.e., they satisfy

$$\int \phi_i^*(x)\phi_j(x) \, dx = \delta_{ij} \tag{A4.6.12}$$

the Kronecker delta δ_{ij} being zero if i is not equal to j and unity if $i = j$. The Slater determinant, properly normalized to unity, can then be shown to have the form

$$\Phi(1, \ldots, N) = \frac{1}{(N!)^{1/2}} \begin{vmatrix} \phi_1(1) & \phi_2(1) & \cdots & \phi_N(1) \\ \phi_1(2) & \phi_2(2) & \cdots & \phi_N(2) \\ \vdots & & & \vdots \\ \phi_1(N) & \phi_2(N) & \cdots & \phi_N(N) \end{vmatrix} \tag{A4.6.13}$$

*$\phi(x) = \psi(\mathbf{r})\alpha$ or $\psi(\mathbf{r})\beta$: x denoting both space $\mathbf{r}$ and spin coordinates.

This expression plainly changes sign when, say, coordinates 1 and 2 are interchanged, as this simply interchanges the first two rows of the Slater determinant (A4.6.13).

Use of Slater Determinants for Open- and Closed-Shell Systems

An open-shell electron configuration of an atom or molecule is representable by a sum of Slater determinants constructed such that the total wave function is an eigenfunction of total spin and orbital angular momentum.

In contrast, a closed-shell configuration can be represented by a single determinant; e.g., for a $1s^2$ closed-shell ground state one can write:*

$$\Phi(1, 2) = \frac{1}{\sqrt{2}} \begin{vmatrix} \psi_{1s}(1)\alpha(1) & \psi_{1s}(1)\beta(1) \\ \psi_{1s}(2)\alpha(2) & \psi_{1s}(2)\beta(2) \end{vmatrix} \qquad \text{(A4.6.14)}$$

For a $1s2s$ (open-shell) case one can write, however,

$$\Phi(1, 2) = \frac{1}{2} \left\{ \begin{vmatrix} \psi_{1s}(1)\alpha(1) & \psi_{2s}(1)\beta(1) \\ \psi_{1s}(2)\alpha(2) & \psi_{2s}(2)\beta(2) \end{vmatrix} \pm \begin{vmatrix} \psi_{1s}(1)\beta(1) & \psi_{2s}(1)\alpha(1) \\ \psi_{1s}(2)\beta(2) & \psi_{2s}(2)\alpha(2) \end{vmatrix} \right\} \qquad \text{(A4.6.15)}$$

The plus sign is for $S = 1$, $M_s = 0$ (i.e., a $1s2s$ $^3\Sigma$ triplet term). The minus sign is for $S = 0$, $M_s = 0$ (i.e., a $1s2s$ $^1\Sigma$ singlet state).

We obtain a generalization of equation (A4.6.6) for the determinantal function (A4.6.13) as [see Blinder (1965)]

$$\left. \begin{aligned} E &= \sum_{i=1}^{N} H_i + \sum_{j>i}^{N} \sum_{j=1}^{N} (J_{ij} - K_{ij}) \\[2mm] H_i &= \int dx\ \phi_i^*(x)(-\tfrac{1}{2}\nabla^2 - Z/r)\phi_i(x) \\[2mm] J_{ij} &= \int dx\,dx'\ \phi_i^*(x)\phi_j^*(x')|\mathbf{r} - \mathbf{r}'|^{-1}\phi_j(x')\phi_i(x) \\[2mm] K_{ij} &= \int dx\,dx'\ \phi_i^*(x)\phi_j^*(x')|\mathbf{r} - \mathbf{r}'|^{-1}\phi_i(x')\phi_j(x) \end{aligned} \right\} \qquad \text{(A4.6.16)}$$

The definitions of H_i and J_{ij} (one-electron and Coulomb integrals, respectively) are equivalent to those in equation (A4.6.6) since the spin parts contribute a factor of unity. The "exchange integral" K_{ij} differs by interchange of the last index i and j. K_{ij} vanishes due to spin orthogonality unless $\phi_i(x)$ and $\phi_j(x)$ have the same spin. The exchange integral is therefore a measure of the energy difference between singlet and triplet states, a feature missing from Hartree's original work.

Inserting the Slater determinant (A4.6.13) into the variational energy (A4.6.3) leads to a new minimization problem for the "best possible" N spin orbitals $\phi_i(x)$,

*The approximate wave function (A4.6.14) is an eigenfunction of the total spin with eigenvalues $S = 0$, $M_S = 0$.

and this leads to a new set of equations which generalize the Hartree equations (A4.6.1) to include spin. These are the Hartree–Fock equations, which again take the form of one-electron Schrödinger equations, but now with an "unconventional" form for the potential energy in these wave equations. Explicitly, these equations can be written, with one-electron energies again introduced, in the variational approach, to take account of the subsidiary conditions summarized in equation (A4.6.12), i.e., as Lagrange multipliers:

$$\mathcal{H}_i^{\text{eff}} \phi_i(x) = \varepsilon_i \phi_i(x), \qquad i = 1, \ldots, N$$

where

$$\mathcal{H}_i^{\text{eff}} = -\tfrac{1}{2}\nabla^2 - Z/r + \sum_{j \neq i} \int dx' \, \phi_j^*(x')|\mathbf{r} - \mathbf{r}'|^{-1}(1 - P_{ij})\phi_j(x') : \tag{A4.6.17}$$

$$P_{ij}\phi_j(x')\phi_i(x) = \phi_i(x')\phi_j(x)$$

The HF equations differ from the Hartree equations by the addition of the terms

$$-\sum_{j \neq i} \left[\int dx' \, \phi_j^*(x')|\mathbf{r} - \mathbf{r}'|^{-1} P_{ij}\phi_j(x') \right] \phi_i(x)$$

$$\equiv -\sum_{j \neq i} \left[\int dx' \, \phi_j^*(x')|\mathbf{r} - \mathbf{r}'|^{-1}\phi_i(x') \right] \phi_j(x) \tag{A4.6.18}$$

The summation in equation (A4.6.18) is interpreted physically in terms of exchange repulsion between electrons of parallel spin. Since the exchange integrals K_{ij} are positive, the total energy is lowered by the presence of these exchange forces where $J_{ij} - K_{ij}$ replaces the Coulomb potential energy J_{ij} for electron pairs with parallel spin.

Considering a many-electron wave function $\Phi(\ldots, x_i x_j, \ldots)$ near a configuration point $x_i = x_j = x'$, the antisymmetric property gives $\Phi(\ldots, x', x', \ldots) = 0$: thus there is zero probability of finding two electrons of parallel spin at the same point in space, in accordance with an intuitive definition of Pauli's exclusion principle (see Appendix A4.7). This is not true for electrons of opposite spin. Moreover, exchange forces, because they tend to keep electrons of the same spin apart, reduce their repulsive energies (i.e., they lower the energy of the system). The region around each electron forbidden to electrons of the same spin is termed the "Fermi hole" (see Figure A.10 for a plot in a uniform electron gas).

Also, if equation (A4.6.17) is rewritten to include $j = i$, equation (A4.6.18) is augmented by that corresponding term which may be interpreted as an electron's exchange interaction with itself, cancelling its Coulombic self-energy, which classically is finite.

The eigenvalues ε_i of the HF equations may be related to the integrals in equation (A4.6.16) evaluated using the spin orbitals. Multiplying equation

(A4.6.17) by $\phi_i^*(x)$ and performing the integration over x we obtain

$$\varepsilon_i = H_i + \sum_{j=1}^{N} (J_{ij} - K_{ij}), \qquad i = 1, \ldots, N \qquad \textbf{(A4.6.19)}$$

This is a generalization of equation (A4.6.10) for the Hartree equations. For closed-shell systems, $-\varepsilon_i$ is equal to the ionization potential for the ith electron minus the total energy of the atom plus the total energy of the ion formed by removing i (see Koopmans' theorem, Appendix A6.2 for more detail).

Summing over the one-electron energies and comparing with the total energy (A4.6.16), one obtains

$$\sum_{i=1}^{N} \varepsilon_i = \sum_{i=1}^{N} H_i + \sum_{j \neq i}^{N} \sum_{i=1}^{N} (J_{ij} - K_{ij})$$

$$= \sum_{i=1}^{N} H_i + 2 \sum_{j>i}^{N} \sum_{i=1}^{N} (J_{ij} - K_{ij})$$

$$= E + \sum_{j>i}^{N} \sum_{i=1} (J_{ij} - K_{ij}) \qquad \textbf{(A4.6.20)}$$

Thus the sum of N ionization energies is greater than the total energy due to the double counting of the electronic repulsive terms $J_{ij} - K_{ij}$. Of course, the HF method provides only approximate solutions to a many-body problem. Physically, the inherent approximation is equivalent to replacing particle-like r_{ij}^{-1} Coulombic interactions with continuous electrostatic interactions among charge clouds, which is a "correlation" error as it implies the need for a more detailed correlation of electron positions than is possible for a charge-cloud description [see Löwdin (1959), Bartlett and Purvis (1978), and Wilson (1984).

The wave functions for atomic systems are separable in polar coordinates (see Appendix A1.1):

$$\phi_{\mathrm{nlm}m_s}(x) = R_{\mathrm{nl}}(r) Y_{\mathrm{lm}}(\theta, \phi) \begin{cases} \alpha \\ \text{or} \\ \beta \end{cases} \qquad \textbf{(A4.6.21)}$$

where $Y_{\mathrm{lm}}(\theta, \phi)$ are spherical harmonics and R is the radial function. In a numerically tractable approach to molecules, the spin orbitals may be represented as linear combinations in a set of basis functions (Roothaan, 1951), i.e.,

$$\phi_i(x) = \sum_{p=1}^{N'} c_{ip} \chi_p(x) \qquad \textbf{(A4.6.22)}$$

The equations for $\phi_i(x)$ are transformed into linear algebraic equations for the coefficients c_{ip}. This form for molecules closely resembles (and is in fact an extension of) LCAO–MO methods. Applied to molecules, Z/r in the Hamiltonian is replaced by the sum $\sum_n (Z_n/r)$. According to equation (A4.6.22), (A4.6.16) becomes:

$$H_i = \sum_{p,q} c_{ip}^* c_{iq}[p|q]$$

$$J_{ij} = \sum_{p,q,r,s} c_{ip}^* c_{iq} c_{jr}^* c_{js}[pq|rs] \qquad \text{(A4.6.23)}$$

$$K_{ij} = \sum_{p,q,r,s} c_{ip}^* c_{iq} c_{js}^* c_{jr}[ps|rq]$$

having defined the one-electron integrals

$$[p|q] \equiv \int dx\ \chi_p^*(x)\left\{-\tfrac{1}{2}\nabla^2 - \sum_n \frac{Z_n}{r_n}\right\}\chi_q(x)$$

and the two-electron integrals

$$[pq|rs] \equiv \int dx\ dx'\ \chi_p^*(x)\chi_q(x)|\mathbf{r} - \mathbf{r}'|^{-1}\chi_r^*(x')\chi_s(x')$$

Substituting equation (A4.6.23) in equation (A4.6.16) we have

$$E(c_{ip},\,c_{ip}^*) = \sum_i \sum_{p,q} c_{ip}^* \left\{ [p|q] \right.$$

$$\left. + \tfrac{1}{2}\sum_j \sum_{rs} c_{jr}^* c_{js}([pq|rs] - [ps|rq]) \right\} c_{iq} \qquad \text{(A4.6.24)}$$

The energy is then minimized with respect to the coefficients c_{ip}, subject to the orthogonalization conditions (imposed since the basis set itself is not orthogonal):

$$\sum_{p,q} c_{ip}^* c_{jq} S_{pq} = \delta_{ij}$$

where $\qquad$ (A4.6.25)

$$S_{pq} \equiv \int \chi_p^*(x)\chi_q(x)\ dx$$

The quantity F is introduced with N^2 Lagrange multipliers λ_{ij}:

$$F \equiv E(c_{ip},\,c_{ip}^*) - \sum_{ij} \lambda_{ij} \sum_{p,q} c_{ip}^* c_{jq} S_{pq}$$

$$= \sum_j \sum_{p,q} c_{ip}^* \left\{ [p|q] + \tfrac{1}{2}\sum_j \sum_{r,s} c_{jr}^* c_{js}([pq|rs] - [ps|rq]) - \varepsilon_i S_{pq} \right\} c_{iq} \quad \text{(A4.6.26)}$$

Evidently the conditions for minimization are:

$$\frac{\partial F}{\partial c_{ip}} = 0, \qquad \frac{\partial F}{\partial c_{ip}^*} = 0, \qquad \begin{array}{l} i = 1, \ldots, N \\ p = 1, \ldots, N' \end{array}$$

and this leads for closed-shell systems ($\lambda_{ij} \rightarrow \varepsilon_i$) to:

$$\sum (H_{pq} - \varepsilon_i S_{pq}) c_{iq} = 0, \qquad p = 1, \ldots, N'$$

where

$$H_{pq} \equiv [p|q] + \sum_j \sum_{rs} c_{jr}^* c_{js}([pq|rs] - [ps|rq]) \qquad \text{(A4.6.27)}$$

The linear equations above (A4.6.27) are the Roothaan–Hall equations and are algebraic approximations to the HF equations.

For nontrivial solutions of equation (A4.6.27), the $N' \times N'$ secular determinant must vanish, i.e.,

$$\det(H_{pq} - \varepsilon S_{pq}) = 0 \qquad \text{(A4.6.28)}$$

providing N' eigenvalues (ε_i) and N' sets of coefficients c_{ip}. This differs from the LCAO–MO secular determinant in that H_{pq} depends on c_{ij}. Values for ε_i are obtained, then H_{pq} is recalculated, etc., until the desired degree of self-consistency is obtained. The lowest eigenvalues are then occupied and pertain to the ground state. The unoccupied orbitals ($i = N + 1, \ldots, N'$ for $N' > N$) are virtual solutions and can be used to construct excited states.

The basis functions $\chi_p(x)$ are often taken to have the form suggested by Slater:

$$\chi_{nlmm_s}(x) = r^{n-1} \exp(-\zeta_{nl} r) Y_{lm}(\theta, \phi) \begin{cases} \alpha \\ \text{or} \\ \beta \end{cases} \qquad \text{(A4.6.29)}$$

The space part is a Slater-type orbital having the appearance of an atomic orbital but without radial nodes (see Appendix A2.4).

HF calculations provide quantitative results for numerous ground-state properties of atomic systems, e.g., a comparison of the radial charge density for Ar calculated by the HF method with experimental electron diffraction data is made in Figure A.9. The agreement is seen to be quite good.

The HF method gives qualitatively good results for excited-state calculations but poorer results for bond energies as they are obtained as a difference between relatively large quantities.

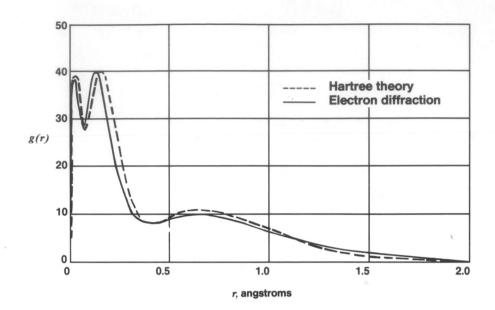

FIGURE A.9. A comparison of the radial charge density for Ar calculated by the Hartree–Fock method with that obtained by experimental electron diffraction data [Bartell and Brockway (1953)].

A4.7. THE DIRAC–SLATER EXCHANGE ENERGY AND EXISTENCE OF A ONE-BODY POTENTIAL INCLUDING BOTH EXCHANGE AND CORRELATION

In this appendix, we shall derive an expression for the exchange energy of a multielectron atom or molecule in terms of its ground state density $\rho(\mathbf{r})$, analogous to equation (4.15) for the kinetic energy. The derivation was first given by Dirac, and the essence of the result was rediscovered and pressed with substantial success by Slater (1951), and later by Slater and his co-workers.

The basic starting point of the derivation of the exchange energy is the single determinantal wave function of occupied orbitals, which may be atomic or molecular, depending on the system considered. The electron density, as written explicitly in equation (4.1), is obtained by integrating out all the electron coordinates but one from the probability density $\Psi^*\Psi$, the total wave function being Ψ.

One can integrate $\Psi^*\Psi$ over all coordinates but, say, $\mathbf{r}_1$ and $\mathbf{r}_2$ of electrons 1 and 2. The result is a pair correlation function, $\Gamma(\mathbf{r}_1\mathbf{r}_2)$, say, giving the probability density of electron 1 being at $\mathbf{r}_1$ and electron 2 at $\mathbf{r}_2$. Quite generally, this pair function can be obtained again for the single Slater determinantal wave function (the antisymmetrized product of occupied one-electron functions, the Pauli exclusion principle thereby being accounted for).

The result when the energy levels are all doubly occupied by electrons of opposed spins is in terms of the "generalized density" given by

$$\sum_{\substack{\text{occupied} \\ \text{states}}} \psi_i^*(\mathbf{r}_1)\psi_i(\mathbf{r}_2) \tag{A4.7.1}$$

which reduces to the electron density $\rho(r)$ when we put $\mathbf{r}_1 = \mathbf{r}_2 = r$, as can be seen by comparing equation (A4.7.1) with equation (4.1), the ψ's in equation (A4.7.1) being the one-electron orbitals in the Slater determinant representing the total wave function.

The result for Γ can then be shown to be

$$\Gamma(\mathbf{r}_1\mathbf{r}_2) = \rho(\mathbf{r}_1)\rho(\mathbf{r}_2) - \frac{1}{2}\left[\sum_{\substack{\text{occupied} \\ \text{states}}} \psi_i^*(\mathbf{r}_1)\psi_i(\mathbf{r}_2)\right]^2 \qquad \text{(A4.7.2)}$$

This result, true for any single Slater determinant with doubly filled orbitals, will be applied below to an electron gas which is uniform. This means that the probability density $\psi^*(\mathbf{r})\psi(\mathbf{r})$ of each orbital is constant, since the density is uniform. The one-electron orbitals are $\exp(i\mathbf{k} \cdot \mathbf{r})$, where $\hbar k$ is readily verified to be the momentum. Thus, in the language used in the phase-space derivation of the electron density theory in Chapter 4, the states will be filled in momentum space out to a maximum momentum $p_m = \hbar k_m$. The generalized density given by (A4.7.1) can be calculated with the ψ's normalized to unity in a volume Ω. Then using the equivalence between energy levels and cells in phase space of volume h^3, the summation over occupied levels in equation (A4.7.1) can be replaced by an integration over phase space, leading to the volume of the system times an integral over momenta from 0 to p_m. The answer, not surprising because of $\exp(i\mathbf{k} \cdot \mathbf{r})$ being basically trigonometric, can be expressed in terms of sines and cosines (and powers of x). If, to be precise, we introduce the quantity $j_1(x)$, known as a first-order spherical Bessel function, and defined by

$$j_1(x) = (\sin x - x \cos x)/x^2 \qquad \text{(A4.7.3)}$$

then for a uniform electron assembly, where $\rho(\mathbf{r}_1 - \mathbf{r}_2)$ can only depend on the distance $|\mathbf{r}_1 - \mathbf{r}_2|$ between electrons 1 and 2, which we shall write as r, the pair function Γ can be expressed in the form of the squared (constant) electron density multiplied by $g(r)$, where

$$g(r) = 1 - \frac{9}{2}\left[\frac{j_1(k_m r)}{k_m r}\right]^2 \qquad \text{(A4.7.4)}$$

This expression, which tends to 1 as the separation between electrons becomes very large, was first derived by Wigner and Seitz (1934) and is plotted in Figure A.10.

It has the important feature that it represents a deficit of charge of precisely one electron, and the hole $g(r) - 1$ in Figure A.10 therefore contains precisely one electron. There is here an electron at the origin $r = 0$, and the plot represents the probability of finding another electron at distance r from the first one. Thus, in the uniform electron gas, as an electron moves around in the gas, it carries with it this spherical hole containing a deficit of one electronic charge. Because this hole is due solely to exchange, there being no Coulomb correlations in this treatment, it is referred to as the exchange (or Fermi) hole around an electron. The existence of such a hole around an electron is due to the Pauli principle—one cannot put

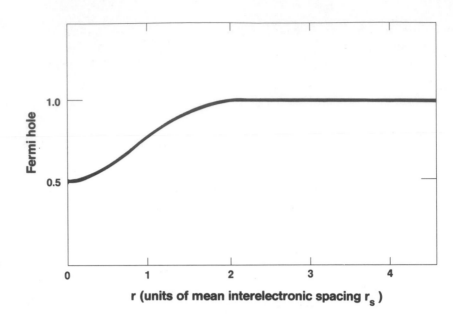

FIGURE A.10. Fermi (or exchange) hole in uniform electron gas. Oscillations exist at large r but the amplitude is too small to be shown graphically.

two electrons with parallel spin at the same point. The reason $g(r)$ is $\frac{1}{2}$ at $r = 0$ and not 0 is that we have been treating half the spins as upward and half as downward. If the electron sitting at the origin has upward spin, say, in the model of a single Slater determinant, it correlates only with the upward spins and not at all with downward-spin electrons in the surrounding gas. Hence, the reason that $g(r) = \frac{1}{2}$ at $r = 0$ should be clear. This model can be used to calculate the exchange energy per electron in a uniform electron gas of density ρ_0 directly.

The principle involved is that there is electrostatic interaction between the electron carrying charge $-e$ at $r = 0$ and the hole containing a total deficit of one electronic charge surrounding that electron. All that is necessary therefore is to calculate the electrostatic potential at the origin due to an electron density $e[g(r) - 1]\rho_0$. This potential then has to be multiplied by the charge of the electron at the origin to convert it to potential (exchange) energy; one must also divide by two to prevent counting electron–electron interactions twice over. This then yields straightforwardly for the potential energy per electron the result

$$\text{Potential energy per electron} = \frac{1}{2}e^2 \int \rho_0 \frac{(g(r) - 1)}{r} \, d\mathbf{r} \qquad \text{(A4.7.5)}$$

To calculate the exchange energy, one merely has to insert the Wigner–Seitz result [equation (A4.7.4)] for $g(r)$ into equation (A4.7.5). The integration is lengthy but can be carried out analytically with the result

$$\text{Exchange energy per electron} = -\text{constant} \times e^2 \rho_0^{1/3}$$

Converting this to exchange energy per unit volume, we find the constant c_e introduced in the main text to be [cf. equation (P4.14)]

$$c_e = \tfrac{3}{4}e^2(3/\pi)^{1/3} \qquad (A4.7.6)$$

Existence of a One-Body Potential Including Exchange

Let us return now to equation (4.6) and note that it can be obtained in the equivalent form

$$\mu = \tfrac{5}{3}c_k\rho^{2/3}(\mathbf{r}) + V_N + V_e \qquad (A4.7.7)$$

where the first term comes from minimizing the kinetic energy c_k with respect to the electron-density variation $\delta\rho(\mathbf{r})$, say, the remaining two terms coming from minimizing the potential energy. To this energy expression, we can now add exchange, and minimizing this with respect to $\rho(\mathbf{r})$ yields [see equation (4.42)]

$$\mu = \tfrac{5}{3}c_k\rho^{2/3} + V_N + V_e - \tfrac{4}{3}c_e\rho^{1/3} \qquad (A4.7.8)$$

This equation, the basis of the so-called Thomas–Fermi–Dirac theory, was interpreted by Slater as equivalent to a total potential energy

$$V(\mathbf{r}) = V_N + V_e + V_X \qquad (A4.7.9)$$

where the exchange potential energy $V_x(\mathbf{r})$ is given by

$$V_x(\mathbf{r}) = -\tfrac{4}{3}c_e[\rho(\mathbf{r})]^{1/3} \qquad (A4.7.10)$$

This exchange potential, which is proportional to the electron density to the one-third power, is called the Dirac Slater exchange potential. It has had considerable success in quantum chemistry and solid state physics [see Callaway and March (1984) for a review and also the book by Dahl and Avery (1984)].

Inclusion of Electron Correlation into the Potential

Though it is a more advanced topic, a correction V_c can be added to equation (A4.7.9) to include electron correlation c. This can be calculated in the same spirit as the $\rho^{4/3}$ exchange energy density for a uniform electron gas. The most accurate results are in the computer simulation of Ceperley and Alder (1980), fitted analytically by Vosko *et al.* (1980) and by Perdew and Zunger (1981). These afford an approximate basis for the calculation of the one-body potential energy $V(\mathbf{r})$ including exchange and correlation

$$V(\mathbf{r}) = V_N + V_e + V_x + V_c \qquad (A4.7.11)$$

Kohn and Sham (1965) have shown, in principle, that such a one-body potential can formally be defined exactly, but at the time of writing its calculation has been based, in a first approximation, on the homogeneous electron assembly.

Nevertheless, in view of the importance for chemistry of one-electron orbital energies, it is of interest that a one-body potential energy $V(\mathbf{r})$ of the form of equation (A4.7.11) including exchange and correlation interactions between electrons does exist.

The final point of this appendix takes us back to the semiclassical phase-space arguments. These can be used to calculate the mean momentum $\bar{p}$ as a function of the atomic number for atoms, as was done independently by Konya (1951) and by Coulson and March (1950). Their result is

$$\bar{p} = \text{constant} \times Z^{2/3} \tag{A4.7.12}$$

and if we calculate the exchange energy with the Thomas–Fermi self-consistent density we find the result [Scott (1952)] that (see also Appendix A4.5)

$$\text{Exchange energy} = \text{constant} \times Z^{5/3} \tag{A4.7.13}$$

for neutral atoms. These results show [see, e.g., Gadre and Pathak (1982)]

$$Z\bar{p} \propto \text{exchange energy} \tag{A4.7.14}$$

The relation (A4.7.14) has been generalized to molecules by Allan and March (1983) and numerical comparison with Hartree–Fock–Roothaan calculations is given by Allan *et al.* (1985). Again the use of high-quality ground-state densities in the inhomogeneous electron gas formula leads to good agreement with wave function calculations derived from the self-consistent field method set out in Appendix A4.6.

A4.8. PROOF THAT THE GROUND-STATE ENERGY OF A MOLECULE IS UNIQUELY DETERMINED BY THE ELECTRON DENSITY

It is of course obvious that the ground-state energy of a molecule is uniquely determined if the many-electron wave function is known for the ground state. However, the electron density description of the ground state in Chapter 4 sets up an admittedly approximate ground-state energy expression in terms of the electron density. For example, the total kinetic energy is approximated by $c_k \int [\rho(\mathbf{r})]^{5/3} \, d\tau$, where ρ is the ground-state electron density. This is a basic approximation of the Thomas–Fermi statistical theory developed in some detail in the main text. That theory, based as it is on statistical arguments, is only rigorously valid (in a nonrelativistic framework as discussed in Chapter 4, although the relativistic generalization is effected in Appendix A5.8), as the number of electrons becomes really large. Doubts therefore remained about the basic foundations of the electron density description for numbers of electrons met in atoms and in many molecules of chemical interest. These doubts were dispelled by the so-called Hohenberg–Kohn theorem (1964), which states that the ground-state energy of an atom, molecule, or solid is uniquely determined by the electron density. Of course, conceptually this is a great simplification because regardless of the number of electrons N the electron density $\rho(\mathbf{r})$ is a function of three variables only, whereas the N-electron wave function already depends on $3N$ spatial variables. However, to date, it has to be said that no formally exact expression for the ground-state energy E in terms of the electron density $\rho(\mathbf{r})$ is known, though there are corrections to the Thomas–Fermi energy to allow for density gradients (Lundqvist and March, 1983).

Electron density theory is now of substantial importance in quantum chemistry and solid state physics [Lundqvist and March (1983), Dahl and Avery (1984), and Callaway and March (1984)], so we shall now give the proof of the Hohenberg–Kohn theorem.

We are concerned with the ground state of an N-electron system, which is assumed to be nondegenerate. The N electrons move under the influence of an external potential energy $V_{\text{ext}}(r)$ (e.g., this might be simply $-Ze^2/r$ for an atom of atomic number Z) and their mutual interelectronic Coulomb repulsion. Let us denote the many-electron ground-state wave function by Ψ.

An essential first step in proving that $V_{\text{ext}}(\mathbf{r})$ is uniquely fixed by $\rho(\mathbf{r})$ is to show that for electrons interacting Coulombically, two different external potential energies cannot lead to the same electron density $\rho(\mathbf{r})$.

Thus, to approach this question, assume that another external potential energy, $V'_{\text{ext}}(\mathbf{r})$, with corresponding ground-state wave function Ψ', leads to the same density $\rho(\mathbf{r})$ as $V_{\text{ext}}(\mathbf{r})$.

Now, clearly, unless we are dealing with the trivial case when V and V' differ only by an additive constant, $\Psi' \neq \Psi$, because they satisfy different Schrödinger equations. Hence, if we denote Hamiltonians and ground-state energies associated with Ψ and Ψ' by H, H' and E, E', respectively, we have by the minimum property of the ground-state energy established in Appendix A2.1:

$$E' = \int \Psi^{*\prime} H' \Psi' \, d\tau < \int \Psi^* H' \Psi \, d\tau$$

$$= \int \Psi^* (H + V'_{\text{ext}} - V_{\text{ext}}) \Psi \, d\tau \qquad \text{(A4.8.1)}$$

and therefore (dropping the subscript 'ext' for notational convenience)

$$E' < E + \int (V' - V)\rho(\mathbf{r}) \, d\tau \qquad \text{(A4.8.2)}$$

Using the same argument for energy E is equivalent to interchanging primed and unprimed quantities, and we have then

$$E < E' + \int (V - V')\rho(\mathbf{r}) \, d\tau \qquad \text{(A4.8.3)}$$

which involves the assumption that the same electron density $\rho(\mathbf{r})$ is generated by both V and V'.

Adding equations (A4.8.2) and (A4.8.3) yields

$$E' + E < E + E'$$

which is plainly incorrect! Thus the initial assumption must be erroneous, and $V_{ext}(\mathbf{r})$ is (to within a constant) uniquely fixed by the electron density $\rho(\mathbf{r})$. Since, in turn, $V_{ext}(\mathbf{r})$ completely determines the Hamiltonian for the given Coulombic electron–electron interactions, we see that the many-electron ground-state energy is uniquely determined by the electron density $\rho(\mathbf{r})$. This is therefore the theorem verifying the validity of the density theory initiated by Thomas (1926) and Fermi (1928).

A4.9. MODELING OF THE CHEMICAL POTENTIAL IN HYDROGEN HALIDES AND MIXED HALIDES

A number of researchers have attempted to model the chemical potential of molecules in which charge transfer occurs [see, e.g., the review by March (1981)]. In this appendix we shall follow the approach of Alonso and March (1983), in which a simple model is developed based on the principle of constant chemical potential, to allow the calculation of the charge transfer in hydrogen halides and mixed halide molecules.

Description of Charge Transfer in Molecules

As discussed in Chapter 4, we can write the chemical potential $\mu(Z, N)$ for an atomic ion of atomic number Z and with N electrons, at least approximately, in the "thermodynamic form" [see equation (4.43)]

$$\mu(Z, N) = \left. \frac{\partial E(Z, N)}{\partial N} \right|_Z \tag{A4.9.1}$$

While this equation does raise some basic questions for quantum mechanics as to how the energy E varies with respect to the number of electrons N, we shall nevertheless use it to treat charge transfer in some halide molecules. If we start with two atoms A and B, with chemical potentials $\mu_A = \mu(Z_A, Z_A)$ and $\mu_B = \mu(Z_B, Z_B)$, then the AB molecule can be regarded as composed, in general, of fractionally charged atomic ions. The fractional charges evidently arise from the process of electron redistribution to bring the chemical potential μ_{AB} of the molecule to its constant equilibrium value. Because of redistribution of charge, applying equation (4.6) to an atomic ion with nuclear charge Z and N^* electrons yields

$$\mu(Z_A, N_A^*) = \mu(Z_B, N_B^*) \tag{A4.9.2}$$

where N_A^* and N_B^* are evidently the (nonintegral) number of electrons of atomic ions A and B comprising the molecule. It is clear from total charge neutrality that the fractional numbers N_A^* and N_B^* must satisfy

$$N_A^* - Z_A = -(N_B^* - Z_B) \tag{A4.9.3}$$

Equations (A4.9.2) and (A4.9.3) will now be exploited to calculate the amount of electronic charge transferred between the two atoms. This charge transfer will then be compared with that derived from the measured electric dipole moments of some halides.

The procedure adopted below is to assume that $\mu(Z, N^*)$ can be usefully Taylor-expanded about the neutral atom limit $N^* = Z$, as

$$\mu(Z, N^*) = \mu(Z, Z) + (N^* - Z)\mu' + \tfrac{1}{2}(N^* - Z)^2\mu'' + \cdots \tag{A4.9.4}$$

where $\mu' = (\partial\mu(Z, N)/\partial N)_{N=Z}$, etc. Neglecting quadratic and higher-order terms, and using equation (A4.9.4) in equation (A4.9.2) one finds

$$\mu(Z_A, Z_A) + \mu_A'Q_A = \mu(Z_B, Z_B) + \mu_B'Q_B \tag{A4.9.5}$$

where $Q_i = N_i^* - Z_i$ (i = A, B). Since $Q_A = -Q_B = Q$, say, then equation (A4.9.5) gives the following result for the charge transfer

$$Q = \frac{\mu(Z_B, Z_B) - \mu(Z_A, Z_A)}{\mu_B' + \mu_A'} \tag{A4.9.6}$$

To be quite specific in calculating the quantities appearing in equation (A4.9.6), a finite difference approximation to μ in equation (A4.9.1) is adopted:

$$-\bar{\mu}(Z, Z) = \frac{E(Z, Z-1) - E(Z, Z+1)}{2} = \frac{I + J}{2} \tag{A4.9.7}$$

where, to avoid notational confusion, we have written J for electron affinity. Evidently, there is immediate contact through equation (A4.9.7) with the definition of electronegativity given by Mulliken (1934, 1935). The differences between $\bar{\mu}$ and μ have been found to be small in the calculations of Balbás *et al.* (1982). Then it follows [Ray *et al.* (1979), Lackner and Zweig (1981)] that

$$\bar{\mu}' = (I - J) \tag{A4.9.8}$$

and if one uses equations (A4.9.7) and (A4.9.8) in equation (A4.9.6) one obtains the result

$$Q = -\frac{1}{2}\left(\frac{I_B + J_B - I_A - J_A}{I_B - J_B + I_A - J_A}\right) \tag{A4.9.9}$$

Below, we shall follow Alonso and March (1983) in relating this result with experiment for some halide molecules.

Comparison with Electric Dipole Moments in Some Halide Molecules

To compare with equation (A4.9.9), Alonso and March computed an experimental value of Q, say, Q_E, defined through the equation

$$D = Q_E R \qquad \text{(A4.9.10)}$$

where R is the experimentally determined equilibrium separation in the molecule. The values of Q_E for the molecules FH, ClH, BrH, IH, BrCl, BrI, ICl, ClF, and BrF obtained from measured dipole moments [see *The Handbook of Chemistry and Physics, 1971–72* and Netercot (1978)] and internuclear separations [Mitchell and Cross (1958)] using equation (A4.9.10) are plotted against Q found by using experimental values of I and J in equation (A4.9.9).

The molecules studied here can be separated into two groups: mixed halides and hydrogen halides. It can be seen from Figure A.11 that a satisfactory linear correlation between the theoretical and the empirical charge transfer exists within each group. The slope of the line corresponding to the mixed halides is close to unity, testifying to the validity of the above simple theory of charge transfer within this group.

Obviously, however, a more complete verification of the model proposed would require that both groups of molecules lie on a single line in Figure A.11. The problem here is not with the theory developed for the charge transfer Q, but rather with the approximation adopted in equation (A4.9.8) for μ'. To support

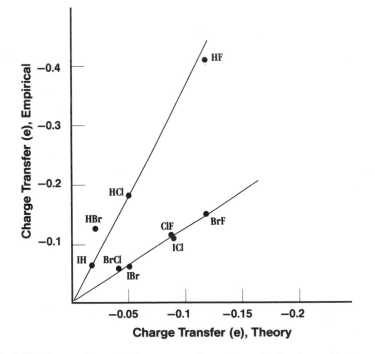

FIGURE A.11. Comparison of charge transfer values obtained empirically *vs.* those obtained by theory.

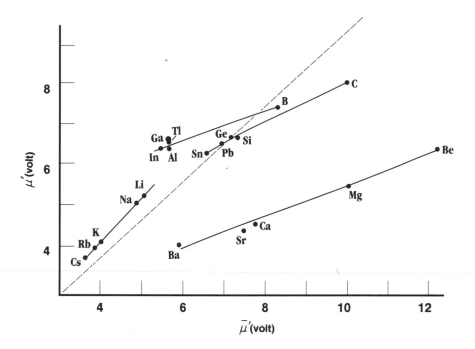

FIGURE A.12. Deviation of the chemical potential calculated by density functional theory *vs.* a finite difference approximation to the same quantity.

this view, we have plotted in Figure A.12 μ' *vs.* $\bar{\mu}'$ for elements of the Li, Be, B, and C groups, where $\bar{\mu}'$ has been computed from equation (A4.9.8) using experimental values of I and J, while μ' has been taken from the results of Balbás and coworkers (1983). These authors have calculated the electronegativity of the neutral atom, and of several of its ions by using the electron density method discussed in the main text. This was implemented via a pseudopotential approach. The results were then fitted to a polynomial and μ' was calculated. Figure A.12 shows that $\bar{\mu} > \mu'$ in cases of large $\bar{\mu}'$. For hydrogen, $\bar{\mu}' = 12.8$ and the trend in Figure A.12 suggests that this value may be an overestimate of the true μ'. A smaller value of $\mu'(H)$ would lower the value of the denominator in equation (A4.9.9) and would then increase the absolute value of Q, thus lowering the slope of the line corresponding to the hydrogen halides in Figure A.11. This explanation must be regarded as somewhat tentative, however, since it is based on Figure A.12, where results for hydrogen and for the halide elements are lacking.

Using equation (A4.9.6) in equation (A4.9.5), one obtains for the chemical potential of the molecule

$$\mu_{AB} = \mu(Z_A, Z_A) + \mu'_A Q$$

$$= \mu(Z_A, Z_A) + \frac{1}{1 + \mu'_B/\mu'_A}[\mu(Z_B, Z_B) - \mu(Z_A, Z_A)] \qquad \text{(A4.9.11)}$$

We have plotted $\bar{\mu}_{AB}$, that is, the value of μ_{AB} obtained using $\bar{\mu}(Z_B, Z_B)$,

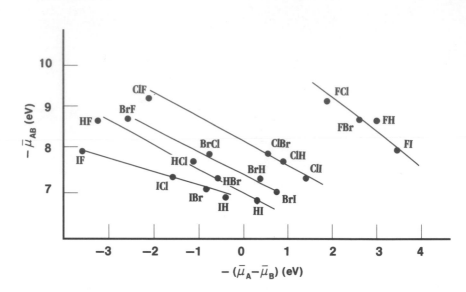

FIGURE A.13. Chemical potential of the molecule AB *vs.* the difference of chemical potential of A and B atoms.

$\bar{\mu}(Z_A, Z_A)$, $\bar{\mu}'_A$, and $\bar{\mu}'_B$ in equation (A4.9.11) *vs.* $[\bar{\mu}(Z_B, Z_B) - \bar{\mu}(Z_A, Z_A)]$ in Figure A.13. The set of linear relations in this figure (each linear relation corresponds to a group of AB molecules, with A fixed and B variable), is reminiscent of a similar set of linear relations between I_{AB} and $(I_A - I_B)$ emphasized by Mucci and March (1981). Here I_A, I_B, and I_{AB} are the experimental first ionization potentials of the A and B atoms and the AB molecule, respectively.

The results in Figure A.13 serve as a theoretical explanation of the empirical correlations developed by Mucci and March (1981) if $I \propto \mu(Z, Z)$. This is approximately true. In Mulliken's approximation, $-\mu(Z, Z) = (I + J)/2$, and since J is generally small compared to I then $-\mu \sim I/2$.

A4.10. X-RAY SCATTERING BY NEON-LIKE MOLECULES

As a further example of the principle of equal chemical potential on all the atomic fragments in the molecule, used in Appendix A4.9 for hydrogen halides and mixed halides, we consider here a very different set of molecules, all with a ten-electron "neon-like" configuration, namely, H_2O, NH_3, and CH_4, following the study of Alonso *et al.* (1983). Comparison with the model can be made by confronting its predictions with measured X-ray scattering data for this series.

The differences between these experimental X-ray scattering functions of the ten-electron molecules Ne, H_2O, NH_3, CH_4 [Thomer (1937)] reflect systematic changes in the spatial distribution of the electronic cloud. Banyard and March (1956) provided a theoretical interpretation based on approximate molecular wave functions obtained by the Hartree–Fock method (Appendix A4.6). The approximation consisted of using a spherically symmetric molecular electron density, obtained by solving the Hartree–Fock equations after averaging the nuclear field over angles about the central nucleus.

In this appendix, we follow Alonso *et al.* (1983) by calculating the scattering factor using a model which consists of building the molecule from atomic fragments, each fragment having a total electronic charge slightly different from that of the neutral atom [Hinze (1970), Parr *et al.* (1978), Ray *et al.* (1979)]. These charges are calculated using the condition of equal chemical potential of the fragments in the molecule [Sanderson (1971)]. Since the electronic charges of the fragments are very similar to the charges in the corresponding neutral atoms, the electron density of a fragment is constructed by simply renormalizing the free-atom electron density (in fact, only the density corresponding to the external electrons is renormalized). The molecular electron density $\rho(\mathbf{r})$ is then formed as

$$\rho(\mathbf{r}) = \sum_i \rho_i(\mathbf{r} - \mathbf{R}_i) \tag{A4.10.1}$$

where $\mathbf{R}_i$ are the nuclear positions and ρ_i are the fragment electron densities. Finally, the scattering function is computed from $\rho^s(r)$, the spherical average of $\rho(\mathbf{r})$ about the "center" of the molecule (see below).

The interest in this work goes beyond the computation of X-ray scattering functions. Since one of the objectives of the theory of molecular binding is to explain the properties of the molecule in terms of the corresponding properties of the atoms building the molecule, there is interest in decomposing the molecular electron density into localized fragments [March (1975), Bader and Nguyen-Dang (1981), Bader and Tang (1982)].

Construction of Molecular Electron Density

The electron density $\rho(\mathbf{r})$ in a molecule XH_m (in the present example, OH_2, NH_3, or CH_4) is constructed using equation (A4.10.1). The nuclear distances are the experimental ones [see Tables of interatomic distances, and Configurations in Molecules and Ions, Chemical Society, London, Special Publication No. 11 (1958)] and the coordinate origin is taken at the nucleus of the X atom. Since the component atoms have different chemical potentials (electronegativities), charge transfer between X and H atoms will occur on molecule formation, leading to effective charges $q(X)$ and $q(H)$ of the atomic fragments in the molecule. Here $q = -(n - Z)$, where Z is the atomic number and n is the number of electrons (in general, fractional after charge transfer). A simple theory for the charge transfer effect [Hinze (1970), Ray *et al.* (1979), Balbas *et al.* (1982), Alonso and March (1983)] gives

$$q(X) = -\frac{\phi_X - \phi_H}{\phi'_X + (\phi'_H/m)} \tag{A4.10.2}$$

where ϕ here is written for the electronegativity and ϕ' is its first derivative with respect to n. The number of electrons in the atomic fragments is then

$$n(X) = n^0(X) - q(X) \tag{A4.10.3}$$

$$n(H) = 1 + q(X)/m \tag{A4.10.4}$$

where $n^0(X)$ is the number of electrons in the neutral X atom. In computing ϕ and ϕ', Mulliken's approximation

$$\phi = (I + A)/2 \tag{A4.10.5}$$

$$\phi' = I - A \tag{A4.10.6}$$

is used, where I is the first ionization potential and A is the electron affinity. Taking experimental values for I and A [Lackner and Zweig (1981)], the following charges $q(X)$ result: $q(O) = -0.022$, $q(N) = 0.011$, and $q(C) = 0.071$, in OH_2, NH_3, and CH_4, respectively. For these XH_m molecules, the electronic configuration of the free X atom (X = C, N, O) is $1s^2 2s^2 2p^{6-m}$, and the corresponding free-atom electron density, separated into shell contributions, is

$$\rho_X^0(\mathbf{r}) = \rho_{1s}(r) + \rho_{2s}(r) + \rho_{2p}(\mathbf{r}) \tag{A4.10.7}$$

$\rho_{1s}(r)$ and $\rho_{2s}(r)$ are spherically symmetric functions. The electron density $\rho_X(\mathbf{r})$ of the fragment X in the molecule is built as

$$\rho_X(\mathbf{r}) = \rho_{1s}(r) + \rho_{2s}(r) + \rho_{2p}^*(\mathbf{r}) \tag{A4.10.8}$$

where $\rho_{1s}(r)$ and $\rho_{2s}(r)$ are the same as in the free X atom, and the $2p$-shell density,

$$\rho_{2p}^*(r) = M_X \tilde{\rho}_{2p}(\mathbf{r}) \tag{A4.10.9}$$

is formed by first averaging $\rho_{2p}(\mathbf{r})$ over angles ($\tilde{\rho}_{2p}(\mathbf{r})$ is the spherical average) and then renormalizing $\tilde{\rho}_{2p}(\mathbf{r})$ to have the correct number of electrons, i.e.,

$$4\pi \int_0^\infty M_X \tilde{\rho}_{2p}(r)\, dr = 6 - m - q(X) \tag{A4.10.10}$$

This equation leads to the normalization constant

$$M_X = 1 - [q(X)/(6 - m)] \tag{A4.10.11}$$

In a similar way, the electron density of each H fragment in the molecule is

$$\rho_H(r) = M_H \rho_H^0(r) \tag{A4.10.12}$$

where $\rho_H^0(r)$ is the electron density of a free H atom and M_H is the normalization constant,

$$M_H = 1 + [q(X)/m] \tag{A4.10.13}$$

The total molecular density is then constructed from equation (A4.10.1), using (A4.10.8) and (A4.10.12) and the experimental internuclear distances. The analytical Hartree–Fock wave functions of Clementi and Roetti (1974) were used to construct ρ_{1s}, ρ_{2s} and ρ_{2p} in equation (A4.10.8).

X-Ray Scattering Factor

Banyard and March (1956) obtained good results for the scattering factors of these molecules by using a spherically symmetric molecular electron density by solving the Hartree–Fock equations after averaging the nuclear field over angles about the central nucleus. To simplify the computation one can spherically average $\rho(\mathbf{r})$ over angles about the nucleus of the X fragment. Since $\rho_X(\mathbf{r})$ is already spherically symmetric, only the H fragment electron densities need to be averaged. Let us call $\rho^s(r)$ the spherically averaged molecular electron density.

For the spherically symmetric charge distribution $\rho^s(r)$, the formula giving the scattering factor is

$$f = \int_0^\infty \rho^s(r) 4\pi r^2 (\sin \kappa r / \kappa r) \, dr \qquad \text{(A4.10.14)}$$

where $\kappa = 4\pi \sin \frac{1}{2}\theta / \lambda$, with λ the wavelength of the incident radiation and θ the angle of scattering. $\rho^s(r)$ can be written

$$\rho^s(r) = \rho_X^s(r) + m\rho_H^s(r) \qquad \text{(A4.10.15)}$$

where $\rho_X^s(r) \equiv \rho_X(r)$ and $\rho_H^s(r)$ is the electron density of an H fragment averaged about the center of the molecule. $\rho_X^s(r)$ and $\rho_H^s(r)$ are analytical functions (see Appendix A1.1), and the integral in equation (A4.10.14) can be calculated in closed form.

Results

The calculated scattering factors of OH_2, NH_3, and CH_4 are plotted together in Figure A.14, where the systematic differences owing to the different spatial distribution of the molecule's ten electrons are clearly seen. The calculated scattering factor of the Ne atom is given, for comparison, in the same figure. In this figure as well as in the earlier work of Banyard and March (1956) (where spherically averaged densities were used), the theoretical curves provide a good description of the experimental results of Thomer (1937). The simple charge transfer model of this appendix proves slightly more accurate than the scattering predictions based on quantum-mechanical calculations of the spherically averaged densities. In both calculations, in fact, the scattering factor is obtained from spherically averaged densities $\rho(\mathbf{r})$. This is known from the work of Banyard and March (1956) on H_2O to introduce negligible error in the scattering factor by neglect of the p and higher angular terms in $\rho(\mathbf{r})$. The spherically averaged densities used to predict the X-ray scattering differ by small amounts, one reason being that Banyard and March (1956) obtained the molecular density by first spherically averaging the field of the nuclear framework. In contrast, the charge transfer

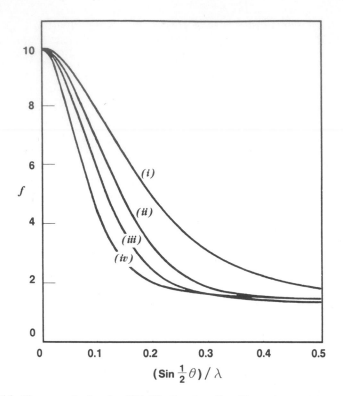

FIGURE A.14. X-ray scattering for "Ne-like" series: (i) self-consistent field results for Ne, (ii) results for H_2O using analytical wave functions, (iii) results for NH_3 using analytical wave functions, and (iv) similar results for CH_4 [redrawn from Banyard and March (1956)].

model calculates the molecular density as a spherical average of a superposition of densities corresponding to atomic-like fragments. It turns out that both procedures for this series of ten-electron molecules explain the X-ray scattering measurements quite satisfactorily.

A4.11. TWO-CENTER CALCULATIONS FROM THE THOMAS–FERMI THEORY

In discussing the problem of density descriptions in molecules, one must eventually seek to solve the Euler equation of the variation problem [i.e., equation (4.6) and its generalization to include electron density gradients] to calculate the electron density $\rho(\mathbf{r})$. Early calculations exist [Hund (1932) for N_2, March (1951) for benzene] for the Thomas–Fermi (TF) equation itself, but such densities are nowadays too crude to be useful without substantial refinements. In what follows we shall consider first the interaction energy between closed-shell atoms and ions, and then present a summary of some more recent calculations involving chemical bonding and the associated charge redistribution.

Following early work by Jensen, Gombas, Wedepohl and numerous other researchers, Gordon and Kim (1972) used the density description to calculate the energy of interaction $E(R)$ of rare-gas atoms as a function of the separation R.

A basic assumption is that the electron density in the two-center problem is simply the superposition of Hartree–Fock (HF) densities for the two isolated atoms involved. It is essential to use such HF rather than TF densities in the calculation. Once the superposition approximation is made, the HF approximation to the (single-center) atomic densities should be satisfactory, since the HF predictions for one-electron properties are known to be accurate to second order in the electron–electron interactions [Möller and Plesset (1934); Appendix AI].

Writing the superposition approximation explicitly for the two-center case of atoms A and B at separation R as

$$\rho(\mathbf{r}) = \rho_a + \rho_b \qquad (A4.11.1)$$

ρ_a and ρ_b being the two isolated atom densities, the Coulombic potential energy may be written

$$E_C = \frac{Z_a Z_b}{R} + \frac{1}{2} \int \frac{[\rho_a(\mathbf{r}_1) + \rho_b(\mathbf{r}_1)][\rho_a(\mathbf{r}_2) + \rho_b(\mathbf{r}_2)]}{\mathbf{r}_{12}} \, d\mathbf{r}_1 \, d\mathbf{r}_2$$

$$- Z_a \int \frac{\rho_a(\mathbf{r}_1) + \rho_b(\mathbf{r}_1)}{\mathbf{r}_{1a}} \, d\mathbf{r}_1 - Z_b \int \frac{\rho_a(\mathbf{r}_1) + \rho_b(\mathbf{r}_1)}{\mathbf{r}_{1b}} \, d\mathbf{r}_1 \qquad (A4.11.2)$$

Evidently the first term on the right-hand side represents the nuclear repulsion energy between atoms A and B with nuclear charges Z_a and Z_b.

From the above expression, one wishes to subtract the Coulomb energies of the isolated atoms to obtain the corresponding contribution to the interaction energy. One has then for the Coulomb energy of atom A the result

$$E_C^{(a)} = \frac{1}{2} \int \frac{\rho_a(\mathbf{r}_1)\rho_a(\mathbf{r}_2)}{\mathbf{r}_{12}} \, d\mathbf{r}_1 \, d\mathbf{r}_2 - Z_a \int \frac{\rho_a(\mathbf{r}_1)}{\mathbf{r}_{1a}} \, d\mathbf{r}_1 \qquad (A4.11.3)$$

with a corresponding expression for atom B. By subtraction, one finds for the Coulomb interaction energy $\Delta E_C(R)$ in this superposition approximation:

$$\Delta E_C(R) = \frac{Z_a Z_b}{R} + \int \frac{\rho_a(\mathbf{r}_1)\rho_b(\mathbf{r}_2)}{\mathbf{r}_{12}} \, d\mathbf{r}_1 \, d\mathbf{r}_2$$

$$- Z_b \int \frac{\rho_a(\mathbf{r}_1)}{\mathbf{r}_{1b}} \, d\mathbf{r}_1 - Z_a \int \frac{\rho_b(\mathbf{r}_2)}{\mathbf{r}_{2a}} \, d\mathbf{r}_2 \qquad (A4.11.4)$$

As Gordon and Kim (1972) pointed out, it is useful in practice to combine the four terms in this expression into a single integrand since there is considerable cancellation between the separate contributions. Thus we may write

$$\Delta E_C(R) = \int \rho_a(\mathbf{r}_1)\rho_b(\mathbf{r}_2)[R^{-1} + r_{12}^{-1} - r_{1b}^{-1} - r_{2a}^{-1}]\, d\mathbf{r}_1\, d\mathbf{r}_2 \qquad \text{(A4.11.5)}$$

using the normalization results for neutral atoms that

$$\int \rho_i(\mathbf{r})\, d\mathbf{r} = Z_i \qquad (i = \text{a or b}) \qquad \text{(A4.11.6)}$$

Other Energy Terms

For the other contributions to the total interaction energy the electron gas formulas discussed in the text and earlier appendixes were employed. In particular, the kinetic and exchange energy densities were the simple expressions proportional to $\rho^{5/3}$ and $\rho^{4/3}$, respectively. An electron-gas approximation to the correlation energy density was also incorporated.

Summary of Main Results

If we take by way of illustration the Ar–Ar interaction [for numerical details see Gordon and Kim (1972)], the theoretical interaction energy tends to zero with increasing R more rapidly than the experimental results.

This could have been anticipated, for only part of the dispersion energy is taken into account in the present form of the inhomogeneous electron-gas model, i.e., that part which comes from the overlap region. One would have to transcend the local density approximation employed above in order to include the long-range dispersion forces which are present even when overlap is absent. Also polarization (induction) forces are not included since the correlation energy density is approximated by a simple electron-gas formula. Therefore the Gordon–Kim potentials are too high at distances at and beyond the minimum in the potential, since their treatment does not account for the negative contributions from long-range dispersion and induction forces.

For these reasons, the calculated well depths for the rare-gas interactions tend to be too small. But this deficiency can be corrected for, since the dispersion and induction forces can be estimated with considerable accuracy at distances somewhat beyond the minimum in the potential curve.

Detailed examination of the Gordon–Kim results shows that the "hard core diameter" is slightly too small for Ar–Ar, and also for Ne–Ne and Kr–Kr interactions. This may be due to underestimating the kinetic energy. Thus, as noted above, gradient corrections to the kinetic energy should raise its value for a given density $\rho(\mathbf{r})$ and thereby increase the predicted values of the hard-core diameter.

Ion–Ion Interaction from Superposition Density

Since many ions that form various ionic compounds and crystals are of closed-shell structure, one can develop the approach outlined above for rare-gas systems to discuss such ion–ion interactions. Kim and Gordon (1974) studied three different types of ion pairs:

1. Ion pairs made of an alkali ion and a halide ion
2. Two alkali ions or two halide ions
3. Combination of an alkaline earth ion and a halide ion

Coulomb Interaction between Ions

Consider the interaction between two ions A and B with ionic charges z_i and atomic number Z_i, $i = $ a, b, respectively. The Coulombic interaction between these two ions is given by

$$\Delta E_C(R) = \frac{Z_a Z_b}{R} + \int \frac{\rho_a(\mathbf{r}_1)\rho_b(\mathbf{r}_2)}{r_{12}} \, d\mathbf{r}_1 \, d\mathbf{r}_2 - Z_b \int \frac{\rho_a(\mathbf{r}_1)}{r_{1b}} \, d\mathbf{r}_1 - Z_a \int \frac{\rho_b(\mathbf{r}_2)}{r_{2a}} \, d\mathbf{r}_2$$

$$(\textbf{A4.11.7})$$

One now makes use of the normalization conditions

$$\int \rho_a(\mathbf{r}) \, d\mathbf{r} = Z_a - z_a \qquad (\textbf{A4.11.8})$$

$$\int \rho_b(\mathbf{r}) \, d\mathbf{r} = Z_b - z_b \qquad (\textbf{A4.11.9})$$

A little rewriting then yields

$$\Delta E_C(R) = \frac{z_a z_b}{R} + \int \rho_a(\mathbf{r}_1)\rho_b(\mathbf{r}_2)\left[\frac{Z_a Z_b - z_a z_b}{(Z_a - z_a)(Z_b - z_b)} \frac{1}{R} + \frac{1}{r_{12}} \right.$$

$$\left. - \frac{Z_b}{Z_b - z_b} \frac{1}{r_{1b}} - \frac{Z_a}{Z_a - z_a} \frac{1}{r_{2a}} \right] d\mathbf{r}_1 \, d\mathbf{r}_2 \qquad (\textbf{A4.11.10})$$

For spherical densities ρ_a and ρ_b, such as obtain in closed-shell ions, the angular integrations can in fact be performed, reducing the calculation of $\Delta E_C(R)$ to a two-dimensional integral over $\mathbf{r}_1$ and $\mathbf{r}_2$.

Again, the other energy contributions are calculated from electron-gas formulas. Kim and Gordon (1974) have created tables for the interactions between the alkali ions (Li^+, Na^+, K^+, and Rb^+) and the halide ions (F^-, Cl^-, and Br^-), between two alkali ions, between two halide ions, and between the alkaline earth ions (Be^{2+}, Mg^{2+}, and Ca^{2+}) and the halide ions (F^- and Cl^-).

Apart from their role in the discussion of the structures of ionic crystals, these potentials are useful, for example, in a computer simulation of superionic conduction in solid CaF_2 [Rahman (1976), Dixon and Gillan (1978)].

Summary of Inhomogeneous Electron-Gas Findings

Because of the statistical character of the assumptions underlying the above approximation employed in the framework of the inhomogeneous electron gas, the method is expected, and indeed found, to improve in accuracy for systems with more and more electrons. This is important because in this same situation the difficulties facing self-consistent field and/or configuration interaction calculations are exacerbated by increasing numbers of electrons. The inhomogeneous electron-gas approach in the form employed by Gordon and Kim with HF atom densities provides a simple, eminently tractable way of obtaining useful estimates of interaction energies between closed-shell atoms.

However, a note of caution needs sounding when one goes to open-shell configurations. It would violate chemical intuition to use a superposition density in such calculations when one knew that strong chemical bonding would be present. For recent references in which local density approximations are used in that context, the reader should refer to Gunnarsson *et al* (1977).

A5.1. ROTATIONAL ENERGY LEVELS OF SOME SIMPLE CLASSES OF MOLECULES: THE SYMMETRIC ROTOR

In the body of the text, we made some use of the rotational energy levels of a rigid rotor, a model for diatomic and triatomic linear molecules.

The purpose of this appendix is to gather together useful results for some of the other classes of molecular rotational energy levels. The proofs can be found in any of the standard works on vibrational properties of molecules, e.g., Wilson *et al.* (1955) or Woodward (1972).

In a diatomic or linear polyatomic molecule, the angular momentum vector **P** due to the rotation of the molecule lies along the axis of rotation. In a prolate symmetric rotor, defined in the main text, the rotational angular momentum vector **P** need not be perpendicular to the top axis, which is the *a*-axis of CH_3I shown in Figure A.15. The molecule rotates (or nutates) about the axis of **P**, which passes through the center of gravity of the molecule and is to be regarded as fixed in space and not fixed to the molecule. In addition, the molecule rotates about the *a*-axis and the vector P_a represents the magnitude and direction of the angular momentum due to this motion. Solution of the Schrödinger equation for this system in any nondegenerate vibrational state gives the term values

$$F_v(J, K) = B_v J(J + 1) + (A_v - B_v)K^2 \qquad \text{(A5.1.1)}$$

This is for the rigid rotor model in which centrifugal distortion has been neglected.

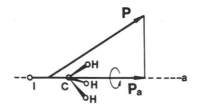

FIGURE A.15. The angular momentum vector P, and the component P_a along the top axis in the prolate symmetric rotor CH_3I.

The rotational constants A and B are related to I_a and I_b, the principal axes moments of inertia, by

$$A = h/8\pi^2 I_a, \qquad B = h/8\pi^2 I_b \qquad \text{(A5.1.2)}$$

In the case we are considering $A > B = C$.

The quantum number K in equation (A5.1.1) can take the values $0, 1, 2, 3, \ldots, J$. The fact that K cannot be greater than J follows from the fact that the magnitude of the vector $\mathbf{P}_a$ cannot be greater than that of $\mathbf{P}$. All rotational levels with $K > 0$ possess a doubly degenerate character. This can be viewed classically as being due to clockwise or anticlockwise rotation about the top axis, resulting in the same magnitude of the angular momentum.

For an oblate symmetric rotor, such as NH_3, the rotational-term values for any nondegenerate vibrational state are

$$F_v(J, K) = B_v J(J + 1) + (C_v - B_v)K^2 \qquad \text{(A5.1.3)}$$

where

$$C = h/8\pi^2 I_c \qquad \text{(A5.1.4)}$$

and $A = B > C$.

The main difference between the levels for a prolate and oblate symmetric rotor is that the levels for a given J and different values of K are more widely spaced with increasing K for a prolate symmetric rotor, because $A - B$ is always positive, than for an oblate situation, for which $C - B$ is always negative.

The selection rules governing pure rotational transitions between levels in a symmetric rotor are

$$\Delta J = \pm 1, \qquad \Delta K = 0 \qquad \text{(A5.1.5)}$$

The result is that transition frequencies or wave numbers for a prolate or oblate symmetric rotor are given by

$$v = F_v(J + 1, K) - F_v(J, K) = 2B_v(J + 1) \qquad \text{(A5.1.6)}$$

This result is the same as that given in equation (5.13) for a diatomic or triatomic linear molecule. The result is a series of equally spaced lines separated by $2B_v$, but since $K = 0, 1, 2, \ldots, J$, each line contains $J + 1$ components. In practice, the centrifugal distortion, discussed in Section 5.2, modifies equation (A5.1.1) somewhat and in fact the spectrum of the symmetric rotor is more complicated than has been indicated here. The interested reader is referred to the book by Hollas (1982) for further details.

A5.2. THE ROTATIONAL PARTITION FUNCTION IN RELATION TO SPECTROSCOPIC INTENSITIES

We discussed in the main text of Chapter 5 the population N_J/N_0 (see, e.g., Figure 5.5) in relation to the intensities of pure rotational transitions. However, strictly speaking, these intensities are proportional to the fraction of the total number of molecules occupying each rotational level, rather than to N_J/N_0. This fraction N_J/N can be obtained as was equation (5.14) and is given by

$$N_J/N = (2J + 1) \exp(-E_r/k_B T)/Q_r \qquad \text{(A5.2.1)}$$

where Q_r is the rotational partition function given by

$$Q_r = \sum_i g_i \exp(-E_{ri}/k_B T) \qquad \text{(A5.2.2)}$$

the g_i's being degeneracy factors.

At high temperatures or low values of the constant B in equation (A5.1.6) for the rotational energy levels and normal temperatures, many rotational levels are significantly populated. Under these circumstances, the distribution of rotational levels can be usefully approximated by a continuum of levels and the summation in equation (A5.2.2) can be replaced by an integration yielding

$$Q_r \simeq \int_0^\infty (2J + 1) \exp(-E_r/k_B T) \, dJ = k_B T/hB \qquad \text{(A5.2.3)}$$

where, as written, B has dimensions of frequency.

In the case of a set of vibrational or electronic energy levels of a molecule, the spacing of the levels is often so large that the summations such as that in equation (A5.2.2) are essentially equal to the first term, namely $g_0 \exp(-E_0/k_B T)$. If $E_0 = 0$ and $g_0 = 1$, then

$$Q_v \simeq 1 \qquad \text{and} \qquad Q_e \simeq 1 \qquad \text{(A5.2.4)}$$

where Q_v and Q_e are the vibrational and electronic partition functions. Only in cases where there are low-energy excited vibrational or electronic states can either of these partition functions differ appreciably from unity.

The intensity of absorption I for a transition between a lower state m and an upper state n is the net result of stimulated absorption, stimulated emission, and spontaneous emission (cf. Appendix A5.5). For the low-frequency transition typical of pure rotation spectra, the population of state n is appreciable compared to that of m, and therefore emission processes are significant.

The rate I_{mn} of stimulated absorption of energy is given by

$$I_{mn} = (N_m/N)B_{nm}\rho(\nu_{nm})h\nu_{nm} \qquad \text{(A5.2.5)}$$

where N_m/N is the fraction of molecules in state m and, as in Appendix A5.5, B_{nm} is the Einstein coefficient, $\rho(\nu_{mn})$ the density of radiation of frequency ν_{nm}, and

$h\nu_{mn}$ the energy difference between the two states. The rate of stimulated emission I_{nm} is given by

$$I_{nm} = (N_n/N)B_{nm}\rho(\nu_{mn})h\nu_{mn} \qquad \text{(A5.2.6)}$$

so that the net absorption is

$$I_{abs} = I_{mn} - I_{nm} = (N_m - N_n)B_{nm}\rho(\nu_{mn})h\nu_{mn}/N \qquad \text{(A5.2.7)}$$

Employing the Boltzmann relation between N_m and N_n this becomes

$$I_{abs} \simeq N_m[1 - \exp(-h\nu_{mn}/k_BT)]B_{nm}\rho(\nu_{nm})h\nu_{mn}/N \qquad \text{(A5.2.8)}$$

In the case where one has $\nu_{mn} \ll k_BT/h$, the exponential can be expanded to yield

$$I_{abs} \simeq N_m h^2 \nu_{mn}^2 B_{nm}\rho(\nu_{nm})/Nk_BT \qquad \text{(A5.2.9)}$$

It turns out on numerical examination that the expansion of the exponential used above in reaching equation (A5.2.9) is reasonable in the microwave, milli-meter wave, and part of the far-infrared spectrum at temperatures around room temperature or lower.

In the rotation spectrum of a diatomic or any linear polyatomic molecule, the square of the transition moment $|\mathbf{R}^{nm}|^2$ varies with J as

$$|\mathbf{R}^{nm}|^2 = \mu^2(J + 1)(2J + 1) \qquad \text{(A5.2.10)}$$

where μ denotes the permanent dipole moment. Therefore one can write [see Hollas (1982) for details]

$$B_{nm} = [8\pi^3/(4\pi\varepsilon_0)3h^2][\mu^2(J + 1)(2J + 1)] \qquad \text{(A5.2.11)}$$

Combining equations (A5.2.9) with equations (A5.2.1), (A5.2.3), and (A5.2.11) yields

$$I_{abs} \simeq \rho(\nu_{J'J})\left[\frac{8\pi^3}{3(4\pi\varepsilon_0)}\right]\left(\frac{hB}{k_BT}\right)\mu^2\nu_{J'J}^2(J+1)\exp\left[-\frac{hBJ(J+1)}{k_BT}\right] \qquad \text{(A5.2.12)}$$

where $\nu_{J'J}$ is the frequency of the transition between the rotational energy levels with quantum numbers J'' and J', and J is understood to stand for J''.

The most important difference between equation (5.14) for N_J/N_0 giving population ratios of rotational energy levels and equation (A5.2.12) is the factor ν^2. This has the consequence of producing a maximum in the absorption intensity at a higher J-value than J_{max} corresponding to the maximum population given by equation (5.14). In addition, the factor $(J + 1)$ in equation (A5.2.12) replaces $(2J + 1)$ in equation (5.14). Consulting Figure 5.5, we can get values of $\tilde{\nu}^2(J + 1)\exp(-hcBJ(J + 1)/k_BT)$ (where $\tilde{\nu}$ here is ν in wave number units) calcu-lated for CO ($B = 1.9314$ cm^{-1}). These values reveal that although the maximum population is at $J = 7$, the most intense transition is that with $J'' = 12$. This result is confirmed by the far-infrared spectrum observed by Fleming and Chamberlain (1974) and reproduced in Figure 5.6, in which the intensity must be taken as the integrated area for each transition.

Intensity measurements of pure rotation spectra in the far-infrared have been used to determine the value of $0.108 \pm 0.005D$ for the dipole moment of CO. This is one good example of the determination of an important molecular parameter by spectroscopy. In this case, the use of the Stark effect on the millimeter wave spectrum affords an alternative approach, and the values obtained from the two methods agree well.

A5.3. NORMAL MODES OF VIBRATION OF MOLECULES

We shall be content here with an outline of the way one sets up and solves the problem of finding the normal modes of vibration of a molecule followed by one simple example. We have, basically, a problem in classical mechanics provided that we are concerned only with the classical vibration wave number of each of the normal modes. It follows that each normal mode is treated in the harmonic oscillator approximation, and it is obvious that if one requires vibrational wave functions, then one must solve the quantum-mechanical Schrödinger equation. However, in this appendix we restrict the discussion to classical mechanics.

It is fair to say that the two most generally useful approaches to normal mode calculations are (i) to start from known vibrational frequencies, and hence to obtain the force constants and the form of the normal modes, or (ii) to assume values of the force constants, which, say, can be transferred with some useful accuracy to structurally similar molecules, and then to obtain the vibrational frequencies and the form of the normal modes.

In both approaches, one must solve Lagrange's equation of classical mechanics

$$(d/dt)(\partial T/\partial \dot{q}_i) + \partial V/\partial q_i = 0 \qquad \text{(A5.3.1)}$$

which involves the kinetic energy T and the potential energy V due to motion along a coordinate q_i, the quantity $\dot{q}_i$ being the time derivative $\partial q_i/\partial t$.

Kinetic Energy

If Cartesian axes are attached to each nucleus, with the origin at the equilibrium position of the nucleus, the kinetic energy is given by

$$T = \tfrac{1}{2} \sum_{j=1}^{N} m_j(\dot{x}_j^2 + \dot{y}_j^2 + \dot{z}_j^2) \qquad \text{(A5.3.2)}$$

where x_j, y_j, and z_j denote the coordinates of the jth nucleus with mass m_j. In the summation, j runs over the total number of nuclei N. The orthogonality of the Cartesian displacement coordinates in equation (A5.3.2) ensures that it contains only squared terms in the coordinates and no cross products.

Translational and rotational, as well as vibrational, motions of the molecule contribute to the kinetic energy of equation (A5.3.2). In order to preclude the possibility of translational motion, one must impose the requirements

$$\sum_{j=1}^{N} m_j \dot{x}_j = 0, \qquad \sum_{j=1}^{N} m_j \dot{y}_j = 0, \qquad \sum_{j=1}^{N} m_j \dot{z}_j = 0 \qquad \text{(A5.3.3)}$$

Similarly, to exclude the possibility of rotational motion we impose the requirement that there shall be no angular momentum:

$$\sum_{j=1}^{N} m_j(y_j \dot{z}_j - z_j \dot{y}_j) = 0$$

$$\sum_{j=1}^{N} m_j(z_j \dot{x}_j - x_j \dot{z}_j) = 0 \qquad \text{(A5.3.4)}$$

$$\sum_{j=1}^{N} m_j(x_j \dot{y}_j - y_j \dot{x}_j) = 0$$

Potential Energy

The potential energy of the vibrating molecule is due to distortion from the equilibrium configuration which can be described in terms of changes of bond lengths and angles. It is therefore natural to express the potential energy in terms of internal displacement coordinates rather than the Cartesian coordinates used for the kinetic energy. One set of internal coordinates consists of the displacements δr_{ij} and δr_{kl} of all pairs of nuclei i, j and k, l. In the harmonic oscillator approximation, the potential energy is then given by

$$V = \tfrac{1}{2} \sum_{ij} \sum_{kl} k_{ij,kl} \delta r_{ij} \delta r_{kl} \qquad \text{(A5.3.5)}$$

where the $k_{ij,kl}$ represent force constants.

As an example, let us consider the application of equation (A5.4.5) to the water molecule. In the notation

$$\underset{H}{\nearrow} \overset{O}{\underset{}{\searrow}} \underset{H'}{}$$

we can write the potential energy, with the abbreviation $k_{OH,OH}$ equals k_{OH} etc., as

$$V = \tfrac{1}{2} k_{OH} (\delta r_{OH})^2 + \tfrac{1}{2} k_{OH'} (\delta r_{OH'})^2 + \tfrac{1}{2} k_{HH'} (\delta r_{HH'})^2 + k_{OH,OH'} \delta r_{OH} \delta r_{OH'}$$

$$+ k_{OH,HH'} \delta r_{OH} \delta r_{HH'} + k_{OH',HH'} \delta r_{OH'} \delta r_{HH'} \qquad \text{(A5.3.6)}$$

and it has been assumed, of course, that $k_{OH} = k_{OH'}$. This expression for the potential energy assumes a quite general force field, taking all possible interactions into account.

The drawback of such a general approach is that while there are four independent force constants in equation (A5.3.6), there are only three normal modes whose fundamental vibration frequencies may be determined experimentally. This problem of having more force constants in a molecule than there are normal modes is a quite general characteristic of this area and the problem gets worse for larger molecules. Even when one deals with small molecules, the problem is exacerbated when the symmetry is lower. As one example, in the bent molecule FOH there are six force constants, compared to the four discussed above for H_2O. Therefore, in the approach in which one wishes to extract force constants from vibrational frequencies, it is plain that use of a general force field is impractical and one must resort to chemical or physical simplifications.

Stretching Vibrations of a Linear Molecule

Motion of the three nuclei of a linear unsymmetrical triatomic molecule ABC, e.g., HCN or OCS, along the axis of the molecule, conventionally labeled the z-axis, represents three degrees of freedom; two of these are vibrational, involving stretching and compression of the bonds, and the other is translational, the molecule as a whole being translated along the z-axis with no change of bond lengths. The molecule is shown in Figure A.16.

The potential energy can be expressed in this example as

$$V = \tfrac{1}{2}k_1R_1^2 + \tfrac{1}{2}k_2R_2^2 \qquad \text{(A5.3.7)}$$

where $R_1 = \delta r_{AB}$, the displacement of the bond A—B from its equilibrium length r_{AB}, and $R_2 = \delta r_{BC}$, k_1 and k_2 being the corresponding stretching force constants.

The kinetic energy T for which the general expression is given by equation (A5.3.2) due to motion of the three nuclei along the space-fixed z-axis is given by

$$T = \tfrac{1}{2}m_A\dot{z}_A^2 + \tfrac{1}{2}m_B\dot{z}_B^2 + \tfrac{1}{2}m_C\dot{z}_C^2 \qquad \text{(A5.3.8)}$$

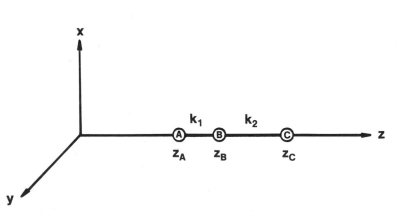

FIGURE A.16. Force constants and nuclear coordinates for the stretching vibrations of a linear molecule ABC.

The internal coordinates R_1 and R_2 can be expressed in terms of the Cartesian coordinates

$$R_1 = \delta r_{AB} = (z_B - z_A) - r_{AB}$$

$$(A5.3.9)$$

$$R_2 = \delta r_{BC} = (z_C - z_B) - r_{BC}$$

Differentiating with respect to time we find

$$\dot{R}_1 = \dot{z}_B - \dot{z}_A, \qquad \dot{R}_2 = \dot{z}_C - \dot{z}_B \qquad (A5.3.10)$$

since r_{AB} and r_{BC} are equilibrium distances. These equations (A5.3.10) are conveniently written in matrix notation as

$$\begin{pmatrix} \dot{R}_1 \\ \dot{R}_2 \end{pmatrix} = \begin{pmatrix} -1 & 1 & 0 \\ 0 & -1 & 1 \end{pmatrix} \begin{pmatrix} \dot{z}_A \\ \dot{z}_B \\ \dot{z}_C \end{pmatrix} \qquad (A5.3.11)$$

However, to change from z's to R's we require a third equation and this is obtained from the requirement that there must be no overall translation along the z-axis and thus no net linear momentum. This condition reads

$$m_A \dot{z}_A + m_B \dot{z}_B + m_C \dot{z}_C = 0 \qquad (A5.3.12)$$

Including equation (A5.3.12) in matrix form, we get

$$\begin{pmatrix} \dot{R}_1 \\ \dot{R}_2 \\ 0 \end{pmatrix} = \begin{pmatrix} -1 & 1 & 0 \\ 0 & -1 & 1 \\ m_A & m_B & m_C \end{pmatrix} \begin{pmatrix} \dot{z}_A \\ \dot{z}_B \\ \dot{z}_C \end{pmatrix} \qquad (A5.3.13)$$

Equation (A5.3.13) is readily inverted to yield

$$\begin{pmatrix} \dot{z}_A \\ \dot{z}_B \\ \dot{z}_C \end{pmatrix} = \begin{pmatrix} -(m_B + m_C)/M & -m_C/M & 1/M \\ m_A/M & -m_C/M & 1/M \\ m_A/M & (m_A + m_B)/M & 1/M \end{pmatrix} \begin{pmatrix} \dot{R}_1 \\ \dot{R}_2 \\ 0 \end{pmatrix} \qquad (A5.3.14)$$

where $M = m_A + m_B + m_C$, and the 3×3 matrix in equation (A5.3.14) is the inverse of that matrix in equation (A5.3.13). We merely note here that provided a matrix A possesses an inverse A^{-1}, then the jith element of A^{-1}, where j and i are row and column labels, respectively, is given by

$$(A^{-1})_{ji} = (-1)^{i+j} |M_{ij}| / |A| \qquad (A5.3.15)$$

where M_{ij} is called the minor of the matrix and is obtained from the matrix A by striking out the ith row and the jth column. Using this result, and calculating the determinants in numerator and denominator, we readily derive equation (A5.3.14).

Substituting the expressions for the $\ddot{z}$'s in terms of $\dot{R}_1$ and $\dot{R}_2$ from equation (A5.3.14) into the kinetic energy T yields

$$T = \frac{m_A}{2M^2}(m_B^2 + 2m_B m_C + m_C^2 + m_A m_B + m_A m_C)\dot{R}_1^2$$

$$+ \frac{m_A m_C}{2M^2}(2m_C + 2m_A + 2m_B)\dot{R}_1\dot{R}_2$$

$$+ \frac{m_C}{2M^2}(m_B^2 + 2m_A m_B + m_A^2 + m_B m_C + m_A m_C)\dot{R}_2^2 \qquad \text{(A5.3.16)}$$

Substituting this expression for T and that of equation (A5.3.7) for V into Lagrange's equation (A5.3.1), for which $q_i = R_1$ or R_2, gives

$$\frac{m_A}{M^2}(m_B^2 + 2m_B m_C + m_C^2 + m_A m_B + m_A m_C)\ddot{R}_1$$

$$+ \frac{m_A}{2M^2}(2m_C^2 + 2m_A m_C + 2m_B m_C)\ddot{R}_2 + k_1 R_1 = 0 \qquad \text{(A5.3.17)}$$

and

$$\frac{m_C}{M^2}(m_B^2 + 2m_A m_B + m_A^2 + m_B m_C + m_A m_C)\ddot{R}_2$$

$$+ \frac{m_A}{2M^2}(2m_C^2 + 2m_A m_C + 2m_B m_C)\ddot{R}_1 + k_2 R_2 = 0 \qquad \text{(A5.3.18)}$$

Solutions of equations (A5.3.17) and (A5.3.18) are now sought having the simple-harmonic time dependence

$$R_1 = A_1 \cos 2\pi \nu t, \qquad R_2 = A_2 \cos 2\pi \nu t \qquad \text{(A5.3.19)}$$

Substituting for the second time derivatives obtained from these two equations gives the simultaneous secular equations

$$-\left[4\pi^2 \nu^2 \frac{m_A}{M^2}(m_B^2 + 2m_B m_C + m_C^2 + m_A m_B + m_A m_C) + k_1\right]A_1$$

$$-\left[4\pi^2 \nu^2 \frac{m_A}{2M^2}(2m_C^2 + 2m_A m_C + 2m_B m_C)\right]A_2 = 0 \qquad \text{(A5.3.20)}$$

and

$$-\left[4\pi^2 v^2 \frac{m_A}{2M^2}(2m_C^2 + 2m_A m_C + 2m_B m_C)\right]A_1$$

$$-\left[4\pi^2 v^2 \frac{m_C}{M^2}(m_B^2 + 2m_A m_B + m_A^2 + m_B m_C + m_A m_C) + k_2\right]A_2 = 0 \qquad \text{(A5.3.21)}$$

If we rewrite these last two equations as

$$aA_1 + bA_2 = 0, \qquad cA_1 + dA_2 = 0 \qquad \text{(A5.3.22)}$$

then the condition that they have nontrivial solutions is that

$$\begin{vmatrix} a & b \\ c & d \end{vmatrix} - 0 \qquad \text{(A5.3.23)}$$

The expansion of this secular determinant gives a quadratic in v^2 from which two values of v^2 and, taking only the positive values of v, two values of v can be obtained.

If one takes the two stretching vibration frequencies of HCN to be 2089 cm^{-1} and 3312 cm^{-1}, then one finds

$$k_{CH} = 5.8 \text{ aJ Å}^{-2}, \qquad k_{CN} = 17.9 \text{ aJ Å}^{-2} \qquad \text{(A5.3.24)}$$

which are "typical" values for this molecule.

The ratio A_1/A_2 of the amplitudes of the modes is also readily obtained from the above analysis. One finds that the frequency $v = 2089$ cm^{-1} is primarily stretching of the C—N bond, while the frequency 3312 cm^{-1} involves stretching of the C—H bonds. These results turn out to be consistent with the approximate transferability of group vibration wave numbers from one molecule to another.

A5.4. THE FRANCK–CONDON PRINCIPLE

Before the proposal of the Schrödinger equation, Franck put forward qualitative arguments to explain the various intensity distributions found in vibronic transitions. His main contribution was to recognize the fact that an electronic transition in a molecule is so much more rapid than a vibrational transition that in a vibronic transition the nuclei have very nearly the same position and velocity before and after the transition.

The possible consequences that stem from this recognition are illustrated in Figure A.17, which shows potential energy curves for the lower state (which is the ground state if an absorption process is being considered) and the upper state. The curves in Figure A.17a are drawn such that $r_e' > r_e''$. When the lower state is the ground state, this is frequently the case because the electron promotion involved is often from a bonding orbital to an orbital which is either less bonding, or even antibonding. As an example, in N_2 (see the discussion in Chapter 2) promotion of an electron from the $\sigma_g 2p$ to the $\pi_g^* 2p$ orbital gives two states, $a\,^1\Pi_g$ and $B\,^3\Pi_g$, in which r_e is 1.2203 Å and 1.2126 Å, respectively, involving a considerable increase from the value of 1.0977 Å in the $X\,^1\Sigma_g^+$ ground state.

In absorption from point A of the ground state in Figure A.17 (zero-point energy can be permissibly neglected in the context of Franck's semiclassical arguments) the transition will be to point B of the upper state. The requirement that the nuclei have the same position before and after the transition means that this is between points which lie on a vertical line in the figure, which means that r remains constant and such a transition is termed "vertical." The second requirement, that the nuclei have the same velocity before and after the transition means that a transition from A, where the nuclei are stationary, must go to B as this is the classical turning-point of a vibration, where the nuclei are also stationary.

A transition from A to C is highly improbable because, although the nuclei are stationary at A and C, there is a large change in r. An A to D transition is also unlikely because, although r is unchanged, the nuclei are in motion at point D.

Figure A.17b illustrates the case where $r_e' \simeq r_e''$. An example of such a transition is the $D\,^1\Sigma_u^+ \rightarrow X\,^1\Sigma_g$ Mulliken band system of C_2 (see Appendix IV). The value of r_e is 1.2380 Å in the D state and 1.2425 Å in the X state. Here the most probable transition is from A to B with no vibrational energy in the upper configuration. The transition from A to C maintains the value of r but the nuclear velocities are increased because they have kinetic energy equivalent to the distance BC.

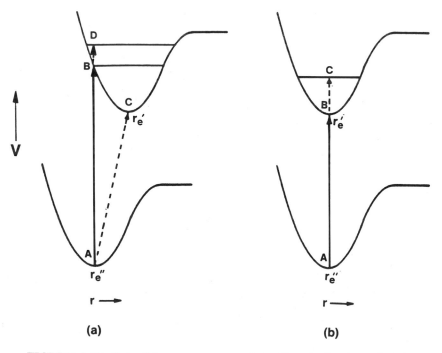

FIGURE A.17. Potential energy curves and the Franck–Condon principle.

Shortly after the advent of Schrödinger's equation, Condon treated the intensities of vibronic transitions quantum-mechanically. The intensity of such a transition is proportional to the square of the transition moment $\mathbf{R}_{ev}$ given by (cf. Appendix 5.2)

$$\mathbf{R}_{ev} = \int \psi_{ev}'^* \mathbf{\mu} \psi_{ev}'' \, d\tau_{ev} \tag{A5.4.1}$$

$\mathbf{\mu}$ being the electric dipole moment operator, while ψ_{ev}' and ψ_{ev}'' are the vibronic wave functions of the upper and lower states, respectively. The integration in equation (A5.4.1) is over electronic and vibrational coordinates. Assuming that the Born–Oppenheimer approximation holds, ψ_{ev} can be factorized into $\psi_e \psi_v$. In addition $\mathbf{\mu}$ can be decomposed into $\mathbf{\mu}_e + \mathbf{\mu}_n$, which are the contributions from electrons and from the nuclei, respectively. Then equation (A5.4.1) becomes

$$\mathbf{R}_{ev} = \int \psi_e'^* \psi_v' (\mathbf{\mu}_e + \mathbf{\mu}_n) \psi_e'' \psi_v'' \, d\tau_e \, dr \tag{A5.4.2}$$

Since the vibrational wave functions are always real, $\psi_v^* = \psi_v$. As $\mathbf{\mu}_e$ is a constant with respect to ψ_v and $\mathbf{\mu}_n$ is a constant with respect to ψ_e we find

$$\mathbf{R}_{ev} = \int \psi_e'^* \psi_e'' \, d\tau_e \int \psi_v' \mathbf{\mu}_n \psi_v'' \, dr + \int \psi_v' \psi_v'' \, dr \int \psi_e'^* \mathbf{\mu}_e \psi_e'' \, d\tau_e \tag{A5.4.3}$$

Because of the orthogonality of electronic states (see Appendix A1.4),

$$\int \psi_e'^* \psi_e'' \, d\tau_e = 0 \tag{A5.4.4}$$

so that

$$\mathbf{R}_{ev} = \mathbf{R}_e \int \psi_v' \psi_v'' \, dr \tag{A5.4.5}$$

where $\mathbf{R}_e$ is the electronic transition moment. In the absence of phenomena such as "intensity stealing," R_e is constant to an approximation that is good enough for the present purposes for all vibronic transitions associated with the electronic transition. The quantity $\int \psi_v' \psi_v'' \, dr$ is termed the vibrational overlap integral for reasons which are clear from analogy with electronic overlap integrals already treated. The square of this vibrational overlap integral is known as the Franck–Condon factor.

In carrying out the integration, the requirement that r remain constant during the transition is necessarily taken into account.

Illustrations of the above discussion are given in some detail by Hollas (1982) to whose account the interested reader is referred.

A5.5. TIME-DEPENDENT PERTURBATION THEORY AND SELECTION RULES FOR ELECTRIC DIPOLE TRANSITIONS IN ATOMS

In Chapter 5, we dealt in some detail with the question of selection rules for electric dipole transitions. These rules were simply quoted there. For the reader wishing to take the matter further, we treat here the simplest case in which selection rules for electric dipole transitions can be derived, namely the atomic as opposed to the molecular problem. The treatment below outlines the main steps in the argument, this paralleling closely that given in Fermi's lecture notes (Fermi 1961).

We are interested in studying a system in which the total Hamiltonian H can be divided into two parts, namely

$$H = H_0 + H_1 \tag{A5.5.1}$$

It will be supposed that H_0 is time-independent, while H_1 is time-dependent, owing, say, to the fluctuating electric field of a light wave.

Then the unperturbed Schrödinger equation in the time-dependent form

$$i\hbar \frac{\partial}{\partial t} \psi_0 = H_0 \psi_0 \tag{A5.5.2}$$

has the solution

$$\psi_0 = \sum_n a_n^{(0)} u_0^{(n)} \exp\left(-\frac{i}{\hbar} E_0^{(n)} t\right) \tag{A5.5.3}$$

where the u's and E's satisfy the stationary state Schrödinger equation

$$H_0 u_0^{(n)} = E_0^{(n)} u_0^{(n)} \tag{A5.5.4}$$

Whereas $a_n^{(0)}$ in equation (A5.5.3) is a constant, independent of time, one now follows Dirac by seeking a solution of the full Schrödinger equation

$$i\hbar \frac{\partial}{\partial t} \psi = (H_0 + H_1)\psi \tag{A5.5.5}$$

by writing

$$\psi = \sum_n a_n(t) u_0^{(n)} \exp\left(-\frac{iE_0^{(n)}}{\hbar} t\right) \tag{A5.5.6}$$

where one has now generalized the a_n's to become functions of time t. Substituting this expansion in equation (A5.5.5), and multiplying the result by $u_0^{(s)}$, say, on the

$$\frac{\partial a_s}{\partial t} = -\frac{i}{\hbar} \sum_n a_n \langle s|H_1|n\rangle \exp\left[\frac{i}{\hbar}(E_0^{(s)} - E_0^{(n)})t\right] \qquad \text{(A5.5.7)}$$

where

$$\langle s|H_1|n\rangle = \int u_0^{(s)*} H_1 u_0^{(n)} \, d\tau \equiv (H_1)_{sn} \qquad \text{(A5.5.8)}$$

Equation (A5.5.7) is, in fact, still exact, but pretty intractable until some approximation is introduced into it. The most obvious starting approximation, which we shall adopt here, is to substitute $a_n(0)$ for $a_n(t)$ on the right-hand side. Then we can integrate with respect to time to get the useful but now approximate equation

$$a_s(t) \doteq a_s(0) - \frac{i}{\hbar} \sum_n a_n(0) \int_0^t (H_1)_{sn} \exp\left[\frac{i}{\hbar}(E_0^{(s)} - E_0^{(n)})t\right] dt \qquad \text{(A5.5.9)}$$

Emission and Absorption of Radiation

Now we turn to apply the above theory, valid when H_1 is a small perturbation, to the emission and absorption of radiation. We represent the perturbation by the form

$$H_1 = eBz \cos \omega t \qquad \text{(A5.5.10)}$$

where B measures the amplitude of the perturbation. Suppose that at time $t = 0$ the atom under consideration is in a pure state n. Then, after time t, to determine the problem of the perturbation having caused a transition to a state m we must evaluate $a_m(t)$ and take its modulus squared. This can be done from equation (A5.5.9), when one finds first that

$$a_m(t) = -\frac{ie}{\hbar} B z_{mn} \int_0^t \cos \omega t \exp(i\omega_{nm}t) \, dt \qquad \text{(A5.5.11)}$$

where we have written ω_{mn} for the transition frequency, which is equal to $(E_m - E_n)/\hbar$. We next decompose $\cos \omega t$ in equation (A5.5.11) into two complex exponentials, $\cos \theta = (e^{i\theta} + e^{-i\theta})/2$, and note that when the applied frequency in equation (A5.5.11) is near to the transition frequency ω_{nm}, only the term $\exp(-i\omega t)$ is important. Then we can clearly write

$$a_m(t) \doteq -\frac{ieB}{2\hbar} z_{mn} \int_0^t \exp[i(\omega_{nm} - \omega)t] \, dt$$

$$= \frac{eB}{2\hbar} z_{mn} \frac{\exp[-i(\omega - \omega_{mn})t] - 1}{\omega - \omega_{mn}} \qquad \text{(A5.5.12)}$$

Thus we can form the squared modulus as

$$|a_m(t)|^2 = \frac{e^2 B^2}{\hbar^2}\, |z_{mn}|^2 \frac{\sin^2(t/2)(\omega - \omega_{mn})}{(\omega - \omega_{mn})^2} \qquad \text{(A5.5.13)}$$

Now with amplitude B, the light intensity is proportional to the square of the amplitude, and with the normalization adopted in equation (A5.5.10) it is given by $cB^2/8\pi$. The transition probability $a_m^2(t)$ is naturally proportional to this as seen from equation (A5.5.13).

However, for our purposes, all that is necessary in an atom to discuss the transitions for electric dipole radiation [ez in equation (A5.5.13) can be regarded as a dipole moment] is to find the conditions under which the matrix element z_{mn} is nonzero, that is when there is a nonvanishing probability of transition under the influence of the perturbation in equation (A5.5.10).

With the atomic wave functions being radial parts times spherical harmonics, as discussed in Appendix A1.1, we want generally to consider terms like $z = r\cos\theta$, and also terms with z replaced by x and y (say, from isotropic radiation). Then we are led, readily, since the spherical harmonics have the property that

$$Y_{11} = \text{const}\ \sin\theta\ \exp(i\phi)$$

$$Y_{10} = \text{const}\ \cos\theta \qquad \text{(A5.5.14)}$$

$$Y_{1,-1} = \sin\theta\ \exp(-i\phi)$$

to integrals that involve the product of three spherical harmonics. While it is true that orthogonality properties only exist for products of two spherical harmonics, the product of the above spherical harmonics with another spherical harmonic can be expressed as a linear combination of spherical harmonics through the identities [see Fermi (1961) for the complete expressions, with all the constants given explicitly]:

$$Y_{11}Y_{l,m-1} = \text{const}\ Y_{l+1,m} + \text{const}\ Y_{l-1,m}$$

$$Y_{10}Y_{l,m} = \text{const}\ Y_{l+1,m} + \text{const}\ Y_{l-1,m} \qquad \text{(A5.5.15)}$$

$$Y_{1,-1}Y_{l,m+1} = \text{const}\ Y_{l+1,m} + \text{const}\ Y_{l-1,m}$$

It can be seen from the fact that only $l+1$ and $l-1$ appear on the left-hand side of equation (A5.5.16) that the orthogonality properties will make the matrix element under examination vanish unless l changes by ± 1 in the transition. Similarly for quantum number m, inspection of equations (A5.5.15) shows that m must change by ± 1 or 0 for a nonvanishing transition probability. This then completes the discussion of the selection rules for electric dipole transitions in atoms.

Absorption from continuous overlapping ω_{nm} can be expressed using

$$\int \frac{(\sin^2 \alpha x)\, dx}{x^2} = \pi \alpha \qquad \text{(A5.5.16)}$$

to show that, by integration, a_m^2 becomes proportional to t, as it must on physical grounds, to yield

$$a_m^2 = t(4\pi^2 e^2/c\hbar^2)|z_{mn}|^2\, dI/d\omega \qquad \text{(A5.5.17)}$$

with ω the angular frequency. Thus the rate of absorption is a constant, independent of time, and is given by

$$\text{Rate of absorption} = (4\pi^2 e^2/c\hbar^2)|z_{mn}|^2\, dI/d\omega \qquad \text{(A5.5.18)}$$

For the case of isotropic radiation of volume energy density $u(\omega)\, d\omega$, used below to derive the Einstein coefficient for spontaneous emission from the rate of absorption, one can finally write

$$\text{Rate of absorption} = (4\pi^2 e^2/3\hbar^2)|\mathbf{x}_{mn}|^2 u(\omega_{mn}) \qquad \text{(A5.5.19)}$$

where the factor of $1/3$ has come from averaging over the direction of polarization. The vector dipole moment $e\mathbf{x}$ has now replaced ez in this step. We now turn to the relationship between emission and absorption.

Einstein Coefficients

Having dealt with selection rules, we shall turn to the second objective of this appendix, namely the derivation of expressions for the so-called Einstein coefficients.

We note with Fermi (1961) that the relation between emission and absorption could be derived from quantum electrodynamics, but that the use of Einstein's A and B coefficients makes the treatment much simpler.

We start by introducing the coefficient B (related to B above), through the rate of transitions from n to m. This is expressed in terms of the energy density $u(\omega)$ of radiation of angular frequency ω and $N(n)$, which denotes the number of atoms in state n. We then define the coefficient B precisely through

$$\text{Rate of transitions from } n \text{ to } m = Bu(\omega)N(n) \qquad \text{(A5.5.20)}$$

From this derivation, it follows that B has the explicit form

$$B = (4\pi^2 e^2/3\hbar^2)|\mathbf{x}_{mn}|^2 \qquad \text{(A5.5.21)}$$

Now we consider the rate of transitions from upper state m to lower state n to have two terms: a spontaneous emission, measured by a coefficient A, and a forced or stimulated emission term, say, $Cu(\omega)$. Then the expression for the rate from m to n may be expressed as

$$\text{Rate from } m \text{ to } n = [A + Cu(\omega)]N(m) \tag{A5.5.22}$$

where evidently the rate is proportional to the number of atoms in state m, denoted $N(m)$.

However, we now note that in thermal equilibrium at temperature T the ratio $N(m)/N(n)$ is determined by the Boltzmann distribution as

$$N(m)/N(n) = \exp[-(E^{(m)} - E^{(n)})/k_B T]$$
$$= \exp(-\hbar\omega_{mn}/k_B T) \tag{A5.5.23}$$

where ω_{mn} is evidently the transition angular frequency corresponding to the spacing between levels m and n.

It is quite clear that at equilibrium the rate $n \to m$ must equal the rate from $m \to n$. Hence we can write the equation

$$A/Bu(\omega) + C/B = N_n/N_m = \exp(\hbar\omega/k_B T) \tag{A5.5.24}$$

However, from Planck's law for the distribution of energy in the spectrum of a blackbody we can write [see, e.g., Roberts and Miller (1960)]

$$u = \frac{\hbar\omega^2/\pi^2 c^3}{\exp(\hbar\omega/k_B T) - 1} \tag{A5.5.25}$$

and we find that by inserting equation (A5.5.25) into equation (A5.5.24), we get the result

$$\frac{\pi^2 c^3}{\hbar\omega^3}\frac{A}{B}\left[\exp\left(\frac{\hbar\omega}{k_B T}\right) - 1\right] + \frac{C}{B} = \exp\left(\frac{\hbar\omega}{k_B T}\right) \tag{A5.5.26}$$

Clearly this equation must hold for all temperatures T and hence

$$(\pi^2 c^3/\hbar\omega^3)A/B = 1 \qquad \text{and} \qquad C/B = 1 \tag{A5.5.27}$$

In summary, therefore, one sees the need for the A coefficient of spontaneous emission, related to B by

$$A = (\hbar\omega^3/\pi^2 c^3)B \tag{A5.5.28}$$

However, B has already been calculated in equation (A5.5.21) and hence we find for spontaneous transitions

$$A = \tfrac{4}{3}(e^2\omega^2/\hbar c^3)|x_{mn}|^2 \tag{A5.5.29}$$

A5.6. SPIN–ORBIT COUPLING

Spin–orbit coupling allows the mixing of energy levels which, in its absence, would be orthogonal. In this appendix the topic will be introduced with reference to atoms followed by a brief account of spin–orbit coupling in diatomic molecules. In the latter area, experimental data are provided by high-resolution spectroscopy (see, e.g., Richards *et al.*, 1981).

Effects in the Atomic Spectra of Alkali Metals

The **LS** (Russell–Saunders) coupling scheme, with **L** the total orbital and **S** the total spin angular momentum, suggests that the ground and first-excited electronic states of the simplest alkali atom, Li, should be ^2S and ^3P arising from electronic configurations $1s^2 2s$ and $1s^2 2p$, respectively. The energy difference between these two states can be usefully viewed as arising from the different spatial distribution of the $2s$ and $2p$ electrons. However, the electronic transition between these states gives rise not to one but to two lines in the atomic spectrum. For the next heaviest alkali, the Na D lines are especially easy to detect experimentally.

In fact, Figure A.18 shows the transitions responsible for the D lines of the Na spectrum. From this figure, it can be seen that the fact that two transitions are observed is due to the splitting of the ^2P level into $^2P_{1/2}$ and $^2P_{3/2}$ sublevels. The origin of this small splitting is spin–orbit coupling.

A simplified model which nevertheless gives the gist of the origin of this effect is to regard the spinning electron as a magnet which then interacts with the magnetic field created by its own orbital motion about the nucleus. Such an effect would obtain in a classical situation. However, at the atomic level, quantum mechanics leads to a limitation on the orientation of the effective magnetic moment, which is related to the total angular momentum of the system.

Rather roughly, the splitting is found to depend on the integral

$$\int \psi (Z_{\text{eff}}/r^3) \psi \, d\tau$$

where ψ is the wave function, Z_{eff} is an effective nuclear charge of the screened nucleus, while r is the distance of the electron from that nucleus. As the average distance of an electron from the nucleus for light atoms is also roughly inversely proportional to the nuclear charge, the actual splitting in the ^2P level, might, on the basis of evaluating the above integral, be expected to be proportional to the atomic number Z to the fourth power. The point about this crude argument is that it shows a marked increase in the importance of spin–orbit coupling with

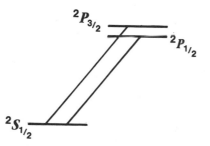

FIGURE A.18. Transitions responsible for the D lines of the Na spectrum.

increasing atomic number. This is qualitatively well borne out, in fact, in the splitting of the $^2P_{3/2}$ and $^2P_{1/2}$ levels in the alkali metal atoms, going from about 17 cm^{-1} in Na to some 550 cm^{-1} in Cs.

Spin in a one-electron atom is most properly treated from the Dirac relativistic wave equation (see Appendix A5.7). The reduction of this equation to nonrelativistic form, and use of atomic units, gives the spin–orbit interaction in this one-electron atom as

$$H_{SO} = \tfrac{1}{2}\alpha^2(1/r)(\partial V/\partial r)\mathbf{l} \cdot \mathbf{s} \qquad \text{(A5.6.1)}$$

where V is the appropriate central field while α is the fine-structure constant $e^2/\hbar c$. This may be written in the conventional form $\zeta \mathbf{l} \cdot \mathbf{s}$ which provides a definition of the spin–orbit coupling constant ζ for a one-electron atom.

Loosely, the form of equation (A5.6.1) can be understood by considering the interaction energy $-\boldsymbol{\mu} \cdot \mathbf{B}$ of the magnetic moment $\boldsymbol{\mu}$ of the electron with magnetic field $\mathbf{B}$. In the absence of an external field, $\mathbf{B}$ from electromagnetic theory is proportional to the vector product of the velocity $\mathbf{v}$ of the electron with the electric field generated according to the central potential $V(r)$;

$$\mathbf{E} = 1/r(\partial V/\partial r)\mathbf{r} \qquad \text{(A5.6.2)}$$

Relating magnetic moment to spin $\mathbf{s}$, and using the usual vector expression $\mathbf{r} \times \mathbf{p}$ for the orbital angular momentum $\mathbf{l}$, $\mathbf{p}$ being $m\mathbf{v}$, the momentum, one obtains the correct form of the result (A5.6.1), though the coefficient is not correctly given for reasons that are discussed, for example, by Richards *et al.* (1981).

Effects in Molecules

Spin–orbit coupling manifests itself in linear molecules in very much the same manner as in atoms. Figure A.19 depicts the multiplet splittings in a $^3\Pi$-state, which can be compared with an analogous diagram to Figure A.18 for a ^3P-state of an atom. The axial symmetry of the molecule, it turns out, causes the sublevels of $^3\Pi$ to be equally spaced.

It is customary to define a spin–orbit coupling constant A by equating the hyperfine spacings to $2A\Lambda\Sigma$, where Λ is the component of the total electronic orbital angular momentum and Σ the component of the total electronic spin angular momentum along the internuclear axis. As is also true in atoms, A is positive for regular states (less than half-filled shells) and negative for inverted states (electronic shells more than half-filled). This means that the sign of the

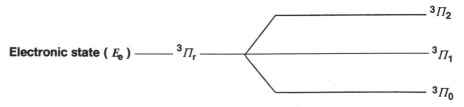

FIGURE A.19. Spin–orbit coupling and the multiplet splittings in $^3\Pi$ state.

spin–orbit coupling constant proves a powerful way of finding out whether a state is regular or inverted. Since spin–orbit coupling constants are usually very different for different electronic states, knowledge of A is important in the identification of states.

It turns out, as a first approximation, that molecular spin–orbit coupling constants may be found by using atomic spin–orbit coupling constants together with Mulliken population analyses.

A5.7. DIRAC'S RELATIVISTIC WAVE EQUATION FOR ONE ELECTRON

So far, we have been concerned solely with the nonrelativistic Schrödinger equation. However, it is rather straightforward to show that for the inner-shell electrons of a heavy atom (e.g., the $1s$ electrons in Pb), and equally so for molecules containing heavy atoms, the electronic velocities are so high that relativistic effects become very important. Though there remain difficulties of principle at the heart of a relativistic theory of a heavy atom [see, e.g., Lindgren and Morrison (1982); March (1983)], useful approximate theories do exist, and we shall consider them briefly in this appendix and also the following one.

We recall first that in classical mechanics the Hamiltonian H (always written in terms of coordinates and momentum; rather than in terms of velocities) is equal to the energy E:

$$H = E \qquad (A5.7.1)$$

and in quantum mechanics, where every dynamical variable has an associated operator, we have instead

$$H\Psi = i\hbar\, \partial\Psi/\partial t \qquad (A5.7.2)$$

where we have used the operator result

$$E \to i\hbar\, \partial/\partial t \qquad (A5.7.3)$$

which embodies the uncertainty relation $|\Delta E\, \Delta t \sim h$ that exists between energy E and time t.

With the Hamiltonian operator H obtained from the classical result for a single particle moving in a potential energy $V(\mathbf{r})$

$$H = (p^2/2m) + V(\mathbf{r}) \qquad (A5.7.4)$$

namely

$$H_{\text{operator}} = -(\hbar^2/2m)\nabla^2 + V(\mathbf{r}) \qquad (A5.7.5)$$

space and time appear on a different footing. To press the point, in equation (A5.7.2), we have a first-order time derivative, whereas in the nonrelativistic Hamiltonian operator (A5.7.5) a second-order spatial operator ∇^2 appears.

Thus the Schrödinger equation obtained by combining equations (A5.7.2) and (A5.7.5) cannot be invariant under a Lorentz transformation [see, e.g., McCrea (1954)] and is not compatible with the special theory of relativity.

Dirac Wave Equation

It should be clear that in order to obtain a relativistic wave equation, one must at least start out from the correct relativistic energy expression for a free electron having momentum p, namely

$$(\text{Energy})^2 = c^2 p^2 + m^2 c^4 \tag{A5.7.6}$$

where m is now the rest mass of the electron, and c is the velocity of light. For small momenta this evidently yields

$$\begin{aligned} \text{Energy} &= mc^2[1 + \tfrac{1}{2}(c^2 p^2 / m^2 c^4) + \cdots] \\ &= mc^2 + p^2/2m + \cdots \end{aligned} \tag{A5.7.7}$$

the first term being the energy associated with the rest mass according to the Einstein equivalence between mass and energy. The second term in the second line of equation (A5.7.7) is simply the nonrelativistic kinetic energy, precisely as in equation (A5.7.4).

Unfortunately, to form a wave equation which is compatible with special relativity, by analogy with equation (A5.7.1), we require not the square of the energy, as in equation (A5.7.6), but the energy itself. Thus, Dirac (1928) proposed writing the quantity $p^2 + m^2 c^2$ appearing on the right-hand side of equation (A5.7.6) as a perfect square; that is, we put

$$\begin{aligned} p^2 + m^2 c^2 &= p_x^2 + p_y^2 + p_z^2 + m^2 c^2 \\ &= (\alpha_x p_x + \alpha_y p_y + \alpha_z p_z + \alpha_m mc)^2 \end{aligned} \tag{A5.7.8}$$

where the vector momentum $\mathbf{p}$ has components (p_x, p_y, p_z) in Cartesian coordinates. It is immediately evident that the α's defined in equation (A5.7.8) cannot be ordinary scalars, and we obtain the rules these must satisfy by multiplying out the right-hand side of equation (A5.7.8) and then equating coefficients. It follows almost immediately that

$$\alpha_x^2 = \alpha_y^2 = \alpha_z^2 = \alpha_m^2 = 1$$

$$\alpha_x \alpha_y = -\alpha_y \alpha_x, \qquad \alpha_x \alpha_z = -\alpha_z \alpha_x, \qquad \alpha_x \alpha_m = -\alpha_m \alpha_x \tag{A5.7.9}$$

$$\alpha_y \alpha_z = -\alpha_z \alpha_y, \qquad \alpha_y \alpha_m = -\alpha_m \alpha_y, \qquad \alpha_z \alpha_m = -\alpha_m \alpha_z$$

It is now clear that the α's must be operators, since they do not commute (i.e., $ab \neq ba$), and provided we satisfy the relations (A5.7.9) we can write the Hamiltonian H for an electron moving in a field corresponding to potential energy $V(\mathbf{r})$ as

$$H = V - c\alpha_x p_x - c\alpha_y p_y - c\alpha_z p_z - \alpha_m mc^2 \qquad \text{(A5.7.10)}$$

Here the negative sign has been chosen in taking the square root of equation (A5.7.8), but this is purely a convention and does not alter the physical content of equation (A5.7.10).

Finally we use the operator substitution $p_x \rightarrow (\hbar/i)\,\partial/\partial x$ etc. to obtain the Dirac Hamiltonian for an electron moving in a potential energy $V(\mathbf{r})$ as

$$H = V + ic\hbar\alpha_x \frac{\partial}{\partial x} + ic\hbar\alpha_y \frac{\partial}{\partial y} + ic\hbar\alpha_z \frac{\partial}{\partial z} - \alpha_m mc^2 \qquad \text{(A5.7.11)}$$

While the theory strictly does not require one to take specific forms for the operators α, it is usually much easier in solving practical problems in chemical physics to do so.

As in the Pauli spin theory (see, e.g., Schiff, 1955) it is convenient to choose the operators to be matrices. It can then be shown that 4×4 matrices are the smallest that can satisfy the relations (A5.7.9).

The set (by no means unique) adopted by Dirac is

$$\alpha_x = \begin{pmatrix} 0 & 0 & 0 & 1 \\ 0 & 0 & 1 & 0 \\ 0 & 1 & 0 & 0 \\ 1 & 0 & 0 & 0 \end{pmatrix}, \qquad \alpha_y = \begin{pmatrix} 0 & 0 & 0 & -i \\ 0 & 0 & i & 0 \\ 0 & -i & 0 & 0 \\ i & 0 & 0 & 0 \end{pmatrix}$$

$$\text{(A5.7.12)}$$

$$\alpha_z = \begin{pmatrix} 0 & 0 & 1 & 0 \\ 0 & 0 & 0 & -1 \\ 1 & 0 & 0 & 0 \\ 0 & -1 & 0 & 0 \end{pmatrix}, \qquad \alpha_m = \begin{pmatrix} 1 & 0 & 0 & 0 \\ 0 & 1 & 0 & 0 \\ 0 & 0 & -1 & 0 \\ 0 & 0 & 0 & -1 \end{pmatrix}$$

With the α's chosen as these 4×4 matrices, the Dirac wave function Ψ in the wave equation (A5.7.2) must be a column vector having four components, namely

$$\Psi = \begin{pmatrix} \psi_1 \\ \psi_2 \\ \psi_3 \\ \psi_4 \end{pmatrix} \qquad \text{(A5.7.13)}$$

We then obtain from equations (A5.7.2), (A5.7.11), and (A5.7.13) four ordinary coupled differential equations to solve for the components ψ_1 to ψ_4 of the column vector (A5.7.13).

Central Field Solutions of the Dirac Equation

As we shall confine ourselves here to the central field solutions of the Dirac equation, with the hydrogen atom especially in mind, the above set of Cartesian differential equations for the components ψ_1 to ψ_4 is required in spherical polar coordinates, directly generalizing the nonrelativistic procedure discussed in detail in Appendix A1.1. To do this, one simply uses the relations

$$\frac{\partial}{\partial x} + i\frac{\partial}{\partial y} = \exp(i\phi)\left(\sin\theta\,\frac{\partial}{\partial r} + \frac{\cos\theta}{r}\frac{\partial}{\partial\theta} + \frac{i}{r\sin\theta}\frac{\partial}{\partial\phi}\right) \tag{A5.7.14}$$

and

$$\frac{\partial}{\partial z} = \cos\theta\,\frac{\partial}{\partial r} - \frac{\sin\theta}{r}\frac{\partial}{\partial\theta} \tag{A5.7.15}$$

while $(\partial/\partial x) - i(\partial/\partial y)$ is obtained from equation (A5.7.14) by replacing i by $-i$ on the right-hand side.

The resulting equations, specialized to the central field case when $V(\mathbf{r}) = V(|\mathbf{r}|) \equiv V(r)$, can then be separated as in the nonrelativistic case. However, as it can be shown that electron spin and magnetic moment are now included in the Dirac equation itself (unlike the Schrödinger equation, where the Pauli spin theory has to be grafted on), the total angular momentum, $\mathbf{J}$, say, plays the role for the Dirac equation that the orbital angular momentum $\mathbf{L}$ played for the Schrödinger equation.

Hence we seek a wave function Ψ satisfying the equations

$$J^2\Psi = j(j+1)\hbar^2\Psi, \qquad j = \tfrac{1}{2}, \tfrac{3}{2}, \ldots \tag{A5.7.16}$$

j being half-integral as spin is now included and

$$J_z\Psi = m\hbar\Psi, \qquad m = \pm\tfrac{1}{2}, \pm\tfrac{3}{2} \tag{A5.7.17}$$

with $j \geq |m|$. The solutions Ψ for a given m and j can then be expressed, as in nonrelativistic theory, in terms of the spherical harmonics Y as

$$\left.\begin{aligned}
\psi_1 &= f(r)\left(\frac{j+1-m}{2j+2}\right)^{1/2} Y_{j+\frac{1}{2},m-\frac{1}{2}}(\theta,\phi) + g(r)\left(\frac{j+m}{2j}\right)^{1/2} Y_{j-\frac{1}{2},m-\frac{1}{2}}(\theta,\phi) \\[2mm]
\psi_2 &= -f(r)\left(\frac{j+1+m}{2j+2}\right)^{1/2} Y_{j+\frac{1}{2},m+\frac{1}{2}}(\theta,\phi) + g(r)\left(\frac{j-m}{2j}\right)^{1/2} Y_{j-\frac{1}{2},m+\frac{1}{2}}(\theta,\phi) \\[2mm]
\psi_3 &= F(r)\left(\frac{j+1-m}{2j+2}\right)^{1/2} Y_{j+\frac{1}{2},m-\frac{1}{2}}(\theta,\phi) + G(r)\left(\frac{j+m}{2j}\right)^{1/2} Y_{j-\frac{1}{2},m-\frac{1}{2}}(\theta,\phi) \\[2mm]
\psi_4 &= -F(r)\left(\frac{j+1+m}{2j+2}\right)^{1/2} Y_{j+\frac{1}{2},m+\frac{1}{2}}(\theta,\phi) + G(r)\left(\frac{j-m}{2j}\right)^{1/2} Y_{j-\frac{1}{2},m+\frac{1}{2}}(\theta,\phi)
\end{aligned}\right\}$$

$$\tag{A5.7.18}$$

We note, for completeness, that the Y's are related explicitly to the associated Legendre functions $P_l^m (\cos \theta)$ by

$$Y_{lm}(\theta, \phi) = (-1)^m \left[\frac{(2l + 1)(l - m)!}{4\pi(l + m)!} \right]^{1/2} P_l^m(\cos \theta) \exp(im\phi) \qquad \text{(A5.7.19)}$$

Substituting these results into the spherical polar forms of the coupled equations for ψ_1 to ψ_4 finally yields the equations to determine the four functions f, g, F, and G:

$$\left. \begin{array}{l} \dfrac{df}{dr} + (j + \tfrac{3}{2}) \dfrac{f}{r} + \dfrac{i}{\hbar c} [E - V(r) - mc^2] G = 0 \\[2ex] \dfrac{dG}{dr} - (j - \tfrac{1}{2}) \dfrac{G}{r} + \dfrac{i}{\hbar c} [E - V(r) + mc^2] f = 0 \\[2ex] \dfrac{dF}{dr} + (j + \tfrac{3}{2}) \dfrac{F}{r} + \dfrac{i}{\hbar c} [E - V(r) + mc^2] g = 0 \\[2ex] \dfrac{dg}{dr} - (j - \tfrac{1}{2}) \dfrac{g}{r} + \dfrac{i}{\hbar c} [E - V(r) - mc^2] F = 0 \end{array} \right\} \qquad \text{(A5.7.20)}$$

When these radial equations are investigated for the hydrogen-like atom with $V(r) = -Ze^2/r$, that is, with one electron moving in the bare Coulomb field of a nucleus of charge Ze, the famous Sommerfeld formula for the energy level spectrum results, namely

$$E = mc^2 \{1 + \zeta^2 / [(n - j - \tfrac{1}{2} + \gamma)^2]\}^{-1/2} \qquad \text{(A5.7.21)}$$

Removing the electron rest mass and using the explicit forms

$$\zeta = Z(e^2/\hbar c) \qquad \text{and} \qquad \gamma = [(j + \tfrac{1}{2})^2 - \zeta^2]^{1/2} \qquad \text{(A5.7.22)}$$

the Sommerfeld formula

$$E' = mc^2 \left\{ \left[1 + \frac{Z^2 e^4 / \hbar^2 c^2}{\{n - j - \tfrac{1}{2} + [(j + \tfrac{1}{2})^2 - Z^2 e^4/\hbar^2 c^2]^{1/2}\}^2} \right]^{-1/2} - 1 \right\} \qquad \text{(A5.7.23)}$$

results. For small values of $Ze^2/\hbar c$, this evidently reduces to the nonrelativistic formula (A1.1.41) for $Z = 1$ or for a general nuclear charge Ze to

$$E'_{\text{nonrel}} = -(Z^2/2n^2)e^2/a_0 \qquad \text{(A5.7.24)}$$

A5.8. RELATIVISTIC ELECTRON DENSITY THEORY FOR MOLECULES COMPOSED OF HEAVY ATOMS

The Dirac relativistic wave equation for one electron was derived in Appendix A5.7, and solutions discussed for the wave functions and energy levels of the hydrogen atom. While a fully satisfactory relativistic theory of multielectron systems is still lacking, the electron density description presented in Chapter 4 can be generalized to conform with the basic requirements of the special theory of relativity. The purpose of the present appendix is to effect this generalization.

To make a start on this, let us return to equation (4.3) for the chemical potential μ of the electron cloud in the molecule. In writing equation (4.3), one has made use of the familiar nonrelativistic expression $E = p^2/2m$ relating energy E and momentum p of a free particle. This is the point then at which to introduce the relation

$$E^2 = c^2 p^2 + m^2 c^4 \tag{A5.8.1}$$

for the energy in terms of momentum p in special relativity, m being the rest mass of the particle and c the velocity of light. Hence, equation (4.3) must be replaced by

$$\mu = [c^2 p_m^2(\mathbf{r}) + m^2 c^4]^{1/2} - mc^2 + V(\mathbf{r}) \tag{A5.8.2}$$

where we have subtracted out the rest-mass energy mc^2. Equation (A5.8.2) can be expanded to lowest order in $1/c$ when equation (4.3) is regained. Using equation (4.5) in equation (A5.8.2), one has the relativistic generalization of the electron density–potential relation in equation (4.6).

This theory, due to Vallarta and Rosen (1932), can be used to determine the self-consistent field in heavy atoms, as well as their binding energies. In contrast to the nonrelativistic case, it is necessary to replace the point-charge model of the nucleus, so useful there, by a nucleus of finite size. This is to allow one to impose normalization on the electron density.

We conclude this appendix by noting: (i) The relativistic electron density theory presented above has been worked out self-consistently for heavy atoms by Hill *et al.* (1985); the interested reader can consult their work for the generalization of equation (A4.5.11) for the binding energy of heavy positive atomic ions. (ii) The above theory has been related [Bettolo and March (1981)], for large numbers of electrons N, to the relativistic generalization of the $1/Z$ expansion presented in Appendix A1.3, the latter being effected by Layzer and Bahcall (1962) [see also Senatore and March (1985)].

A6.1. THE JAHN–TELLER EFFECT

The Jahn–Teller (1937) theorem states that for an electronically degenerate molecule, there will exist a distortion of the nuclear framework that lowers the energy and reduces the symmetry of the system.

The only exceptions are linear systems, which are stable against bending, and Kramers' degeneracy. This is embodied in Kramers' (1929) theorem, which states that for a system with an odd number of electrons in the absence of a magnetic field, all the states have even degeneracy. This theorem is valid in the presence of an arbitrary electrostatic field and of spin–orbit coupling. A pair of degenerate levels exhibiting this degeneracy is called a Kramers doublet. The simplest illustration is a single electron in an orbitally nondegenerate state, where the twofold degeneracy is simply the spin degeneracy. Clearly, therefore, such a Kramers' degeneracy can only be removed by a magnetic field.

Electronically degenerate systems are intrinsically unstable against such asymmetric Jahn–Teller distortions. The instability manifests itself in several ways, and it is possible to detect the lowering of symmetry in spin resonance (cf. Section 5.11).

The asymmetric state, having lower symmetry than the Hamiltonian, can be attained by several equivalent but distinct distortions. Indeed, many of the novel features of Jahn–Teller systems are a consequence of these equivalent distinct distortions. Useful reviews of the Jahn–Teller instability are those of Sturge (1967), Ham (1972), and Englman (1972).

A6.2. KOOPMANS' THEOREM AND ITS USE IN INTERPRETING PHOTOELECTRON SPECTRA

As was discussed in Section 5.9 of the main text, photoelectron spectroscopy measures the ionization potentials of individual electrons from molecules. Molecular orbital calculations should provide similar data and have been widely used in the interpretation of the spectra, aided by Koopmans' theorem. There are, however, some problems inherent in this approach. Here, we follow the discussion of Richards (1969), who exposes these for both closed-shell molecules and open-shell cases where additional complications arise.

Closed-Shell Molecules

For a closed-shell molecule, the orbital energy computed by, say, the Roothaan self-consistent field (SCF) method, is approximately equal to the ionization potential of an electron from that orbital. This is, in fact, what is known as Koopmans' theorem.

Richards stresses that three approximations underly Koopmans' idea:

1. No redistribution of electronic orbitals in space is taken into account on removing an electron, i.e., the one-electron orbitals of the neutral atom are also assumed appropriate for the ion.

2. Self-consistent Hartree–Fock–Roothaan theory is nonrelativistic. In essence, while not troublesome for outer electrons, which are moving slowly, in heavier molecules the inner electrons move fast. Koopmans' theorem, in essence, will still be valid if the relativistic energy is the same in both molecule and ion. This may be reasonable if only outer electrons are removed, and it would be surprising if it were also true for the removal of core electrons. As an example quoted also by Richards in this context, the following numbers give some indication of the magnitude of the relativistic contributions of various subshells to the total energy of Ar: $1s$: 1.226, $2s$: 0.235, $2p$: 0.257, $3s$: 0.025, and $3p$: 0.022, these being measured in atomic units (1 a.u. = 27.2 eV).

3. Correlation energy is neglected in Hartree–Fock–Roothaan theory. This error arises from the fact that electrons with antiparallel spins are essentially uncorrelated in the Hartree–Fock theory, parallel-spin electrons having only Pauli principle correlations (exchange) between them. Naturally, antiparallel-spin electrons also avoid one another to some lesser extent, owing solely to electrostatic Coulombic repulsion. Koopmans' theorem, in taking the difference between two relatively simple energy expressions makes the assumption, at least implicitly, that the correlation energy will be accurately the same in a neutral molecule and a positive ion. This is clearly not going to be precisely true from the foregoing discussion of the origin of correlation energy; it must be expected to be different in the parent molecule and in the ion.

Open-Shell Molecules

For open-shell molecules, the three objections leveled above against Koopmans' theorem apply with equal force. In addition there are two more sources of difficulty in its noncritical application [Richards (1969)], which we now consider.

First of all, the orbital energies, $\varepsilon_i^{\text{SCF}}$, say, found from Hartree–Fock calculations, can be viewed as constants introduced to avoid awkward normalization constraints in applying the variational method (the analogue of the important chemical potential μ in the electron density description, which takes care of the normalization condition on the electron density). However, new difficulties arise for the open-shell case. Roughly speaking the "constant" eigenvalues representing constraints in the closed-shell case, namely ε_i, remain in the matrix form ε_{ij} in the open-shell case, and then no clear physical interpretation is possible. Koopmans' theorem does not apply in any clear-cut way, even though it may still give approximate numbers of reasonable magnitude. If one uses different orbitals for different spins, it may be possible to give a clearer interpretation, though this is not generally true, as discussed, for instance, by Laidlaw and Birss (1964).

The second difficulty encountered with open-shell molecules may also be present in the case of some excited states of ions derived from closed-shell molecules. For open-shell states, the wave functions may not be usefully written as single Slater determinants. In the energy expression for such multideterminant states, there will be cross terms due to matrix elements between the component determinants. These cross terms come in when taking energy differences and may result in an energy difference which cannot be represented by an orbital energy $\varepsilon_i^{\text{SCF}}$, even supposing that all the other assumptions of Koopmans' theorem are valid.

Richards has pointed out that if sufficient care is taken, MO calculations can still be used to resolve problems in the interpretation of photoelectron spectra, and notes the following:

1. The energy of the molecule and each separate state of the ion should be computed as closely as possible to the Hartree–Fock limit. The variation of the SCF eigenvalues with the basis set is thereby removed. Estimates should then be made of the differences in correlation and relativistic energy among the various energy states. If this is done with care, even using semiempirical notions, the energy spacings, and, with more certainty, the order of the ionization potentials should be given correctly.

2. The computed potential curves and surfaces of the levels are frequently more realistic than the absolute energies so that the fine structure of the photoelectron spectrum should be amenable to interpretation. Work exists, for instance, in calculating the vibrational spacing of various states of the ions (Richards and Wilson, 1968). It is now also feasible to compute accurate values of fine-structure effects such as spin–orbit coupling (see Appendix A5.6) and spin–spin splittings in excited levels.

A7.1. SYMMETRY ARGUMENTS FOR ELECTROCYCLIC REACTIONS

We summarize here an alternative approach to electrocyclic reactions, due originally to Longuet-Higgins and Abrahamson (1965). Only orbitals, either occupied or unoccupied, that are involved in bonds being made or broken during the course of the reaction are involved. For butadiene, there are four π-MOs; and for cyclobutene there are two π-MOs associated with the isolated two-center π-bond and two σ-MOs associated with the new C—C σ-bond. These orbitals and their energies are shown in Figure A.20.

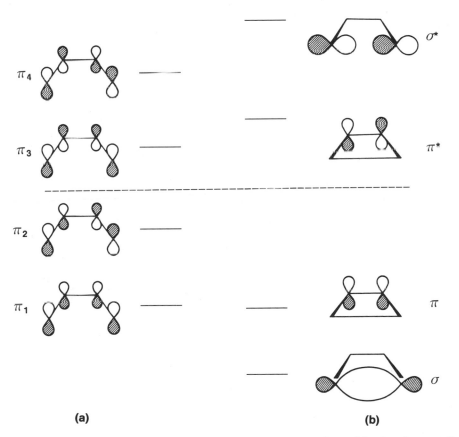

(a)   **(b)**

FIGURE A.20. MOs associated with bonds being broken or formed in the electrocyclic closure of (a) *cis*-1,3-butadiene to (b) cyclobutene [after Lowe (1978)].

The σ- and σ^*-MOs in cyclobutene are more widely separated than the π- and π^*-states because the $p\sigma$-AOs overlap more strongly. Also the σ-MO is assumed to be lower than π_1 of butadiene. These details, however, are not absolutely essential: what one must be certain of is that one has correctly divided the occupied from the unoccupied MOs on the two sides. The dashed line in Figure A.20 separates occupied and unoccupied states.

The next step in the argument is to decide which symmetry elements are preserved throughout the idealized reactions we wish to treat. Let us consider first the reactants and products. These have C_{2v} symmetry, that is, a twofold rotational axis C_2 and two reflection planes σ_1 and σ_2 containing the C_2 axis (cf. Figure A.21). A conrotatory twist preserves C_2 but σ_1 and σ_2 are lost as symmetry operations during the intermediate stages between reactant and product. A disrotatory twist preserves σ_1 but destroys C_2 and σ_2. Therefore, when energy levels are connected together for the disrotatory mode, one must connect levels of the same symmetry for σ_1, but for the conrotatory mode they must agree in symmetry for C_2. The σ_2-plane does not apply to either mode and is therefore not involved in the arguments. The symmetries of each MO are readily determined by reference to Figure A.20 and are collected in Table A.1. Those assignments lead to two different correlation diagrams, one for each mode, and these are shown in Figure A.22.

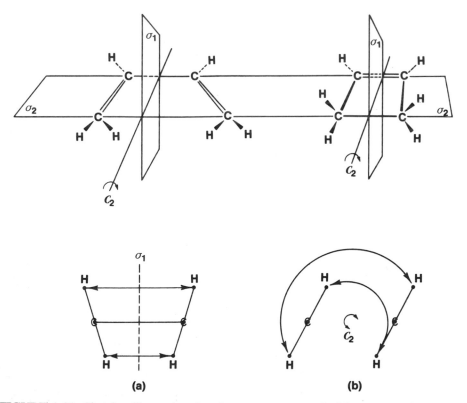

FIGURE A.21. Sketches illustrating that the conrotatory mode (b) preserves the C_2 axis while the disrotatory mode (a) preserves the reflected plane σ_1 [after Lowe (1978)].

TABLE A.1
Symmetries for C_{2v} MOs[a]

	Butadiene			Cyclobutene	
MO	σ_1	C_2	MO	σ_1	C_2
π_1	S[b]	A	σ	S	S
π_2	A	S	π	S	A
π_3	S	A	π^*	A	S
π_4	A	S	σ^*	A	A

[a]From Lowe (1978).
[b]S is symmetric; A antisymmetric.

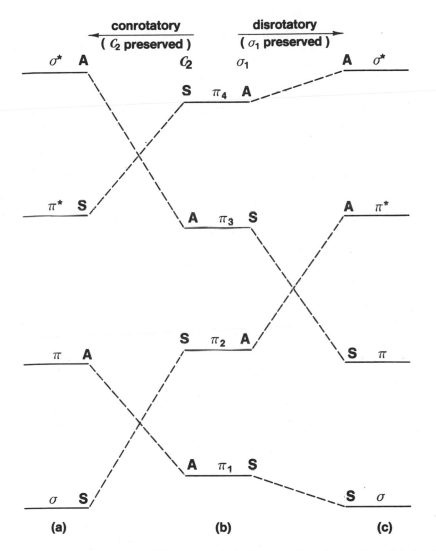

FIGURE A.22. A pair of two-sided correlation diagrams (one for each mode) for the electrocyclic reactions of *cis*-1,3-butadiene: (a) cyclobutene, (b) butadiene, (c) cyclobutene [after Lowe (1978)].

One sees that a curve crossing occurs in these correlation diagrams, but as it is always lines of different symmetry that cross, there is no violation of the noncrossing rule.

If we are considering a thermal reaction, the lowest two π-MOs of butadiene are occupied. These correlate with the lowest two MOs of cyclobutene if the conrotatory mode is followed, and the thermal conversion of *cis*-butadiene to cyclobutene by a conrotatory closure is said to be symmetry-allowed. The other mode correlates π_2 with the empty π^* cyclobutene MO. Following this route moves the reactant toward doubly excited cyclobutene. Even though one might anticipate de-excitation along the route, the energy required in the early stages would still be much higher than would be needed for the symmetry-allowed mode. This is said to be a symmetry-forbidden reaction.

If we now envisage photoexcitation of *cis*-butadiene to have generated a state associated with the configuration $\pi_1^2\pi_2\pi_3$ and follow the development of the species for the two modes of reaction, we note that the disrotatory route leads to cyclobutene in the configuration $\sigma^2\pi\pi^*$ while the conrotatory mode gives $\sigma\pi^2\sigma^*$. Both of these are excited, but the first corresponds to the lowest excited configuration $\pi \rightarrow \pi^*$, while the second corresponds to a very high-energy excitation $\sigma \rightarrow \sigma^*$. Therefore the former is allowed, since it goes from the lowest excited reactant to the lowest excited product and the latter is forbidden.

The correlation diagrams of Figure A.22 lead us to the same predictions as the argument given in the main text, but it is worth stressing the difference between the two procedures:

1. The frontier orbital approach involves considering the highest occupied MO and then judging overlap changes as the nuclei move, by qualitative MO arguments.

2. The correlation diagram approach requires sketching all the MOs, both occupied and unoccupied, of both reactant and product involved in bonds breaking or forming, ordering the corresponding energy levels, and finding symmetry elements preserved throughout the reaction. Once this has been done, the levels are connected by correlation lines without relying on MO-type reasoning. Thus, the correlation diagram method is the more rigorous approach and should be used whenever the system under discussion has enough symmetry to make the approach feasible. For processes with lower symmetry, the other method offers a fruitful way forward [see Lowe (1978)].

A7.2. CHEMICAL REACTIONS: ARRHENIUS' EMPIRICAL WORK, THE COLLISION THEORY, AND THE ABSOLUTE RATE OR TRANSITION STATE THEORY

Arrhenius discovered in 1889 that the temperature dependence of many reactions follow the equation

$$k = A\,e^{-E_a/RT} \qquad\qquad (A7.2.1)$$

where k is the rate constant, E_a is the activation energy, and A is a constant representing the total frequency of encounters between reactants. We can write equation (A7.2.1) above as

$$\ln k = \ln A - E_a/RT \qquad \text{(A7.2.2)}$$

A plot of $\ln k$ *vs.* $1/T$ gives a straight line with a negative slope equal to E_a/R and the intercept of the ordinate yields A. It would be useful to be able to calculate both A and E_a. The collision theory and the transition state theory have both attempted to do this. We will now consider both of these theories briefly.

Collision Theory

The fundamental principle of the collision theory is that the reactant molecules must collide before reaction can occur. Hence, the rate of the reaction will be proportional to the number of collisions per unit time. Let us consider, for example, a gas-phase bimolecular, second-order reaction such as $A + B \rightarrow$ Products. The number of collisions between A and B per second at unit concentration, Z_{AB}^0, can be calculated from the kinetic theory of gases, but for our purposes it will suffice to note that

$$Z = \text{constant} \sqrt{T} \qquad \text{(A7.2.3)}$$

where T is the absolute temperature. The constant is not essential to the present consideration.

The number of collisions Z_{AB}^0 per unit time depends on the concentration. Therefore, the reaction rate will be given by the product of a proportionality constant $\times Z_{AB}^0 \times$ actual concentrations of A and B in moles per liter, so that we must develop an expression for the proportionality constant. We can use the Maxwell–Boltzmann distribution of energies to advantage in this theory. For example, in an assembly of gas molecules at a particular temperature, say, T_1, there is a distribution of energies, and at a higher temperature T_2 the distribution is shifted to higher energies (with the average energy moved higher of course).

From the Maxwell–Boltzmann distribution, the fraction of molecules in a particular assembly having greater than the average energy by an amount equal to or greater than E^* is given by

$$\text{Fraction of molecules with } E \geq (E_{av} + E^*) = e^{-E^*/RT} \qquad \text{(A7.2.4)}$$

We note immediately the identity of this expression with that of the empirical Arrhenius equation, where E^* is the activation energy E_a. Hence, equation (A7.2.4) gives the fraction of all collisions involving enough energy to produce a reaction. This fraction is the proportionality factor expressing the dependence of the reaction rate on the actual number of collisions occurring per unit time and volume. It appears, therefore, that we can express the rate of reaction as follows:

Rate = Number of collisions per unit concentration per unit time
    $\times$ fraction of molecules with $E \geq (E_{av} + E_a) \times$ concentrations
    of second-order reaction $A + B \rightarrow$ Products

or

$$\text{Rate} = Z^0_{AB}\, e^{-E_a/RT}[A][B] \tag{A7.2.5}$$

If we compare (A7.2.5) with the second-order reaction expression Rate = k[A][B], we note that in the collision theory

$$k = Z^0_{AB}\, e^{-E_a/RT} \tag{A7.2.6}$$

Equation (A7.2.6) has been found to need adjustment to conform to experimental values. It is common to include an additional pre-exponential term p, the steric factor. The value of p is usually less than unity. However, even the use of the p-factor does not always yield good results by use of the collision theory when compared with the experimental values (see Table A.2). We now turn to the transition state theory, which is an improvement over the collision theory.

Absolute Rate or Transition State Theory

In absolute rate or transition state theory (TST) we have the following consideration

$$\underbrace{A + B \underset{k_2}{\overset{K^{\ddagger}}{\rightleftharpoons}} AB^{\ddagger} \overset{k_d}{\longrightarrow} \text{Products}} \tag{A7.2.7}$$

for a second-order reaction A + B → Products. That is, (A7.2.7) shows the formation of an activated complex $AB^{\ddagger}$ which is in thermodynamic equilibrium with A and B. The rate of decomposition of the complex $AB^{\ddagger}$ can be represented by

$$dx/dt = k_d[AB^{\ddagger}] \tag{A7.2.8}$$

and the overall rate by

$$dx/dt = k_2[A][B] \tag{A7.2.9}$$

If we combine equations (A7.2.8) and (A7.2.9) we have

$$k_2[A][B] = k_d[AB^{\ddagger}] \tag{A7.2.10}$$

or

$$k_2 = \text{Second order rate constant} = k_d[AB^{\ddagger}]/[A][B] \tag{A7.2.11}$$

Now the equilibrium constant $K^{\ddagger}$ as shown in equation (A7.2.7) can be written as

$$K^{\ddagger} = [AB^{\ddagger}]/[A][B] \tag{A7.2.12}$$

Hence from equations (A7.2.11) and (A7.2.12) we have that

$$k_2 = k_d K^{\ddagger} \qquad \text{(A7.2.13)}$$

At this point we note that the rate constant for the decomposition of the transition state is directly related to the frequency of vibration of the atoms in the "molecule" AB and, hence, using classical vibrational theory we can write that

$$k_d = v_d = RT/Nh \qquad \text{(A7.2.14)}$$

where v_d is the frequency of vibrations necessary to decompose $AB^{\ddagger}$, h is Planck's constant, R is the gas constant, and N is Avogadro's number. Using equation (A7.2.14) we have

$$k_2 = RTK^{\ddagger}/Nh \qquad \text{(A7.2.15)}$$

However, from thermodynamics we know that

$$\Delta G^{0\ddagger} = -2.303RT \log K^{\ddagger} \qquad \text{(A7.2.16)}$$

where*

$$\Delta G^{\ddagger} = \Delta H^{\ddagger} - T\Delta S^{\ddagger} \qquad \text{(A7.2.17)}$$

Hence

$$\ln K^{\ddagger} = -\Delta G^{\ddagger}/RT = \Delta S^{\ddagger}/R - \Delta H^{\ddagger}/RT \qquad \text{(A7.2.18)}$$

and

$$K^{\ddagger} = e^{\Delta S^{\ddagger}/R} e^{-\Delta H^{\ddagger}/RT} \qquad \text{(A7.2.19)}$$

and further that [from equation (A7.2.15)]

$$k_2 = (RT/Nh) e^{\Delta S^{\ddagger}/R} e^{-\Delta H^{\ddagger}/RT} \qquad \text{(A7.2.20)}$$

We now have an expression for the rate constant, k_2, using TST.

To illustrate the improvement of TST over the collision theory, we compare the pre-exponential factor in both theories with that observed experimentally for

---

*The superscript zero in $\Delta G^{0\ddagger}$ is often dropped; $\Delta G^{\ddagger}$ is then referred to as the free energy of activation; however, strictly speaking it is a *standard* free energy change that is involved.

TABLE A.2

Comparison of Pre-Exponential Factor A from Collision Theory and TST[a]

Reaction	$\log_{10} A^b$ (dm^3/mole $\cdot$ s)		
	Observed	TST	Collision theory
$NO + O_3 \rightarrow NO_2 + O_2$	8.9	8.6	10.7
$NO + O_3 \rightarrow NO_3 + O$	9.8	8.1	10.8
$NO_2 + F_2 \rightarrow NO_2F + F$	9.2	8.1	10.8
$NO_2 + CO \rightarrow NO + CO_2$	10.1	9.8	10.6
$2NO_2 \rightarrow 2NO + O_2$	9.3	9.7	10.6
$NO + NO_2Cl \rightarrow NOCl + NO_2$	8.9	8.9	10.9
$2NOCl \rightarrow 2NO + Cl_2$	10.0	8.6	10.8
$NO + Cl_2 \rightarrow NOCl + Cl$	9.6	9.1	11.0
$F_2 + ClO_2 \rightarrow FClO_2 + F$	7.5	7.9	10.7
$2ClO \rightarrow Cl_2 + O_2$	7.8	7.0	10.4

[a]Data are taken from D. R. Herschbach *et al.* (1956).
[b]A is the pre-exponential factor from both theories.

several reactions. Table A.2 shows clearly the good results for the pre-exponential factor in TST.

ADVANCED PROBLEMS

A.1. In electron-density theory discussed in Chapter 4, the ground-state energy of an atom is a function of the electron density $\rho(\mathbf{r})$. However, for atomic number Z and N electrons, this ground-state energy eventually reduces to a function $E(Z, N)$. In the so-called $1/Z$ expansion of atomic theory, this is approximated by

$$E(Z, N) = Z^2\left[\varepsilon_0(N) + \frac{1}{Z}\varepsilon_1(N) + \frac{1}{Z^2}\varepsilon_2(N) + \cdots\right] \qquad \textbf{(PA.1)}$$

(a) Using the "thermodynamic" definition of the chemical potential μ,

$$\mu = (\partial E/\partial N)_Z \qquad \textbf{(PA.2)}$$

show that, restricting consideration only to the terms explicitly displayed in equation (PA.1), the chemical potential is quadratic (and lower order) in Z.
(b) By eliminating Z between μ and E, show that the chemical potential can be written in the form

$$\mu = \alpha E + (\text{constant} + \beta E)^{1/2} \qquad \textbf{(PA.3)}$$

(c) Show also that the quantities α and β can be obtained, at least in principle, from the functions $\varepsilon_n(N)$ in the $1/Z$ expansion (PA.1).

A.2. Dispersion energies can be calculated statically, at least in principle, from a knowledge, in, say, a diatomic, of the total electron density, this necessitating solution of a two-center problem. However, there is an alternative formulation which can reduce the problem to knowledge of solely one-center properties, leading to a formula for c_6 in the dispersion energy $\phi_{disp}(r) \sim -c_6/r^6$, as r tends to infinity. Following, e.g., the account of Murrell (1974), one makes use of the mathematical identity

$$\frac{1}{a+b} = \frac{2}{\pi} \int_0^\infty \frac{ab}{(a^2 + u^2)(b^2 + u^2)} \, du \qquad (a, b > 0) \qquad \text{(PA.4)}$$

(a) Using this equation, transform the given result of perturbation theory:

$$c_6 = \sum_r{}' \sum_s{}' \tfrac{2}{3} (\mu_{0r}^A \mu_{0s}^B)^2 (E_{rs}^{(0)} - E_{00}^{(0)})^{-1} \qquad \text{(PA.5)}$$

(b). Explicitly, for two atoms A and B, the energy denominator characteristic of perturbation theory that appears in c_6 immediately above can be expressed as $1/[(E_r^A - E_0^A) + (E_s^B - E_0^B)]$, which, in turn, is to be written as an integral over u as in the above equation for $1/(a + b)$. Use that expression to show that c_6 can be transformed from the double summation over r and s into the product of two single summations, namely

$$c_6 = \frac{4}{3\pi} \int_0^\infty \left\{ \sum_r{}' \frac{(\mu_{0r}^A)^2 (E_r^A - E_0^A)}{(E_r^A - E_0^A)^2 + u^2} \right\} \left\{ \sum_s{}' \frac{(\mu_{0s}^B)^2 (E_s^B - E_0^B)}{(E_s^B - E_0^B)^2 + u^2} \right\} du \qquad \text{(PA.6)}$$

This is compactly written in terms of the quantity $\alpha^A(\omega)$, and a similar quantity for atom B, $\alpha^B(\omega)$ as

$$c_6 = \frac{3}{\pi} \int_0^\infty \alpha^A(iu) \alpha^B(iu) \, du \qquad \text{(PA.7)}$$

where explicitly

$$\alpha^A(\omega) = 2 \sum_r{}' \frac{(E_r^A - E_0^A)(\mu_{0r}^A)^2}{(E_r^A - E_0^A)^2 - \omega^2} \qquad \text{(PA.8)}$$

Note that $\alpha^A(iu)$ appearing in c_6 is therefore a real quantity.
(c) Following early ideas of London again, one can represent $\alpha^A(iu)$ by the approximate form

$$\alpha^A(iu) \simeq \alpha^A(0)/(1 + u^2/E_A^2) \qquad \text{(PA.9)}$$

where E_A is a characteristic energy of atom A. Prove that inserting this into equation (PA.7) for c_6 yields, using the formula for $1/a + b$ to do the integration,

$$c_6^{AA} \simeq \tfrac{3}{4} \{\alpha^A(0)\}^2 E_A \qquad \text{(PA.10)}$$

One thus recovers the result (3.19) if E_A is identified with the ionization potential I_A. For the A—B pair, the corresponding result should now be shown to be

$$c_6^{AB} \simeq \tfrac{3}{2} \alpha^A(0) \alpha^B(0) E_A E_B / (E_A + E_B) \qquad \text{(PA.11)}$$

which is to be compared with equation (3.20) of the main text. Dalgarno has given integral formulas for c_8 and c_{10} in terms of frequency-dependent polarizabilities. **(d)** Finally, given the perturbative form of the static polarizability $\alpha(0)$, which has the shape

$$\alpha^A(0) = 2 \sum_r [(\mu_{r0}^A)^2 / E_r^A - E_0^A] \qquad \text{(PA.12)}$$

explain why the result $\alpha^A(\omega)$ can be termed a frequency-dependent polarizability. Note here that, given these frequency-dependent polarizabilities $\alpha^A(\omega)$ and $\alpha^B(\omega)$, one can get c_6 for the pairs AA, BB, and AB (compare c_6^{AB} from the London expression above). However, it must also be emphasized that in the actual formula for c_6, there is the strange feature that α appears in the form $\alpha(iu)$, where $i = \sqrt{-1}$, i.e., the polarizability of the atoms involved is needed at pure imaginary frequency. Nevertheless, this problem illustrates that c_6 can be obtained without any knowledge of the two-center solutions for A—A, B—B, and A—B. Needless to say, much more information than just ground-state properties is then required concerning the atoms, and, at least in principle, full knowledge of excited states is needed [see Bishop and Pipin (1992) for $\alpha(iu)$ for H and He].

A.3. The so-called Poschl–Teller potential for a diatomic molecule is given in equation (2b) on p. 147 of their original paper *in Zeitschrift fur Physik*, **83**, 143, 1933). In their notation, write a short computer program to: **(a)** Generate $(8\pi^2 m/\alpha h^2)$ *vs.* $\alpha(r - r_0)$ for the parameter set $\mu = 16$, $\nu = 1$. **(b)** Calculate (i) the corresponding kinetic energy T and (2) the potential energy V, using the virial theorem for a diatomic molecule. **(c)** From (b1), show that there is a minimum in the kinetic energy when $\alpha(r - r_0)$ is approximately 1.3, and briefly comment on its significance for chemical bonding. Similarly, locate the position of a maximum in (b2).

AI. SOME ADVANCED ASPECTS OF QUANTUM-MECHANICAL PERTURBATION THEORY*

In this appendix, a compact, if somewhat formal, presentation will be given of quantum-mechanical perturbation theory. The notation in Dirac's book will be employed, and only toward the end of the appendix will integrals over configuration space be introduced explicitly.

The unperturbed Hamiltonian is denoted here by H^0, while the perturbation is H^1. To keep track readily of the order of terms, a coupling parameter will be

---

*The material presented in these 6 General Appendices is intended for the reader who wishes to go well beyond the level of the main text. Inevitably, they are less self-contained and could form the basis for a set of advanced projects.

introduced, so that the total Hamiltonian H is given by

$$H = H^0 + \lambda H^1 \qquad \text{(AI.1)}$$

The idea now is to solve the Schrödinger equation

$$H\psi = E\psi \qquad \text{(AI.2)}$$

by expanding ψ and E in a power series in λ according to

$$E = E^{(0)} + \lambda E^{(1)} + \lambda^2 E^{(2)} + \lambda^3 E^{(3)} + \lambda^4 E^{(4)} + \cdots \qquad \text{(AI.3)}$$

and

$$\psi = \psi^{(0)} + \lambda \psi^{(1)} + \lambda^2 \psi^{(2)} + \lambda^3 \psi^{(3)} + \lambda^4 \psi^{(4)} + \cdots \qquad \text{(AI.4)}$$

We shall next assume normalization such that the overlap integral $\langle \psi | \psi^{(0)} \rangle$ between ψ and $\psi^{(0)}$ is unity. It then follows that

$$\langle \psi^{(1)} | \psi^{(0)} \rangle = \langle \psi^{(2)} | \psi^{(0)} \rangle = \langle \psi^{(3)} | \psi^{(0)} \rangle = \langle \psi^{(4)} | \psi^{(0)} \rangle = 0 \qquad \text{(AI.5)}$$

The next step is to explicitly insert equations (AI.3) and (AI.4) into the Schrödinger equation (AI.2) and equate powers of λ^n on both sides. The results for $n = 0$ and 1 are readily obtained as

$$H^0 \psi^{(0)} = E^{(0)} \psi^{(0)} \qquad \text{(AI.6)}$$

and

$$(H^1 - E^{(1)}) \psi^{(0)} + (H^0 - E^{(0)}) \psi^{(1)} = 0 \qquad \text{(AI.7)}$$

We now multiply equation (AI.7) on the left by the unperturbed wave function $\psi^{(0)}$, and integrate over configuration space. The result is then formally written as

$$\langle \psi^{(0)} | H^1 | \psi^{(0)} \rangle - E^{(1)} + \langle \psi^{(0)} | H^0 - E^{(0)} | \psi^{(1)} \rangle = 0 \qquad \text{(AI.8)}$$

where the normalization condition on the unperturbed wave function $\psi^{(0)}$, namely $\langle \psi^{(0)} | \psi^{(0)} \rangle = 1$ has been used. Utilizing the result

$$\langle \psi^{(1)} | \psi^{(0)} \rangle = 0 \quad \text{and} \quad \langle \psi^{(0)} | H^0 | \psi^{(1)} \rangle = E^{(0)} \langle \psi^{(0)} | \psi^{(1)} \rangle \qquad \text{(AI.9)}$$

the first-order perturbation energy $E^{(1)}$ is found as

$$E^{(1)} = \langle \psi^{(0)} | H^1 | \psi^{(0)} \rangle \qquad \text{(AI.10)}$$

This, in a somewhat formal guise, is merely the expectation value of the perturbation H^1 with respect to the unperturbed wave function $\psi^{(0)}$, the coupling parameter λ corresponding to unity for the real problem.

Second- and Higher-Order Results

At order λ^2, one can readily generalize equations (AI.7) to read

$$(H^1 - E^{(1)})\psi^{(1)} + (H^0 - E^{(0)})\psi^{(2)} - E^{(2)}\psi^{(0)} = 0 \qquad \text{(AI.11)}$$

The same procedure of multiplying on the left by $\psi^{(0)}$ then leads to

$$E^{(2)} = \langle\psi^{(0)}|H^1 - E^{(1)}|\psi^{(1)}\rangle = \langle\psi^{(0)}|H^1|\psi^{(1)}\rangle \qquad \text{(AI.12)}$$

Similarly at order λ^3 and λ^4 one finds

$$(H^1 - E^{(1)})\psi^{(2)} + (H^{(0)} - E^{(0)})\psi^{(3)} - E^{(2)}\psi^{(1)} - E^{(3)}\psi^{(0)} = 0 \qquad \text{(AI.13)}$$

$$E^{(3)} = \langle\psi^{(0)}|H^1|\psi^{(2)}\rangle \qquad \text{(AI.14)}$$

and

$$E^{(4)} = \langle\psi^{(0)}|H^1|\psi^{(3)}\rangle \qquad \text{(AI.15)}$$

A way of viewing equations for $E^{(n)}$ above is as special cases of the exact "level shift" formula [see, e.g., March *et al.* (1967)]:

$$E - E^{(0)} = \langle\psi^{(0)}|H^1|\psi\rangle \qquad \text{(AI.16)}$$

Evidently, equation (AI.10) follows immediately from equation (AI.16) by replacing the (usually unknown!) wave function ψ by the zeroth-order wave function $\psi^{(0)}$.

Of course, in order to use results (AI.12), (AI.14), and (AI.15) for the higher-order energy "corrections," one must have useful expressions for the higher-order wave-function terms $\psi^{(1)}$, $\psi^{(2)}$, and $\psi^{(3)}$.

Calculation of Higher-Order Wave Function "Corrections"

The aim here is to express higher-order wave-function terms by means of the complete set of solutions of the unperturbed Schrödinger equation:

$$H^0\phi_n = E_n\phi_n, \qquad n = 0, 1, 2, \ldots \qquad \text{(AI.17)}$$

A relationship is established by noting that

$$\psi^{(0)} \equiv \phi_0 \qquad \text{and} \qquad E^{(0)} = E_0 \qquad \text{(AI.18)}$$

Using equation (AI.10) one has

$$E^{(0)} + E^{(1)} = E_0 + \langle\phi_0|H^1|\phi_0\rangle = \langle\phi_0|H^0 + H^1|\phi_0\rangle = \langle\phi_0|H|\phi_0\rangle \qquad \text{(AI.19)}$$

To calculate $\psi^{(1)}$, let us expand in the (complete) set of unperturbed wave functions ϕ_n as

$$\psi^{(1)} = \sum c_n^{(1)} \phi_n \tag{AI.20}$$

(compare the expansion in a Fourier series, say, as a simple analogy). One now substitutes this expansion into equation (AI.7), multiplies on the left by ϕ_m, and readily obtains

$$\langle \phi_m | H^1 - E^{(1)} | \phi_0 \rangle + \langle \phi_m | H^0 - E_0 | \sum c_n^{(1)} \phi_n \rangle = 0$$

from which it follows that the coefficients c in equation (AI.20) are

$$c_m^{(1)} = - \frac{\langle \phi_m | H^1 | \phi_0 \rangle}{E_m - E_0} \tag{AI.21}$$

Thus we have achieved the objective of expressing the first-order wave function term $\psi^{(1)}$ in terms of the perturbation H^1 and the unperturbed wave functions and energy levels.

We now substitute the result for $\psi^{(1)}$ into equation (AI.12) to find

$$E^{(2)} = -\langle \phi_0 | H^1 | c_m^{(1)} \phi_m \rangle \tag{AI.22}$$

and using equation (AI.21), we get

$$E^{(2)} = \frac{\langle \phi_0 | H^1 | \phi_m \rangle \langle \phi_m | H^1 | \phi_0 \rangle}{E_m - E_0} \tag{AI.23}$$

If one wishes, H^1 can be replaced in this equation by H since $\langle \phi_0 | H^0 | \phi_m \rangle = 0$ for $m \neq 0$; the term $m = 0$ is excluded from the sum in equation (AI.23). Hence the second-order energy correction $E^{(2)}$ can be written explicitly as

$$E^{(2)} = - \sum \frac{\langle \phi_0 | H | \phi_m \rangle^2}{E_m - E_0} \tag{AI.24}$$

Second-Order Wave Function. To find $\psi^{(2)}$, one extends the above approach by expanding as

$$\psi^{(2)} = \sum c_n^{(2)} \phi_n \tag{AI.25}$$

Substituting in equation (AI.11) leads to the result for the coefficients $c^{(2)}$:

$$c_m^{(2)} = \frac{-\langle \phi_m | H^1 - E^{(1)} | \psi^{(1)} \rangle}{E_m - E_0} \tag{AI.26}$$

From equation (AI.14), using equation (AI.21), it then follows that

$$E^{(3)} = - \sum_{nm} \langle \phi_0 | H | \phi_m \rangle \frac{1}{E_m - E_0} \langle \phi_m | H^1 - E^{(1)} | \phi_n \rangle \frac{1}{E_n - E_0} \langle \phi_n | H | \phi_0 \rangle \qquad \text{(AI.27)}$$

Third-Order Wave Function. Finally the third-order wave function can be found as

$$\psi^{(3)} = \sum c_n^{(3)} \phi_n \qquad \text{(AI.28)}$$

Substituting into equation (AI.13) we find after some manipulation

$$c_m^{(3)} = - \langle \phi_m | H^1 - E^{(1)} | \psi^{(2)} \rangle + E^{(2)} c_m^{(1)} \qquad \text{(AI.29)}$$

From equation (AI.15) the fourth-order energy follows as

$$E^{(4)} = - \sum_{mnp} \langle \phi_0 | H | \phi_m \rangle \frac{1}{E_n - E_0} \langle \phi_n | H^1 - E^{(1)} | \phi_n \rangle \frac{1}{E_m - E_0} \langle \phi_n | H^1 - E^{(1)} | \phi_p \rangle$$

$$\times \frac{1}{E_p - E_0} \langle \phi_p | H | \phi_0 \rangle - \sum_m E^{(2)} \frac{\langle \phi_0 | H | \phi_n \rangle^2}{E_n - E_0} \qquad \text{(AI.30)}$$

Example of Möller–Plesset Perturbation Theory

Möller–Plesset perturbation theory, in which the unperturbed problem H^0 is taken to be the Hartree–Fock solution, is important in present-day computational quantum chemistry [see Krishnan *et al.* (1980)] and will therefore be given here as an example of the above treatment. Thus one writes (cf. Appendix A4.6)

$$H^0 = \sum F(i), \qquad H^1 = H - H^0 \qquad \text{(AI.31)}$$

where H is the full Hamiltonian for the atom or molecule (with fixed nuclei) under consideration. The SCF orbitals and energies are introduced through

$$F\phi_j = \varepsilon_j \phi_j \qquad \text{(AI.32)}$$

We next denote by Φ_n the set of all determinants made from the ϕ_j's. E_n is the sum of all the ε_j in the determinant ϕ_n.

Specifically, with $\mathscr{A}$ denoting the antisymmetrized product (determinant)

$$\Phi_0 = \mathscr{A}(\phi_1 \phi_2 \cdots \phi_N) \qquad \text{(AI.33)}$$

while

$$E^{(0)} = E_0 = \sum_{j=1}^{N} \varepsilon_j \qquad \text{(AI.34)}$$

Also

$$E^{(0)} + E^{(1)} = \langle \Phi_0 | H^0 + H^1 | \Phi_0 \rangle = \langle \Phi_0 | H | \Phi_0 \rangle \qquad \text{(AI.35)}$$

which is the SCF energy of Φ_0.

From equation (AI.24), it is important to note that if Φ_m is a "single replacement" determinant Φ_i^a of Φ^0, then

$$\langle \Phi_0 | H | \Phi_m \rangle = 0 \qquad \text{(AI.36)}$$

a result known to quantum chemists as Brillouin's theorem. If Φ_m is a three-replacement determinant of Φ_0 ($\equiv \Phi_{ijk}^{abc}$) then $\langle \Phi_0 | H | \Phi_m \rangle$ is again zero. Thus only the double replacements ($\equiv \Phi_{ij}^{ab}$) enter the summation in equation (AI.24) and one has

$$E^{(2)} = -\sum \frac{\langle \Phi_0 | H | \Phi_{ij}^{ab} \rangle^2}{\varepsilon_a + \varepsilon_b - \varepsilon_i - \varepsilon_j} \qquad \text{(AI.37)}$$

the denominator coming because the energy of each determinant is the sum of the ε's.

Finally, using matrix elements of determinants, one can write

$$E^{(2)} = -\tfrac{1}{4} \sum_{\substack{ij \\ ab}} \frac{[(ia|jb) - (ib|ja)]^2}{\varepsilon_a + \varepsilon_b - \varepsilon_i - \varepsilon_j} \qquad \text{(AI.38)}$$

where the factor $1/4$ has been inserted because Φ_{ij}^{ab}, Φ_{ij}^{ba}, Φ_{ji}^{ab}, and Φ_{ji}^{ba} each represent the same determinant. As already indicated, equation (AI.38) has found widespread application in modern computational quantum chemistry.

AII. THE FORMATION OF ACETYLENE FROM TWO CH FRAGMENTS

In this appendix, we present a theoretical study of the reaction between two CH radicals to form acetylene:

$$\text{CH}(X\,^2\Pi) + \text{CH}(X\,^2\Pi) \rightarrow \text{C}_2\text{H}_2(X\,^1\Sigma_g^+)$$

This process is of considerable interest: one source of motivation being to elucidate the possible mechanisms of the formation of acetylene in interstellar gases.

The reaction can be described in simple terms as starting from two CH radicals in their $^2\Pi$ ground states. As the molecules approach they are both excited into the low-lying $a\,^4\Sigma^-$ state, as a result of which they subsequently form a very strong C—C triple bond (see Section 6.6 of the main text).

In order to study this reaction over the whole range of $r(\text{CC})$ distances Gerratt *et al.* (see Raimondi *et al.* (1981)) have constructed a type of description which is a hybrid of MO and VB theories. According to this, the individual CH fragments, referred to as A or B, are described by MO theory but the relevant A—B interaction is described by VB theory. Thus one forms VB structures of the type [A(1)B(2) + B(1)A(2)] (where 1 and 2 now stand for one or more electron coordinates stemming from systems A or B), using the molecular orbitals of the individual fragments. The treatment may therefore be regarded as an *ab initio* "molecules-in-molecules" method. In the case of acetylene, the problem reduces to one analogous to the VB description of the isoelectronic N_2 molecule.

A feature of the interaction between the two CH radicals is the extensive reorganization of the electron spins which occurs. At intermediate C—C distances the two fragments each possess three unpaired spins which are coupled to form a resultant $S = \frac{3}{2}$ in a $^4\Sigma^-$ state. As A and B approach, the electrons of one fragment recouple with partners from the other to form the final C—C triple bond. The particular approach adopted by Gerratt *et al.* enables one to follow this process and to correlate it with features of the calculated acetylene potential energy surface.

The spin functions for the interacting system are constructed by the VB approach [actually according to the Rumer scheme: see McWeeny (1970)]. A useful alternative basis, which corresponds to a definite spin state for each fragment, is obtained by coupling the resultant spins of each CH radical, S_A and S_B to form the required overall spin S. Thus we use the set of spin functions

$$|S_A S_B k_A k_B ; SM\rangle = \sum_{\substack{M_A, M_B \\ M_A + M_B = M}} \langle S_A S_B M_A M_B | SM \rangle | S_A M_A k_A \rangle | S_B M_B k_B \rangle \quad \text{(AII.1)}$$

where k_A, k_B describe the modes of coupling the individual spins within each fragment, and $\langle S_A S_B M_A M_B | SM \rangle$ is a (known) vector coupling coefficient [see Pauncz (1979)]. In the present case $S = 0$ and hence $S_A = S_B$. Analysis of the final eigenvectors for the VB calculation in both the Rumer basis and equation (AII.1) enables one to follow the recoupling process. For this purpose it is necessary to know the transformation matrix connecting the two types of spin function [see Raimondi (1981)].

A description of the CH—CH interaction in terms of the more common MO–CI (configuration interaction) theory does not, in this system, appear to yield as much insight into the mechanism of the reaction. However, the approach adopted here may be regarded as being equivalent to a CI calculation using a multiconfiguration SCF function.

Theory and Results

Each CH fragment is described by the MO configurations $(1\sigma^2 2\sigma^2 3\sigma^2 1\pi)$ in the $X\,^2\Pi$ state, and by $\{1\sigma^2 2\sigma^2 3\sigma 1\pi_x 1\pi_y\}$ in the $a\,^4\Sigma^-$ state. However, this simple description leads to an incorrect ordering, the $a\,^4\Sigma^-$ apparently forming the ground state. This situation is remedied [see Raimondi *et al.* (1981)] by adding two doubly excited configurations to the $^2\Pi$ state so that this is now described as $\{1\sigma^2 2\sigma^2 3\sigma^2 1\pi\} + \{1\sigma^2 2\sigma^2 1\pi^3\} + \{1\sigma^2 3\sigma^2 1\pi^3\}$.

The MOs are expanded in terms of the basis set of Slater orbitals (see Appendix A2.4) shown in Table A.3.

TABLE A.3
C—H Basis Set[a]

Basis set	Orbital exponent
$1s_C$	5.00904
$1s_C$	9.04883
$2s_C$	1.29799
$2s_C$	2.06820
$2p_C$	1.03933
$2p_C$	1.72601
$2p_C$	2.74247
$1s_H$	1.34188
$2s_H$	2.11216
$2p_H$	1.44660

[a]From Raimondi *et al.* (1981).

The expansion coefficients and total energies were obtained from restricted Hartree–Fock (RHF)SCF calculations. The three-configuration $^2\Pi$ state yields a minimum energy of -38.267082 a.u. at $r(CH) = 2.069$ a.u., and this state remains below the $a\,^4\Sigma^-$ state in the vicinity of the C—H equilibrium separation (see Table A.4). It is worth pointing out that in order to describe the dissociation of $CH(X\,^2\Pi)$ to $C(^3P) + H(^2S)$ correctly it would be necessary to add a fourth configuration of the form $(1\sigma^2 2\sigma^2 4\sigma^2 1\pi)$. The equilibrium C—H distance in the $a\,^4\Sigma^-$ state is the same as that of the ground state.

The occupied MOs of CH can be characterized as core (1σ), C—H bond (2σ), and lone-pair (3σ) with valence orbitals largely localized upon C.

Electronic configurations for CH—CH were formed from configurations of the individual fragments corresponding to CH, CH^+, CH^-, CH^{2+}, and CH^{2-}, and these are listed in Table A.5. The number of unpaired electrons in the CH—CH system constructed in this way ranges from two to six. By including all the possible modes of coupling the unpaired spins to give the required overall singlet state, one correctly allows for the spin recoupling discussed above. This procedure, applied to the list of configurations in Table A.5, gives rise to 59 VB structures of which 22 belong to $^1\Sigma_g^+$ states in the linear configuration.

Calculations were carried out by Gerratt *et al.* for the linear approach of the CH radicals keeping the $r(CH)$ distance fixed at 2.069 a.u. (the calculated equilibrium value for isolated CH $^2\Pi$ and $^4\Sigma^-$). The resulting energies as a function

TABLE A.4
Valence Bond Energy of CH Using
MOs from Spin Restricted SCF
Calculations on CH in the $^2\Pi$ State[a]

R (au)	$E\,(^4\Sigma)$	$E\,(^2\Pi)$
2.124	-38.258034	-38.266314
2.069	-38.260229	-38.267082
2.0145	-38.261310	-38.266824
1.959	-38.261098	-38.265328

[a]From Raimondi *et al.* (1981).

TABLE A.5

CH_3—CH_3 Configurations Formed from the Individual Fragments[a]

Fragment states included			Resulting CH—CH configurations
CH:	$1\sigma^2 2\sigma^2 3\sigma^2 1\pi$		$(2\sigma^2 3\sigma^2 1\pi;\ 2\sigma^2 3\sigma^2 1\pi),\ (2\sigma^2 1\pi^3;\ 2\sigma^2 1\pi^3),$
	$1\sigma^2 2\sigma^2 1\pi^3$	2Π	$(3\sigma^2 1\pi^3;\ 3\sigma^2 1\pi^3),\ (2\sigma^2 3\sigma^2 1\pi;\ 2\sigma^2 1\pi^3),$
	$1\sigma^2 3\sigma^2 1\pi^3$		$(2\sigma^2 3\sigma^2 1\pi;\ 3\sigma^2 1\pi^3),\ (2\sigma^2 1\pi^3;\ 3\sigma^2 1\pi^3)$
	$1\sigma^2 2\sigma^2 3\sigma^1 \pi^2\ {}^4\Sigma^-,\ {}^2\Delta,\ {}^2\Sigma,\ {}^2\Sigma^+$		$(2\sigma^2 3\sigma 1\pi^2;\ 2\sigma^2 3\sigma 1\pi^2)$
CH^+:	$1\sigma^2 2\sigma^2 1\pi^2$		
CH^-:	$1\sigma^2 2\sigma^2 3\sigma 1\pi^2$	$\left.\right\}\ {}^3\Sigma^-,\ {}^1\Delta,\ {}^1\Sigma^+$	$(2\sigma^2 1\pi^2;\ 2\sigma^2 3\sigma 1\pi^2)$
CH^+:	$1\sigma^2 2\sigma^2 3\sigma 1\pi$		
CH^-:	$1\sigma^2 2\sigma^2 3\sigma 1\pi^3$	$\left.\right\}\ {}^3\Pi,\ {}^1\Pi$	$(2\sigma^2 3\sigma 1\pi;\ 2\sigma^2 3\sigma 1\pi^3)$
CH^{2+}:	$1\sigma^2 2\sigma^2 3\sigma$		
CH^{2-}:	$1\sigma^2 2\sigma^2 3\sigma 1\pi^4$	$\left.\right\}\ {}^2\Sigma^+$	$(2\sigma^2 3\sigma;\ 2\sigma^2 3\sigma 1\pi^4)$

[a]From Raimondi *et al.* (1981).

of $r(CC)$ are plotted in Figure A.23. The minimum energy is found to occur at $r(CC) = 2.5$ a.u., and the dissociation energy for decomposition into two $CH(X\ {}^2\Pi)$ fragments is calculated to be 6.301 eV. This is 64% of the experimental value and was computed by Gerratt *et al.* as the difference between the VB energy at $r(CC) = 2.5$ a.u. and at $r(CC) = 20$ a.u., but a potential energy barrier, 0.33 eV in height, is found at a distance $r(CC) = 6$ a.u. (see Figure A.23).

Inspection of the eigenvectors from the VB calculation reveals several interesting features of the reaction. Figure A.24 shows the variation with $r(CC)$ of several structure-occupation numbers. In particular it can be seen that there is a sudden

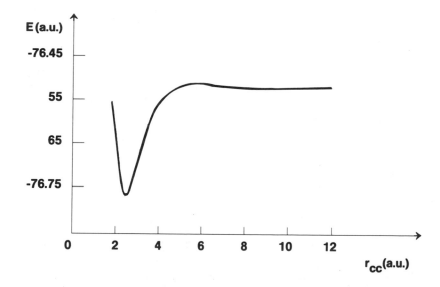

FIGURE A.23. Variation of energies as a function of $r(CC)$ [redrawn from Raimondi *et al.* (1981)].

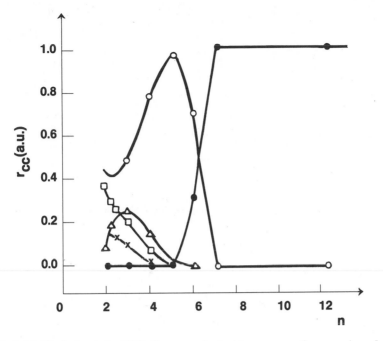

FIGURE A.24. Variation of $r(CC)$ for several structure-occupation numbers [redrawn from Raimondi *et al.* (1981)].

alteration in the relative occupation numbers of the structures

$$(\ldots 3\sigma^2 1\pi, \ldots, 3\sigma^2 1\pi)$$

and

$$(\ldots 3\sigma 1\pi_x 1\pi_y, \ldots, 3\sigma 1\pi_x 1\pi_y)$$

in the region 5 a.u. $< r(CC) < 7$ a.u. The change from $[CH(^2\Pi) - CH(^2\Pi)]$ to $[CH(^4\Sigma^-) - CH(^4\Sigma^-)]$ thus apparently occurs quite suddenly within a very narrow range of C—C separations, and moreover this is precisely where we find the energy barrier. A similar phenomenon is encountered in the formation of BeH $X\,^2\Sigma^+$ from Be(S^2, ^1S) and H(^2S) [see Raimondi *et al.* (1981)].

   Further light is shed upon this process by examining the eigenvectors (both in the Rumer and in the *NSG* basis (AII.1) of spin functions). The members of the two sets are displayed graphically in Table A.6 for the case N = 6, S = 0 assuming that the orbitals are ordered $(3\sigma_A 3\sigma_B 1\pi_{x_A} 1\pi_{x_B} 1\pi_{y_A} 1\pi_{y_B})$ in the Rumer basis and as $(3\sigma_A 1\pi_{x_A} 1\pi_{y_A}; 3\sigma_B 1\pi_{x_B} 1\pi_{y_B})$ in the basis (AII.1). In Table A.7 the variation with $r(CC)$ of the occupation numbers for the different spin-couplings associated with this particular electron configuration is shown, their total contribution being normalized to unity. The relative weights of these structures in the ground-state wave function remain at zero throughout the range 20 a.u. $> r(CC) > 7$ a.u., increasing to 70% of the total at $r(CC) = 6$ a.u. and to 96% at $r(CC) = 5$ a.u. In the region 5 a.u. $< r(CC) < 6$ a.u., the most important coupling is that due to spin function NSG_1 in basis (AII.1) (see Table A.6) corresponding to the direct interaction $CH(^4\Sigma^-) - CH(^4\Sigma^-)$. As $r(CC)$ decreases

TABLE A.6

Spin Functions for Different S and N (number of electrons)[a]

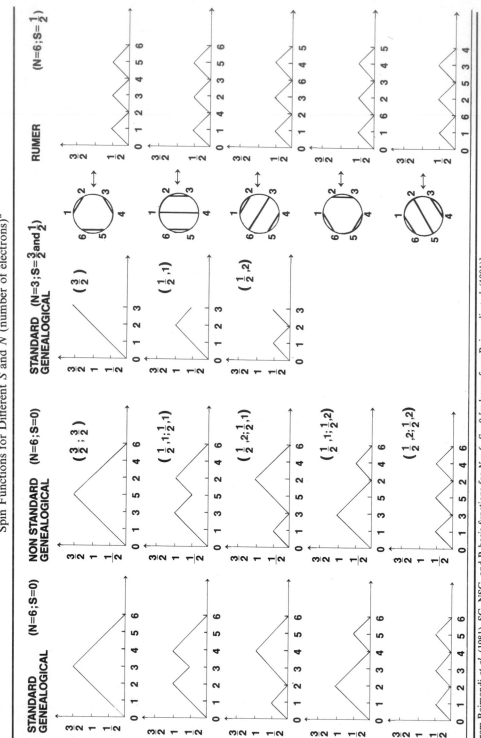

[a]From Raimondi *et al.* (1981). SG, NSG, and R basis functions for $N = 6$, $S = 0$ [redrawn from Raimondi *et al.* (1981)].

TABLE A.7

Renormalized Occupation Numbers of Structures with Configurations
$(3\sigma\pi_x\pi_y, 3\sigma\pi_x\pi_y)$ in the Rumer and NSG Bases[a]

$r(CC)$ (au)	Rumer basis					Non-standard genealogical basis				
	R_1	R_2	R_3	R_4	R_5	NSG_1	NSG_2	NSG_3	NSG_4	NSG_5
2.0	0.72	0.09	0.09	0.0	0.09	0.68	0.04	0.13	0.13	0.02
2.295	0.65	0.11	0.12	0.0	0.11	0.70	0.04	0.11	0.11	0.04
2.5	0.61	0.13	0.14	0.0	0.13	0.73	0.03	0.10	0.10	0.03
4.0	0.47	0.16	0.20	0.0	0.16	0.97	0.0	0.0	0.02	0.0
5.0	0.03	0.31	0.34	0.0	0.31	1.0	0.0	0.0	0.0	0.0
6.0	0.0	0.33	0.33	0.0	0.33	1.0	0.0	0.0	0.0	0.0

[a]From Raimondi *et al.* (1981).

TABLE A.8

Experimental Location of States
of CH with Respect to the
Ground State[a]

State	Separation (eV)
C $^2\Sigma^+$	3.94272
B $^2\Sigma^+$	3.22896
A $^2\Delta$	2.87509
a $^4\Sigma^-$	0.72454
X $^2\Pi_r$	0.0

[a]After Raimondi *et al.* (1981).

further, other couplings assume significant values, but NSG_1 remains dominant. This is to be expected since states of CH above $a\,^4\Sigma^-$ are far removed in energy (see Table A.8). Regarded from the point of view of the Rumer basis it is clear that as $r(CC)$ approaches its equilibrium value, the coupling R_1 corresponding to the C—C triple bond [$\ldots (3\sigma_A, 3\sigma_B)(1\pi_{x_A}, 1\pi_{x_B})(1\pi_{y_A}, 1\pi_{y_B})$] rapidly becomes the most important. However this does not become overwhelmingly so until very small C—C separations when presumably the $3\sigma_A$–$3\sigma_B$ and $1\pi_A$–$1\pi_B$ overlaps approach unity.

For C—C separations greater than 7 a.u. the ground-state wave function is given entirely by the single structure $(\ldots 3\sigma^2 1\pi, \ldots, 3\sigma^2 1\pi)$ and therefore furnishes a correct description of this particular dissociation process. At $r(CC)$ values below about 5 a.u., "charge-rearrangement" structures such as

$$(\ldots 3\sigma^2 1\pi_x, \ldots, 1\pi_x 1\pi_y^2)$$

increase in importance. Figure A.24 shows the contribution to the ground-state wave function of structures of the type $(CH^+ - CH^-)$ and $(CH^{2+} - CH^{2-})$.

Inclusion of local double excitations into the virtual MOs of each fragment would certainly improve the description of ground-state acetylene in the region of the equilibrium C—C separation by allowing for relaxation of the occupied MOs as $r(CC)$ is varied. Nevertheless by using a description intermediate between VB and MO treatments, Gerratt *et al.* have been able to elucidate a number of aspects of an important chemical reaction.

AIII. MACDONALD'S THEOREM

In quantum chemistry, we are interested in ground and excited states. The following theorem gives the mathematical basis for most of the useful approximate methods.

Given a finite set of expansion functions $\phi_1\phi_2 \cdots \phi_m$, solve the *secular equations*

$$\Sigma\langle\phi_r|H - W|\phi_s\rangle Y_s = 0 \qquad \text{(AIII.1)}$$

and denote the eigenvalues in ascending order by $W_1^{(m)}, W_2^{(m)} \cdots W_m^{(m)}$. Then

$$W_{i-1}^{(m)} < W_i^{(m+1)} < W_i^{(m)} \qquad \text{and} \qquad W_i^{(m)} > W_i^{(\infty)} \qquad \text{(AIII.2)}$$

Proof. Let the $(m - 1) \times (m - 1)$ problem have eigensolutions W_t, u_t. In the $m \times m$ problem, by suitable adding of rows and columns, the basis functions can be $u_1, u_2, \ldots, u_{m-1}, \phi_m$, and the secular determinant will have the form

$$D_m = \begin{vmatrix} d_{mm} & d_{mm-1} & \cdots & d_{m1} \\ & \ddots & & 0 \\ \vdots & & d_{rr} & & \vdots \\ & 0 & & \ddots \\ d_{m1} & & \cdots \end{vmatrix}$$

$$d_{rs} = \langle u_r|H - W|u_s\rangle\delta_{rs} \qquad r, s \leq m - 1$$

$$d_{ms} = \langle\phi_m|H - W|u_s\rangle$$

$$d_{mm} = \langle\phi_m|H - W|\phi_m\rangle$$

We can add rows and columns again to form $d_{ms} = \langle\bar{\phi}_m|H - W|u_s\rangle$ where

$$\bar{\phi}_m = \phi_m + \sum_r^{m-1} c_r u_r = \phi_m - \sum_r^{m-1} \langle\phi_m|u_r\rangle u_r$$

so that $\langle u_s|\bar{\phi}_m\rangle = 0$. Hence $d_{ms} = \langle\bar{\phi}_m|H|u_s\rangle$ and

$$D_m = \left(\prod_r^{m-1} d_{rr}\right)\left(d_{mm} - \sum_{s=1}^{m-1} d_{ms} d_{sm}/d_{ss}\right) \qquad \text{(AIII.3)}$$

In the interval $W_{r-1} < W < W_r$, $\Pi_r d_{rr}$ will not change sign, and so if we examine the sign of $D_m/\Pi_r^{m-1} d_{rr}$ in these intervals,

$$D_m/\prod_r d_{rr} = H_{mm} - W - \sum_s^{m-1} |\langle\bar{\phi}_s|H|u_s\rangle|^2/(W_s - W) \qquad \text{(AIII.4)}$$

examination shows that D_m must change sign in each interval, and therefore must have a zero in each interval; there are m such intervals. This proves the theorem that the zeros of the m problem divide the zeros of the $m - 1$ problem.

It is important to note that we must assume that all ϕ_i have the same symmetry, i.e., $\langle \phi_i | H | \phi_j \rangle \neq 0$. If there are several symmetries, then each symmetry is, of course, to be treated separately. The above results were first derived by MacDonald (1933). To summarize, the content of this result for excited states is that if an upper bound of the pth lowest energy level is required it is sufficient to use at least p independent functions in the variation process.

AIV. SPECTROSCOPIC NOMENCLATURE

There is a convention, which is common but not universal, about the labeling of electronic states in diatomic molecules. The ground state is labeled X and higher states of the same multiplicity are labeled $A, B, C, \ldots$ in order of increasing energy. States of multiplicity different from that of the ground state are labeled $a, b, c, \ldots$ in order of increasing energy. As an example, the a state of CO is the first excited triplet state.

In polyatomic molecules the same convention is used except that a tilde is placed over the label, as in $\tilde{A}$ and $\tilde{a}$ states, to remove any confusion with symmetry species. This convention applies even to linear polyatomics since they may be nonlinear in excited electronic states. We will now give some illustrative examples of the labeling of electronic states in diatomic and simple polyatomic molecules that is utilized in the main text.

Configuration and States of the C_2 Molecule

Table A.9 lists the configuration and states of the diatomic molecule C_2, to establish a nomenclature.

The ground configuration, denoted 1 in this table, is obtained by filling up the MO levels with the twelve electrons, as shown. All occupied orbitals are filled and the ground state is $X\,^1\Sigma_g^+$.

The excited configuration 2 in the table promotes one electron from the $\pi_u 2p$ MO into the $\sigma_g 2p$, to give the results shown. This excited configuration 2 gives rise

TABLE A.9
States of the C_2 Molecule

	Configurations	States
1	$(\sigma_g 1s)^2 (\sigma_u^* 1s)^2 (\sigma_g 2s)^2 [\sigma_u^*(2s)]^2 (\pi_u 2p)^4$	$X\,^1\Sigma_g^+$
2	$\ldots (\sigma_u^* 2s)^2 (\pi_u 2p)^3 (\sigma_g 2p)^1$	$a\,^3\Pi_u,\, A\,^1\Pi_u$
3	$\ldots (\sigma_u^* 2s)^2 (\pi_u 2p)^2 (\sigma_g 2p)^2$	$b\,^3\Sigma_g^-,\, E\,^1\Sigma_g^+,\, ^1\Delta_g$
4	$\ldots (\sigma_u^* 2s)^1 (\pi_u 2p)^4 (\sigma_g 2p)^1$	$c\,^3\Sigma_u^+,\, D\,^1\Sigma_u^+$
5	$\ldots (\sigma_u^* 2s)^1 (\pi_u 2p)^3 (\sigma_g 2p)^2$	$d\,^3\Pi_g,\, C\,^1\Pi_g$

to the $a\,^3\Pi_u$ and $A\,^1\Pi_u$ excited states. The three states arising from the configuration labeled 3 in the table are such that the $b\,^3\Sigma_g^-$ state is low-lying, but the $E\,^1\Sigma_g^+$ and $^1\Delta_g$ states are much higher. The latter do not seem to have been observed.

Some Polyatomic (AH₂) Molecules

As an illustrative example for polyatomic molecules, particularly with reference to the labeling of states, Figure A.25 shows the valence electron MO configuration of some AH₂ molecules in ground and low-lying excited electronic states.

In CH₂ there are six electrons to be filled into the MOs of Figure 6.2. Four of them fill the two lowest MOs and one low-energy configuration arises from placing the other two in the $3a_1$ orbital, with antiparallel spins:

$$(2a_1)^2(1b_2)^2(3a_1)^2$$

giving only one state, $\tilde{a}\,^1A_1$, in which the molecule is bent.

A second low-energy configuration is

$$(2a_1)^2(1b_2)^2(3a_1)^1(1b_1)^1$$

which gives rise to two states $\tilde{X}\,^3B_1$ and $\tilde{b}\,^1B_1$ with parallel and antiparallel spins, respectively. The $\tilde{X}\,^3B_1$ state lies lower in energy than $\tilde{b}\,^1B_1$, which follows from one of Hund's rules for two electrons in a degenerate orbital. In CH₂ the $3a_1$ and $1b_1$ orbitals are strictly degenerate only in the linear molecule but the rule is still expected to apply.

It is not clear from the Walsh diagram of Figure 6.2 whether the $\tilde{X}\,^3B_1$ or the $\tilde{a}\,^1A_1$ state is lower in energy, and therefore the true ground state, but there is

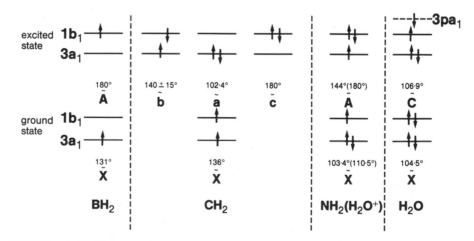

FIGURE A.25. Valence electron MO configurations of some AH₂ molecules [redrawn from Hollas (1982)].

now no doubt that it is $\tilde{X}\,^3B_1$, and that $\tilde{a}\,^1A_1$ is a very low-lying excited state. A quantitative value of the $\tilde{a}\,^1A_1$–$\tilde{X}\,^3B_1$ separation has not been easy to establish as the $\tilde{a} \to \tilde{X}$ transition is forbidden by the $\Delta S = 0$ selection rule and has not been observed. Experiments using chemical, photochemical, and thermochemical methods [see Hollas (1982)] point to a value around 3000 cm^{-1}.

Owing to the fact that the $\tilde{a} \to \tilde{X}$ transition is spin-forbidden, CH_2 in the $\tilde{a}$ state is metastable and, under favorable conditions in the gas phase, has a lifetime sufficiently long that the kinetics of its reactions with other species can be followed before it returns to the ground state.

In the absorption spectrum of CH_2, produced by the flash photolysis of, for example, diazomethane, $CH_2{=}N{=}N$, a band system has been observed in the red region of the spectrum, which arises from the $\tilde{b}\,^1B_1 \to \tilde{a}\,^1A_1$ allowed transition. The HCH angle is 102.4° in the $\tilde{a}$ state and 140° ± 15° in the $\tilde{b}$ state.

In Figure A.25 the electron configurations and bond angles in the $\tilde{X}$, $\tilde{a}$, $\tilde{b}$, and $\tilde{c}$ states are summarized. This reveals how similar is the angle in the $\tilde{X}$ state of BH_2 and the $\tilde{X}$ and $\tilde{b}$ states of CH_2, in all of which there is just one electron in the $3a_1$ orbital. In the $\tilde{A}$ state of CH_2, there is a greatly decreased angle because there are now two electrons in $3a_1$.

AV. THE WENTZEL–KRAMERS–BRILLOUIN SEMICLASSICAL METHOD FOR CALCULATING EIGENVALUES FOR CENTRAL FIELDS

In connection especially with the electron density description of Chapter 4, it is of interest to set out briefly a further semiclassical method, this time to calculate, usually approximately, the eigenvalues of given, central potentials $V(r)$. We shall first state the formula to be used, and then apply it to neutral atoms described by the SCF approach set up in Chapter 4. The method has also been applied to the "almost spherical" molecule GeH_4, within the same density framework [see March and Deb (1987)] by Egorov et al. (1992).

The Wentzel–Kramers–Brillouin (WKB) condition for determining the one-electron eigenvalues in a potential energy $V(r)$ takes the form [March and Plaskett (1956); see also Jeffreys and Jeffreys (1956)]

$$\pi(s + \tfrac{1}{2}) = \frac{1}{\hbar} \int_{r_1}^{r_2} \left\{ 2m \left[\varepsilon - V(r) - \frac{\hbar^2}{2m} \frac{(l + \tfrac{1}{2})^2}{r^2} \right] \right\}^{1/2} dr \qquad \text{(AV.1)}$$

Here r_1 and r_2 are the turning points of the classical motion, which are the roots of

$$V(r) + \frac{\hbar^2}{2m} \left(\frac{l + \tfrac{1}{2}}{r} \right)^2 = \varepsilon \qquad \text{(AV.2)}$$

One wishes to determine the eigenvalues ε as a function of the integers s and l.

Thomas–Fermi Potential Energy for Neutral Atoms

In the TF approximation set out in Chapter 4, it is a straightforward matter to solve the Euler equation (4.2) to obtain the density in the form

$$\rho(\mathbf{r}) = \begin{cases} (8\pi/3h^3)[\mu - V(\mathbf{r})]^{3/2}(2m)^{3/2}: & \mu - V \geq 0 \\ 0, & \text{otherwise} \end{cases} \qquad \text{(AV.3)}$$

For neutral atoms, it is not difficult to show that in the SCF theory obtained by combining equation (AV.2) with the Poisson equation of electrostatics, the chemical potential μ is identically zero.

One of the merits of this SCF problem for neutral atoms is that one can make simple scaling with atomic number Z by writing

$$V(r) = (-Ze^2/r)\phi(x) \qquad \text{(AV.4)}$$

where $\phi(x)$ is obviously a screening function ≤ 1 which reduces the attractive potential $-Ze^2/r$ of the bare nucleus. The quantity x is a dimensionless measure of the distance r from the nucleus through

$$r = bx = \alpha Z^{-1/3}a_0, \qquad \alpha = (9\pi^2/2)^{1/3}/4 \qquad \text{(AV.5)}$$

The Poisson equation combined with equation (AV.3) leads to ϕ satisfying

$$d^2\phi/dx^2 = \phi^{3/2}/x^{1/2} \qquad \text{(AV.6)}$$

and for neutral atoms one knows numerically the solution of equation (AV.6), once for all, which satisfies the boundary conditions

$$\phi(x = 0) = 1, \qquad \phi(x) \to 0 \text{ at infinity} \qquad \text{(AV.7)}$$

It is easy to verify that an exact solution of equation (AV.6) is $144/x^3$ by substitution in the differential equation. Though this obviously does not satisfy the neutral atom boundary condition at the origin $x = 0$ the neutral atom solution does tend to this form for sufficiently large x. In terms of this solution $\phi(x)$ the WKB condition (AV.1) is readily written in the form

$$\pi(s + \tfrac{1}{2}) = Z^{1/3} \int_{x_1}^{x_2} dx \left[2\alpha \frac{\phi(x)}{x} - \frac{(l + \tfrac{1}{2})^2}{x^2} Z^{-2/3} + \varepsilon\alpha^2 Z^{-4/3} \right]^{1/2}$$

$$\equiv Z^{1/3}I(Z, l, \varepsilon) \qquad \text{(AV.8)}$$

In the above equation the scaled length x has evidently been employed, x_1 and x_2 being the turning points already referred to in equation (AV.2); the energy level ε is in Rydbergs. For a given atomic number Z, equation (AV.8) is seen to give ε as a function of the integers s and l.

With use of the known numerical solution for $\phi(x)$ [see, e.g., March 1975], it is a relatively straightforward matter to calculate the eigenvalues numerically.

However, it is to be noted that since the WKB method is semiclassical, it is the energy levels for high quantum numbers that are most accurately given by the WKB approximation set out here: this following from Bohr's correspondence principle, which states that the predictions of quantum and classical mechanics agree in the limit of large quantum numbers.

Some analysis for the case when Z is very large has been given by Senatore and March (1985) and the interested reader is referred to their paper for details. As already noted, the molecule GeH_4 in a spherically symmetrical model has also been treated by the above method.

AVI. ELECTRON CORRELATION IN THE HELIUM ATOM AND THE SLOW CONVERGENCE OF CONFIGURATION INTERACTION

For the He atom ground state, it is helpful to use r_1, r_2 and the interelectronic separation r_{12} in order to rewrite the Hamiltonian

$$H = -\tfrac{1}{2}\nabla_1^2 - \tfrac{1}{2}\nabla_2^2 - \frac{Z}{r_1} - \frac{Z}{r_2} + \frac{1}{r_{12}} \qquad (Z = 2 \text{ for neutral He atom}) \quad \textbf{(AVI.1)}$$

as

$$H = -\frac{1}{2}\frac{\partial^2}{\partial r_1^2} - \frac{1}{2}\frac{\partial^2}{\partial r_2^2} - \frac{\partial^2}{\partial r_{12}^2} - \frac{1}{r_1}\frac{\partial}{\partial r_1} - \frac{1}{r_2}\frac{\partial}{\partial r_2} - \frac{2}{r_{12}}\frac{\partial}{\partial r_{12}}$$
$$- \frac{\mathbf{r}_1 \cdot \mathbf{r}_{12}}{r_1 r_{12}}\frac{\partial^2}{\partial r_1 \partial r_{12}} - \frac{\mathbf{r}_2 \cdot \mathbf{r}_{21}}{r_2 r_{12}}\frac{\partial^2}{\partial r_2 \partial r_{12}} - \frac{Z}{r_1} - \frac{Z}{r_2} + \frac{1}{r_{12}} \qquad \textbf{(AVI.2)}$$

One now makes use of the fact that the "local energy" $H\Psi/\Psi$ is constant everywhere. We utilize this as the two electrons come very close together, i.e., as r_{12} tends to zero. We then assume that Ψ near this limit has the expansion

$$\Psi = a + br_{12} + \cdots \qquad \textbf{(AVI.3)}$$

Forming $H\Psi/\Psi$ using equations (AVI.2) and (AVI.3) is readily shown to yield $b/a = \tfrac{1}{2}$, and hence

$$\Psi = a(1 + \tfrac{1}{2}r_{12}) + \text{higher order corrections} \qquad \textbf{(AVI.4)}$$

This means that Ψ as a function of r_{12} exhibits a cusp at $r_{12} = 0$.

A linear combination of determinants can also introduce electron correlation effects, as discussed briefly in the main text of Chapter 2. Consider the explicit form

$$\Psi = \sum_{i=1}^{4} \phi_i(1)\phi_i(2)(\alpha\beta - \beta\alpha), \qquad \alpha\beta - \beta\alpha \equiv \alpha(1)\beta(2) - \beta(1)\alpha(2) \qquad \textbf{(AVI.5)}$$

where the antisymmetric singlet spin function has now been formally introduced. The choice

$$\phi_1 = \exp(-\alpha r), \qquad \phi_2 = \exp(-\alpha r)x, \qquad \phi_3 = \exp(-\alpha r)y, \qquad \phi_4 = \exp(-\alpha r)z$$
$$\text{(AVI.6)}$$

with $c_1 = 1$, $c_2 = c_3 = c_4 = 2c$, yields then the approximate wave function

$$\Psi = \exp[-\alpha(r_1 + r_2)(1 + 2c\mathbf{r}_1 \cdot \mathbf{r}_2)(\alpha\beta - \beta\alpha)]$$
$$= \exp\{-\alpha(r_1 + r_2)[1 + c(r_1^2 + r_2^2 - r_{12}^2)](\alpha\beta - \beta\alpha)\} \qquad \text{(AVI.7)}$$

i.e., an r_{12}^2 factor has been introduced.

However as was demonstrated above from the local energy calculation as r_{12} tends to zero, there must be an r_{12} term in the wave function [see equation (AVI.4)]. The above is therefore a simple way to understand why configuration interaction is a slowly convergent method.

FURTHER PROBLEMS

1. Following pioneering work of Dirac, it is important to reformulate valence bond theory in terms of electron spins. The main tool that emerges is the so-called Heisenberg Hamiltonian, to be discussed below. (This Hamiltonian is widely used, for example, in microscopic theories of the high-temperature superconducting ceramic oxides, where strong electron–electron interactions are known to be of major importance, and therefore molecular orbital theory is inappropriate.)

(a) As a starting point, give the arguments why, for the H_2 molecule: (i) A singlet spin function $\alpha(1)\beta(2) - \alpha(2)\beta(1)$ demands a symmetric space function

$$\Psi_{\text{symmetric}}^{(1,2)} = \psi_a(1)\psi_b(2) + \psi_b(1)\psi_a(2) \qquad (1)$$

and (ii) Triplet (symmetric) spin functions

$$\alpha(1)\alpha(2)$$
$$\beta(1)\beta(2) \qquad (2)$$
$$\alpha(1)\beta(2) + \alpha(2)\beta(1)$$

demand an antisymmetric space function

$$\Psi_{\text{antisymmetric}}^{(1,2)} = \psi_a(1)\psi_b(2) - \psi_b(1)\psi_a(2) \qquad (3)$$

In other words, the corresponding orbital parts of the wave functions are symmetrical for total spin $S = 0$ (ground state of H_2) and antisymmetrical for $S = 1$. The aim of the problem outlined below is to demonstrate that the above situation can be represented as a strong coupling between the spins s_1 and s_2 of the two individual electrons.

(b) Pressing this point, next calculate the scalar product $s_1 \cdot s_2$ and show its dependence on the total spin S by deriving its value (use units in which one measures the spin angular momentum vectors s_1 and s_2 of the two electrons in units of $\hbar$) as

$$s_1 \cdot s_2 = -\tfrac{3}{4}: S = 0 \tag{4}$$

$$s_1 \cdot s_2 = \tfrac{1}{4}: S = 1 \tag{5}$$

This difference between spin coupling $s_1 \cdot s_2$ for singlet and triplet cases can be used, as sketched below, to characterize a (spin) Hamiltonian representing the two-electron system. The next step in achieving this is to use the space wave functions in the variational principle

$$\varepsilon = \frac{\int \Psi H \Psi \, d\tau_1 \, d\tau_2}{\int \Psi^2 \, d\tau_1 \, d\tau_2} \tag{6}$$

where H is written as

$$H = H_0 + V_{12} \tag{7}$$

Here V_{12} is simply the electron–electron interaction. Show first, without writing out K and J explicitly, that the expectation value of V_{12} is

(i) $K_{12} + J_{12}$, for the symmetric wave function $\Psi_{\text{symmetric}}$

and

(ii) $K_{12} - J_{12}$, for the antisymmetric wave function $\Psi_{\text{antisymmetric}}$.

(c) Show further, using the results for $s_1 \cdot s_2$ derived above for $S = 0$ and 1 that one can write the expectation value of V_{12} in the form

$$\langle V_{12} \rangle = K_{12} - \tfrac{1}{2}J_{12} - 2J_{12}s_1 \cdot s_2 \tag{8}$$

The conclusion to be drawn from this problem is that the two electrons behave as though there were a strong coupling between their two spins which (apart from an unimportant additive constant) is proportional to the scalar product of these spin angular momenta, or, equivalently, to the cosine of the angle between the two spin vectors (compare $\mu_1 \cdot \mu_2/r^3$ discussed in Chapter 3, which appears in the mutual potential energy between two magnetic dipoles). This is not to say, of course, that there is real magnetic coupling; the strong coupling exhibited in this problem is due solely to the (generalized) Exclusion Principle, requiring one type of orbital solution when the spins are parallel and another when they are anti-parallel. One formalizes this into the Heisenberg Hamiltonian

$$H_{\text{Heisenberg}} = -2 \sum J_{kl} \mathbf{s}_k \cdot \mathbf{s}_l \tag{9}$$

where the coupling constant J_{kl} (exchange integral) will depend on the states assumed to be occupied by these two electrons k and l. This Hamiltonian is today a major tool in valence bond theory.

The reader interested in following up this problem is referred to the work of L. F. Mattheiss (1961), who has shown that the low-lying excited states of six hydrogen atoms placed on a "benzene" ring are excellently described by this Hamiltonian, with appropriate choice of J (in other words, this is used as a single adjustable parameter). The high-temperature superconductors referred to above are discussed by P. W. Anderson (1987), within the present context. [Hint: to calculate the numerical values of $\mathbf{s}_1 \cdot \mathbf{s}_2$ for total spins $S = 0$ and $S = 1$, utilize first the identity

$$(\mathbf{s}_1 + \mathbf{s}_2)^2 = \mathbf{s}_1^2 + \mathbf{s}_2^2 + 2\mathbf{s}_1 \cdot \mathbf{s}_2 \tag{10}$$

and secondly the fact that the allowed values of $(\mathbf{s}_1 + \mathbf{s}_2)^2$ are $S(S + 1)$.]

2. A hydrogen atom, say, c, is in a linear configuration with a hydrogen molecule a—b and c is nearer to b. Denoting Coulomb and exchange energies of interaction between atoms i and j as C_{ij} and A_{ij}, respectively, London essentially extended the Heitler–London theory of H_2 to calculate the ground state energy E as

$$E = 3E_H + C_{ab} + C_{ac} + C_{bc}$$
$$+ (A_{ab}^2 + A_{ac}^2 + A_{bc}^2 - A_{ab}A_{ac} - A_{ab}A_{bc} - A_{ac}A_{bc})^{1/2} \tag{11}$$

where E_H is the energy of one atom. With this energy as starting point, follow London's approach to calculate the energy of activation, with c initially far from the molecule a—b (and finally with a molecule b—c and an atom a). This procedure can be summarized as follows: **(a)** Give plausibility arguments why, when c is at a rather large distance R_{bc}, say, from b, and before it interacts appreciably with a, the ground state energy can be replaced by

$$E = 3E_H + C_{ab} + C_{bc} + (A_{ab}^2 + A_{bc}^2 - A_{ab}A_{bc})^{1/2} \tag{12}$$

and that E is greater, for finite large R_{bc} than when c is located at infinite separation from the molecule a—b.

(b) Assuming that the a—b molecular distance remains unchanged on approach of atom c, and that the dominant dependence of energy E on R_{bc} is via A_{bc}, equate $(\partial E/\partial A_{bc})_{A_{ab}}$ to zero to show that E has a maximum value when

$$A_{bc} = A_{ab}/2 \tag{13}$$

(Incidentally, it may be noted that A_{bc} increases with decrease of the distance R_{bc} until this distance is identical with the a—b bond length.)

(c) Show that the energy at this maximum is, apart from C terms:

$$E = 3E_H + \tfrac{1}{2}3^{1/2}A_{ab} \tag{14}$$

(d) Subtract the energy when R_{bc} is infinite to obtain the energy of activation.

3. This problem is about the linear polyenes $C_{2N}H_{2N+2}$. It was first noted by Lennard-Jones that there was interest in ascertaining their electronic properties in the limit as the number $2N$ of carbon atoms tended to infinity. This is evidently then the prototype of a one-dimensional solid.

(a) To obtain a first orientation this problem of $2N$ C atoms with one π-electron per atom, start from the free-electron model (cf. Chapter 4). Show that for the π-electrons enclosed in a box of length l the free-electron model gives the discrete energy levels, ε_r say, as

$$\varepsilon_r = r^2 h^2/8ml^2, \ r = 1, 2, 3, \ldots \tag{15}$$

where m is the electron mass. Show that, if we have $2N$ conjugated C atoms with bond length d and we put infinite barriers at distance d from the end C atoms, then $l = (2N + 1)d$. (In the model, there is a certain degree of flexibility in choosing l; the above choice is one reasonable way only.)

(b) Consider next the absorption spectra of these linear polyenes. Evidently, in this free-electron model, the N lowest levels ε_r given above are doubly occupied by π-electrons with opposed spins in the ground state.

Construct the first excited state by exciting an electron from the level $r = N$, which is the highest occupied level in the above ground state, to the lowest unoccupied level with $r = N + 1$, and in particular show that the excitation energy, $\Delta\varepsilon_N$ say, is given by

$$\Delta\varepsilon_N = \frac{h^2}{8m(2N + 1)d^2} \sim \frac{20}{2N + 1}[\text{eV}] \tag{16}$$

when d is taken as 1.4 Å, which is a reasonable average bond length for conjugated polyenes.

(c) Show finally that the absorption wavelength λ increases proportionally to N. (This, in fact, disagrees with observations on the absorption spectra; we therefore consider the "refinement" of Hückel theory in the next part of the problem.)

(d) Turning from the free electron model above to Hückel theory (HMO; see Chapter 6), it is shown in the book by L. Salem (1966) that with equal bond lengths in the polyene chain, the energy levels ε_r are given by

$$\varepsilon_r = \alpha + \beta \cos(r\pi/2N + 1) \tag{17}$$

Here we remind the reader that α is a property of a single C atom, while β is the resonance integral determined by the properties of two adjacent atoms. Using the fact that the excitation energy is again $\varepsilon_{N+1} - \varepsilon_N$ evaluate this difference $\Delta\varepsilon_N$ to verify the result

$$\Delta\varepsilon_N = -4\beta \, \sin\left[\frac{\pi}{2(2N+1)}\right] \qquad (18)$$

which, for large N, gives $\Delta\varepsilon_N = -2\beta\pi/(2N+1) \propto N^{-1}$. (Note that the discrepancy with experiment for the free-electron model is therefore not removed by the use of Hückel theory.)

(e) Show finally that if in the Hückel formula for ε_r above we employ the approximation $\cos x \simeq 1 - x^2/2$, then the free-electron result can be regained by appropriate choice of the resonance integral β.

The conclusion from this problem [see, for example, J. N. Murrell (1963)], is that to resolve the disagreement with experiment alternating bond lengths, not equal values, must be assumed. In the solid state context referred to above, this situation is formalized in Peierls theorem, which states that a one-dimensional metal cannot exist, the distortion to alternating bond lengths being termed a "Peierls distortion." This leads to an energy gap as $N \to \infty$ and absorption wavelength λ then tends to a finite value as $N \to \infty$.

4.   The ground-state potential energy curve of the H_2^+ ion is denoted below by $E^+(R)$. This can be expressed from the Schrödinger equation as

$$E^+(R) = \frac{H\psi(\mathbf{r}, R)}{\psi(\mathbf{r}, R)} \qquad (19)$$

where the Hamiltonian H is given by

$$H_{\frac{1}{2}} = -\frac{\hbar^2}{2m}\,\nabla^2 - \frac{Ze^2}{|\mathbf{r} - \mathbf{R}/2|}\,\frac{-Ze^2}{|\mathbf{r} + \mathbf{R}/2|} + \frac{(Ze)^2}{R} \qquad (20)$$

the origin $\mathbf{r} = 0$ having been chosen at the bond midpoint, with the nuclei carrying charge Ze at positions $\mathbf{R}/2$ and $-\mathbf{R}/2$ ($Z = 1$, of course, for H_2^+).

(a) Given that the ground state electron density is $\rho(\mathbf{r}, R)$, and that the wave function $\psi(\mathbf{r}, R)$ of the ground state is real, express $E^+(R)$ solely in terms of the electron density. Evaluate the resulting equation at the bond midpoint $\mathbf{r} = 0$ (the answer in terms of $\rho(\mathbf{r}, R)$ is, of course, valid at any value of $\mathbf{r}$, for a given internuclear separation R) [see Amovilli and March (1990)].

   Near the united ion limit R tends to zero, Byers-Brown and Steiner (1962) have shown that $E^+(R)$ has an expansion of the form

$$E^+(R) = Z^2[e^2/R + 4\varepsilon(2ZR)] \qquad (21)$$

where, with $s = 2ZR$, $\varepsilon(s)$ has the small s expansion

$$\varepsilon(s) = -\tfrac{1}{2} + \tfrac{1}{6}s^2 - \tfrac{1}{6}s^3 + \tfrac{11}{270}s^4 - \tfrac{1}{18}s^5 \ln s + O(s^5) \tag{22}$$

(b) Use the virial theorem in the form

$$T = -E - R\, dE/dR \tag{23}$$

to verify that the corresponding small R expansion of the kinetic energy T is

$$T = 4Z^2[\tfrac{1}{2} - \tfrac{1}{2}s^2 + \tfrac{2}{3}s^3 - \tfrac{11}{54}s^4 + \tfrac{1}{3}s^5 \ln s + O(s^5)] \tag{24}$$

5. An almost spherical molecule YH_n (like the tetrahedral GeH_4 example) can be usefully treated by smearing the charge of the protons uniformly over the surface of a sphere of radius R, equal to the Y—H bond length. Writing the self-consistent field as [see equations (AV.4) and (AV.5)]

$$V(r) = -(Ze^2/r)\,\phi(x), \qquad r = bx, \qquad b = Z^{-1/3}[(9\pi^2/2)^{1/3}/4]$$

where Z is the atomic number of the central atom Y, the "screening function" $\phi(x)$ satisfies the differential equation (AV.6) but with boundary conditions now modified from those for the neutral isolated atom Y.

(a) Writing $R = bX$, sketch the form of $\phi(x)$ schematically, labeling the solution ϕ_1 inside X and ϕ_2 outside. What is the discontinuity in slope of $\phi(x)$ at $x = X$, i.e., across the surface charge distribution? [Hint: One must utilize the result of electrostatics for the discontinuity of the electric field across a surface charge distribution.]

(b) Now attempt the following more advanced questions: (1) Denoting by $Z_H e$ the total surface charge ($4e$ for GeH_4, say), (1) use the Hellmann–Feynman theorem of Appendix A2.6 to show that the total molecular energy $E(R)$ has the slope given by [March (1952)]

$$\frac{dE}{dR} = \frac{ZZ_H e^2}{R^2}\left[X\left(\frac{d\phi_1}{dx}\right)_X - \phi_1(X) - \frac{cZ_H}{Z}\right]$$

where c is determined by the geometry through the nuclear–nuclear potential energy $U_{nn} = (Z_H e^2/R)(Z + cZ_H)$, c being $3\sqrt{6}/32$ for tetrahedral geometry. (2) Calculate c for $n = 6$ with assumed octahedral geometry. (3) How many electrons lie outside a sphere of equilibrium radius R_e, where evidently $(dE/dR)_{R_e} = 0$? (4) How should the above model be modified to treat the π-electrons in C_{60} (having the shape of a European football)? How many π-electrons lie outside the surface of the "football" at the equilibrium bond length [see Siringo *et al.* (1992)]? (5) Into the C_{60} molecule discussed above, one can put atoms which become trapped in the fullerene cage. First study Van Cleef *et al.* (1992) and J. Cioslowski and E. D. Fleischmann (1991). Then write an account of (a) why K atoms, when introduced, give up one electron per atom to the carbon cage, and (b) why the minimum energy position of ionic guests for closed-shell endohedral fullerenes is not at the center of the cage.

REFERENCES

A. C. Aitken, *Determinants and Matrices*, Oliver and Boyd, Edinburgh (1954).

N. L. Allan, D. L. Cooper, C. G. West, P. J. Grout, and N. H. March, *J. Chem. Phys.* **85**, 239 (1985).

N. L. Allan and N. H. March, *Int. J. Quant. Chem. Symposium* **17**, 227 (1983).

J. A. Alonso and N. H. March, *Chem. Phys.* **76**, 121 (1983).

J. A. Alonso, L. C. Balbás, and N. H. March, *Mol. Phys.* **50**, 789 (1983).

C. Amovilli and N. H. March, *Chem. Phys.* **146**, 207 (1990).

P. W. Anderson, *Science* **235**, 1196 (1987).

P. W. Atkins, *Molecular Quantum Mechanics, 2nd Ed.*, Oxford University Press, Oxford (1983).

R. F. W. Bader and T. T. Nguyen-Dang, *Adv. Quantum Chem.* **14**, 63 (1981).

R. F. W. Bader, T. T. Nguyen-Dang, and Y. Tal, *Rep. Progr. Phys.* **44**, 893 (1981).

R. F. W. Bader and T. H. Tang, *J. Amer. Chem. Soc.* **104**, 946 (1982).

L. C. Balbás, A. Rubio, J. A. Alonso, N. H. March, and G. Borstel, *J. Phys. Chem. Solids* **49**, 1013 (1982).

L. C. Balbás, J. A. Alonso, and E. Las Heras, *Mol. Phys.* **48**, 981 (1982).

K. E. Banyard and N. H. March, *Acta. Cryst.* **9**, 385 (1956).

L. S. Bartell and L. O. Brockway, *Phys. Rev.* **90**, 833 (1953).

R. J. Bartlett and G. D. Purvis, *Int. J. Quantum Chem.* **14**, 56 (1978).

H. A. Bethe and R. Jackiw, *Intermediate Quantum Mechanics*, 2nd Ed., Benjamin, New York (1968).

U. M. Bettolo and N. H. March *Int. J. Quant. Chem.* **20**, 693 (1981).

D. M. Bishop and J. Pipin, *J. Chem. Phys.* **97**, 3375 (1992).

S. M. Blinder, *Amer. J. Phys.* **33**, 431 (1965).

S. F. Boys, *Proc. Roy. Soc.* **A200**, 542 (1950).

S. F. Boys and N. C. Handy, *Proc. Roy. Soc.* **A311**, 309 (1969).

W. Byers-Brown and E. Steiner, *J. Chem. Phys.* **37**, 461 (1962).

J. Callaway and N. H. March, *Solid State Physics Volume* 38, p. 136 H. Ehrenreich, F. Seitz and D. Turnbull, eds., Academic, New York (1984).

D. M. Ceperley and B. J. Alder, *Phys. Rev. Lett.* **45**, 566 (1980).

J. Cioslowski and E. D. Fleischmann, *J. Chem. Phys.* **94**, 3730 (1991).

E. Clementi and C. Rosetti, *At. Nucl. Data Tables* **14**, 177 (1974).

E. U. Condon and G. H. Shortley, *The Theory of Atomic Spectra* (Cambridge University Press) (1951).

C. A. Coulson and I. Fischer, *Phil. Mag. (London)* **40**, 386 (1949).

C. A. Coulson and N. H. March, *Proc. Phys. Soc. (London)* **A63**, 367 (1950).

J. P. Dahl and J. Avery, *Local Density Approximations in Quantum Chemistry and Solid State Physics*, Plenum, New York (1984).

P. A. M. Dirac, *Proc. Roy. Soc.* **A117**, 610 (1928); **A118**, 351 (1928).

M. Dixon and M. Gillan, *J. Phys.* **C11**, 1165 (1978).

S. A. Egorov, N. H. March, and R. Santamaria, *Int. J. Quantum Chem.* **42**, 1641 (1992).

R. Englman, *The Jahn-Teller Effect in Molecules and Crystals*, Wiley–Interscience, New York (1972).

H. Eyring, J. Walter, and G. E. Kimball, *Quantum Chemistry*, John Wiley and Sons, New York (1944).

E. Fermi, *Z. für Phys.* **48**, 73 (1928).

E. Fermi, *Rev. Mod. Phys.* **4**, 87 (1932).

E. Fermi, *Notes on Quantum Mechanics*, The University of Chicago Press (1961).

J. W. Fleming and J. Chamberlain, *Infrared Phys.* **14**, 277 (1974).

V. Fock, *Z. Physik* **61**, 126 (1930).

S. V. Fomin, *Calculus of Variations*, Prentice-Hall, Englewood Cliffs, N.J. (1963).

D. Fox, *J. Chem. Phys.* **24**, 1103 (1956).

S. R. Gadre and R. K. Pathak, *J. Chem. Phys.* **77**, 1073 (1982).

R. G. Gordon and Y. S. Kim, *J. Chem. Phys.* **56**, 3122 (1972).

P. J. Grout, Y. Tal, and N. H. March, *J. Chem. Phys.* **79**, 331 (1983).

O. Gunnarson, J. Harris, and R. O. Jones, *J. Chem. Phys.* **67**, 3970 (1977).

F. S. Ham, in: *Electron Paramagnetic Resonance*, S. Geschwind, ed., Plenum, New York (1972).

N. C. Handy *et al.*, see, e.g., *Mol. Phys.* **23**, 1 (1972).

D. R. Hartree, *Proc. Camb. Phil. Soc.* **24**, 111, 426 (1928).

D. R. Hartree, *Calculations of Atomic Structures*, John Wiley and Sons, New York (1957).

D. R. Herschbach, H. J. Johnson, K. S. Pitzer, and R. E. Powell, *J. Chem. Phys.* **25**, 7336 (1956).

S. H. Hill, P. J. Grout, and N. H. March, *J. Phys.* **B17**, 4819 (1985).

J. Hinze, *Physical Chemistry*, Vol. 5, H. Eyring, D. Henderson, and W. Jost, eds., Academic, New York (1970).

P. C. Hohenberg and W. Kohn, *Phys. Rev.* **136**, 8864 (1964).

J. M. Hollas, *High Resolution Spectroscopy*, Butterworths, London (1982).

F. Hund, *Z. Physik* **77**, 12 (1932).

H. A. Jahn and E. Teller, *Proc. Roy. Soc. (London)*, **A161**, 220 (1937).

H. A. Jahn and E. Teller, *Proc. Roy. Soc. (London)*, **A164**, 117 (1938).

H. Jeffreys and B. S. Jeffreys, *Methods of Mathematical Physics*, Cambridge University Press, Cambridge (1956).

T. Kato, *Commun. Pure Appl. Math.* **10**, 151 (1957).

Y. S. Kim and R. G. Gordon, *J. Chem. Phys.* **60**, 4332 (1974).

W. Kohn and L. J. Sham, *Phys. Rev.* **A140**, 1133 (1965).

A. Konya, *Acta Physica Hungarica* **1**, 12 (1951).

H. A. Kramers, *Z. Physik* **53**, 422 (1929); **53**, 429 (1929).

R. Krishnan, M. J. Frisch, and J. A. Pople, *J. Chem. Phys.* **72**, 4244 (1980).

K. S. Lackner and G. Zweig, California Institute of Technology, Report No. CALT 68-865 (1981).

W. G. Laidlaw and R. Birss, *J. Chem. Phys.* **48**, 63 (1964).

D. Layzer, *Ann. Phys.* **8**, 271 (1959).

D. Layzer and A. Bahcall, *Ann. Phys.* **17**, 177 (1962).

R. LeSar, *J. Chem. Phys.* **86**, 1485 (1987).

I. Lindgren and J. Morrison, *Atomic Many-Body Theory*, Springer-Verlag, Berlin (1982).

H. C. Longuet-Higgins, *Proc. Roy. Soc.* **A235**, 537 (1956).

H. C. Longuet-Higgins and E. W. Abrahamson, *J. Am. Chem. Soc.* **87**, 2045 (1965).

P. O. Löwdin, *Adv. Chem. Phys.* **2**, 207 (1959).

J. P. Lowe, *Quantum Chemistry*, Academic, New York (1978).

S. Lundquist and N. H. March, eds: *Theory of the Inhomogeneous Electron Gas*, Plenum, London (1983).

J. K. L. MacDonald, *Phys. Rev.* **43**, 830 (1933).

N. H. March, *Acta Cryst.* **5**, 187 (1951).

N. H. March, *Proc. Camb. Phil. Soc.* **48**, 665 (1952).

N. H. March, *J. Chem. Phys.* **86**, 2262 (1974).

N. H. March, *Self-Consistent Fields in Atoms*, Pergamon, Oxford (1975).

N. H. March, *Localization and Delocalization in Quantum Chemistry*, Vol. I, D. Chalvet *et al.*, eds., Reidel, Dordrecht (1975).

N. H. March, *Special Periodical Reports Chemical Society-Theoretical Chemistry*, Vol. 4 (Chem. Soc. London) (1981).

N. H. March and B. M. Deb, *The Single-Particle Density in Physics and Chemistry*, Academic, New York (1987).

N. H. March and R. G. Parr, *Proc. Natl. Acad. Sci. USA* **77**, 6285 (1980).

N. H. March and J. S. Plaskett, *Proc. Roy. Soc. (London)*, **A235**, 419 (1956).

N. H. March, W. H. Young, and S. Sampanthar, *The Many-Body Problem in Quantum Mechanics*, Cambridge University Press (1967).

L. F. Mattheiss, *Phys. Rev.* **123**, 1219 (1961).

W. H. McCrea, *Relativity Physics*, Methuen, London (1954).

R. McWeeny, *Spins in Chemistry*, Academic Press, New York (1970).

C. Möller and M. S. Plesset, *Phys. Rev.* **46**, 618 (1934).

J. F. Mucci and N. H. March, *J. Chem. Phys.* **75**, 5789 (1981).

R. S. Mulliken, *J. Chem. Phys.* **2**, 782 (1934); **3**, 573 (1935).

J. N. Murrell, *Theory of the Electronic Spectra of Organic Molecules*, John Wiley and Sons, New York (1963).

J. N. Murrell, in: *Orbital Theories of Molecules and Solids*, N. H. March, ed., Clarendon, Oxford (1974), p. 311.

A. H. Netercot, *Chem. Phys. Lett.* **59**, 346 (1978).

R. G. Parr, R. A. Donnelly, M. Levy, and W. E. Palke, *J. Chem. Phys.* **68**, 3801 (1978).

L. Pauling and E. B. Wilson, *Introduction to Quantum Mechanics*, McGraw-Hill, New York (1935).

R. Pauncz, *Spin Eigenfunctions*, Plenum, New York (1979).

J. P. Perdew and A. Zunger, *Phys. Rev.* **B23**, 5048 (1981).

A. Rahman, *J. Chem. Phys.* **65**, 4845 (1976).

M. Raimondi, M. Simonetta, and J. Gerratt, *Chem. Phys. Letts.* **77**, 12 (1981).

N. K. Ray, L. Samuels, and R. G. Parr, *J. Chem. Phys.* **70**, 3680 (1979).

W. G. Richards, *Int. J. Mass Spectrometry and Ion Physics* **2**, 419 (1969).

W. G. Richards, H. P. Trivedi, and D. L. Cooper, *Spin-Orbit Coupling in Molecules*, Clarendon Press, Oxford (1981).

W. G. Richards and R. C. Wilson, *Trans. Faraday Soc.* **64**, 1729 (1968).

J. K. Roberts and A. R. Miller, *Heat and Thermodynamics*, Blackie, London (1965).

C. C. J. Roothaan, *J. Chem. Phys.* **19**, 445 (1951); *Rev. Mod. Phys.* **23**, 69 (1951).

K. Ruedenberg, *Rev. Mod. Phys.* **34**, 326 (1962).

G. S. Rushbrooke, *Introduction to Statistical Mechanics*, Oxford University Press, New York (1949).

D. E. Rutherford, *Vector Methods*, Oliver and Boyd, Edinburgh (1946).

L. Salem, *Molecular Orbital Theory of Conjugated Systems*, Benjamin, New York (1966).

R. T. Sanderson, *Chemical Bonds and Bond Energy*, Academic, New York (1971).

L. I. Schiff, *Quantum Mechanics*, 2nd Ed., McGraw-Hill, New York (1955).

J. M. C. Scott, *Phil. Mag.* **43**, 859 (1952).

G. Senatore and N. H. March, *Phys. Rev.* **A32**, 1322 (1985).

F. Siringo, G. Piccitto, and R. Pucci, *Phys. Rev.* **A46**, 4048 (1992).

J. C. Slater, *J. Chem. Phys.* **1**, 687 (1933); *Phys. Rev.* **81**, 385 (1951).

I. N. Sneddon, *Fourier Transforms*, McGraw-Hill, New York (1951).

M. D. Sturge, in: *Solid State Physics*, *Vol. 20*, F. Seitz, D. Turnbull, and H. Ehrenreich, eds., Academic, New York (1967), p. 91.

L. H. Thomas, *Proc. Camb. Phil. Soc.* **23**, 542 (1926).

G. Thomer, *Phys. Zeit* **38**, 48 (1937).

J. R. Townsend and G. S. Handler, *J. Chem. Phys.* **36**, 3325 (1962).

M. S. Vallarta and N. Rosen, *Phys. Rev.* **41**, 708 (1932).

G. W. Van Cleef, G. D. Renkes, and J. V. Coe, *J. Chem. Phys.* **98**, 860 (1992).

S. H. Vosko, L. Wilk, and M. Nusair, *Can. J. Phys.* **58**, 1200 (1980).

E. P. Wigner and F. Seitz, *Phys. Rev.* **46**, 509 (1934).

E. B. Wilson Jr., J. C. Decius, and P. C. Cross, *Molecular Vibrations*, McGraw-Hill, New York (1955).

S. Wilson, *Electron Correlation in Molecules*, Clarendon, Oxford (1984).

W. S. Wilson and R. B. Lindsay, *Phys. Rev.* **47**, 681 (1935).

L. A. Woodward, *Introduction to the Theory of Molecular Vibrations and Vibrational Spectroscopy*, Clarendon Press, Oxford (1972).

INDEX

Ab initio calculations, 207, 217, 257
Absorption, 24
Acetylene, 175, 178, 209, 210, 367
Activated complex, 227, 228, 233, 235, 358
Activation energy, 225, 227, 235, 356, 357
Active medium, 24
Air, 91
Alkali halide, 15, 17, 18
Ammonia molecule, 167
Angular momentum, 55, 56, 136, 264, 268, 326
Angular wave function, 170, 171, 261
Anharmonic constants, 135
Anharmonic oscillator, 136
Anharmonicity
  electrical, 135
  mechanical, 135
Antibonding states, 41, 54, 57, 183
Anti-Stokes Raman scattering, 27
Antisymmetric wave function, 49
Arithmetic mean, 9
Associated Laguerre function, 270
Associated Legendre function, 266, 349
Asymmetric rotor, 127, 132
Atomic beam, 93
Atomic electron affinities, 6, 10, 11, 17
Atomic ground-state configurations, 271, 272
Atomic mass, 13
Aufbau principle, 56
Auger electron spectroscopy (AES), 18, 20
Avogadro's number, 87
Azomethane, 167
Azulene, 32, 33

Barrier, 226, 227
Basis set, 179, 194, 281, 307
Bending mode, 234
Bending vibration, 234
Benzene, 110, 145, 177, 192, 193, 195
Bimolecular reaction, 357
Binding energy, 106
Blackbody radiation, 342
Bohr's correspondence principle, 294
Bohr energy levels, 119, 270
Bohr magneton, 149
Bohr radius, 35

Boltzmann distribution, 22, 25, 131, 357
Bond angles, 24
Bond breaking, 4, 5, 224, 225, 227, 238
Bond dissociation, 4, 9, 24, 67
Bond energies, 4, 9, 14, 47
Bond forming, 224, 227, 238
Bond length, 13, 14, 109
Bond order, 13, 14, 134, 201
Bond strength, 4, 133
Born–Haber cycle, 15, 17, 18, 34
Born–Oppenheimer approximation, 42, 161, 278, 279
BrCl, 4
Building blocks, 37
Butadiene, 184, 185, 197, 213, 353, 356
Butanol, 84, 85

Ca_2, 29
Catalyst, 219, 227
C_2, 143
CCl_4, 84
Central field, 262, 263
Centrifugal distortion, 132, 137
Chapman–Enskog theory, 92
Charge cloud, 16, 101, 119
Charge density, 197, 199
Charge exchange reactions, 256
Charge transfer, 314
Chemical lasers, 24, 26, 248
Chemical potential, 101, 115, 116, 119, 295, 314
Chemical reactivity, 36
Chemical shift, 24, 84, 85, 149, 153, 154
Chemisorption, 201
Classical fluids, 97
CNDO method, 204
CO, 91, 129, 130, 131
Coherence, of laser radiation, 23, 24
Collision diameter, 98
Collisional energy transfer, 31
Collisional relaxation, 30
Collisions, 28, 93, 94, 357
Column vector, 196
Compton line, 207
Configuration interaction, 51, 326, 379
Conjugated hydrocarbons, 175, 180
Correlation diagram, 52, 56, 108

Correlation energy, 299, 352
Correspondence Principle, 294
Coulomb integral, 45, 81, 205
Coulomb operator, 111
Coulson–Fischer wave function, 51, 282
Covalent bond, 2, 9, 49
Cross section, 94
Cyclobutadiene, 189, 353, 356
Cyclopentadienyl, 214
Cyclopropenyl, 188, 189
Cylindrical symmetry, 53

Degeneracy, 56, 59
Degrees of freedom, 123
Density gradient, 121
Differential scattering cross section, 94
Diffusion, 90
Dimensional analysis, 104
Dipole–dipole interaction, 70, 72
Dipole moment, 13, 24, 65, 66, 71, 340
Dirac equation, 345
Dirac–Slater exchange
    energy, 308
    potential, 311
Dispersion
    energy, 71, 96
    force, 71, 74, 98
Dissociation, 5, 9, 11, 60
Double bond, 1, 176, 184
Drude model, 71

Eigenfunction, 111, 265, 267
Eigenvalue, 265
Einstein coefficients, 25, 341
Electric dipole moment, 128
Electric dipole selection rules, 130
Electric dipole transitions, 128, 338
Electric field strength, 27, 73
Electrocyclic reactions, 353
Electromagnetic radiation, 124
Electromagnetic spectrum, 124
Electron affinity, 5, 6, 118
Electron correlation, 311
Electron density, 1, 35, 61, 64, 98, 101, 115, 119,
    214, 312
Electron diffraction, 308
Electronegativity, 9, 12, 34, 60, 65, 66, 83, 115
Electron exchange, 121
Electron gas model, 214
Electron impact spectroscopy, 20
Electron–nuclear potential energy, 103
Electron repulsions, 45, 104
Electron spin, 22, 348
Electron spin resonance (ESR), 24, 123, 124,
    154–157
Electron volt, 19
Electronic selection rules, 140–142, 154–157

Electronic spectra, 125, 138
Electronic transitions, 24, 140
Emission of radiation, 24
Energy scaling relations, 106, 114
Equilibrium constant, 233
ESCA, 20, 36
Ethanol, 84
Ethylene, 174, 176, 180, 181, 183
Exchange energy, 113, 114, 310, 312
Exchange energy density, 121
Exchange hole, 309, 310
Exchange integral, 81
Exchange operator, 111
Excited states, 1, 142
Excluded volume, 74
Exponential-6 potential, 79, 80, 98

Fermi–Dirac statistics, 102
Fermi hole, 309, 310
Fixed nucleus approximation, 42, 133
Flash photolysis, 142
Fluorescence, 28, 29, 32, 35
Fluorine F_2, 59, 133
Force constants, 95, 132, 133, 134, 233
Force field, 331
Fourier transform, 98
Franck–Condon factor, 337
Franck–Condon principle, 28, 335, 336
Free radicals, 201
Free valence, 201, 202
Frontier orbital, 241

Gaussian orbital, 280, 281
GeH_4, 377
Geometric mean, 9
Gerade state, 39, 163
Germanium atom, 377

Hamiltonian, 39, 42, 147, 288, 313, 351
Hard-sphere model, 76, 87, 90
Harmonic oscillator, 293
Hartree-Fock method, 110, 113
HBr, 37
HCl, 30, 37, 91
HCN, 126, 127
Heisenberg uncertainty principle, 102
Heitler–London theory, 37, 47, 230
Helium atom, 43, 296
Hellmann–Feynman theorem, 282, 286, 289
Helmholtz free energy, 87
Heteronuclear diatomics, 37
Hexatriene, 215
HF, 37, 130
Hohenberg–Kohn theorem, 117, 312
Homonuclear diatomics, 41, 106, 295
HOMO orbital, 244
Hückel MO method, 175, 177, 180, 194, 205
Hund's coupling cases, 139, 140, 142

Hund's rules, 59
Hybridization, 168, 169, 173
Hydrocarbons, 175
Hydrogen atom, 1, 83, 207, 208, 278
Hydrogen bond, 82
Hydrogen halides, 314
Hydrogen molecule, 41, 42
Hydrogen molecule ion, 38, 39, 41, 277
Hyper-Raman effect, 28

Impact parameter, 92
Imperfect gases, 85, 98
Indistinguishability, of electrons, 47
Induced dipole–dipole interaction, 69, 70
Induced emission, 22
Inert gases (*see* Noble or Rare Gases)
Infrared fluorescence, 31
Inhomogeneous electron gas, 104, 294
Intensity stealing, 142
Intermolecular forces, 69, 77, 98, 99, 289
Inverse power law potential, 79
Iodine negative ions, 31
Ionic resonance energy, 10
Ionization potential, 34, 35, 74, 118, 145, 181, 195
Isotopes, 251

Jahn–Teller effect, 167

Kato's theorem, 213
Kinetic energy, 93, 101, 110, 112, 113
Kinetic theory, 99
Koopmans' theorem, 18, 21, 118, 146, 351
Krypton, 91

Lagrange equation, 330
Lagrange multiplier, 117
Laguerre polynomial, 270
Laser excitation, 31
Lasers, 22, 27, 28, 30–32, 251
Lattice energy, 16, 17, 188
LCAO method, 41, 47, 62, 111, 112
Legendre polynomials, 266
Lennard-Jones potential, 79, 87, 98
Lewis concept, 3
Local density approximation, 122, 326

Madelung constant, 16
Magnetic field, 128, 147
Magnetic quantum number, 264
Magnetogyric ratio, 147
Maxwell–Boltzmann distribution, 89, 357
Mayer function, 86, 97
Mean free path, 88
Mean momentum, 113, 312
Methane, 127, 232
Microwave spectroscopy, 28, 35, 83
Mie potential, 290

Molar volume, 98
Molecular beams, 93, 98, 290
Molecular orbital, 18, 19, 48, 123, 175
Molecular properties, 22, 202
Molecular shape, 107, 108
Molecular spectroscopy, 22
Molecule-fixed axes, 125
Momental ellipsoid, 125, 126
Moments of inertia, 123, 125, 327
Momentum density, 120
Monochromatic light source, 30
Mulliken electronegativity, 10, 11, 13, 117, 118, 315, 318
Multicenter integrals, 45
Multiplicity, 30

NaCl, 16
Negative ion, 117, 118
Neutron diffraction, 96, 97
Newton equation of motion, 133, 263, 283
N_2, 21, 130
NO, 21, 130
Noble gases, 2
Nonuniform gases, 99
Normal vibrations, 167, 332
Normalized wave function, 47
Nuclear magnetic resonance (NMR), 24, 123, 147–153

Octahedral molecules, 127
Octet formation, 4
Octet rule, 3
O_2, 21, 60, 92
Open-shell molecules, 352
Operator, 104, 111, 261, 265, 268
Orbital angular momentum, 139, 263, 348
Orbital energy, 19, 112, 194, 195
Orbital energy sum, 106, 181
Orthogonality, 275
Overlap integral, 40, 81, 112, 174, 179

Pair correlation function, 97, 98
Pair potential, 97
Paramagnetic molecules, 18, 23, 60
Pariser–Parr–Pople method, 205
Partition function, 87, 132, 328
Pauli exclusion principle, 309
Penning ionization spectroscopy, 18, 20
Perfect gas, 86
Periodic Table, 41, 271, 272, 298
Phase space, 102, 104, 119, 121, 214, 290, 291, 293
Photochemistry, 36
Photoelectric effect, 19
Photon cascade, 24
Photons, 24, 25
Picosecond spectroscopy, 31
Planck's constant, 105, 291
Polarizability, 70, 75, 76, 134

Polyatomic molecules, 145, 161, 203
Population inversion, 22, 26
Potential energy curves, 48, 103, 142
Pressure, 87, 128
Principal quantum number, 270
Progressions, 144, 181
p-type orbitals, 52, 54, 62, 109, 162, 270
Pumping, 22
Pyrolysis, 167

Quadrupole moment, 289

Radial distribution function, 289
Radial wave functions, 269
Radiation density, 25, 35
Radiationless decay, 29, 30
Radiationless transitions, 36
Raman scattering, 33, 34
Raman spectroscopy, 27, 28, 83
Rare gases, 2–4
Rayleigh frequency, 28
Rayleigh line, 27, 28
Rayleigh scattering, 27
Reaction coordinate, 232, 235
Reaction dynamics, 36, 219, 251
Relaxation processes, 32
Relaxation time, 32
Renner–Teller effect, 211, 212
Resonance energy, 185, 186
Resonance integral, 205
Rest mass, 349, 350
Ring currents, 150, 216
Roothaan–Hall equations, 110, 204
Rotational degrees of freedom, 123
Rotational process, 31
Rotational spectroscopy, 123, 128, 133
Russell–Saunders coupling, 138, 343
Rydberg, Klein, and Rees (RKR) procedure, 95
Rydberg orbitals, 138

Scattering cross section, 94
Schrödinger equation, 72, 101, 102, 261, 263, 279, 326, 338
Secular determinant, 213, 214, 215
Selection rules, 30, 129, 135, 340
Self-consistent field, 104, 107, 110, 114, 120, 194, 195, 299, 351
Semiclassical radius, 116
Separation constant, 265
Sequences, 144
Shape, of spectral lines, 24
Shock tube, 81
Single bond, 2, 37
Singlet–triplet transition, 30, 141
Slater determinant, 121, 309
Slater orbitals, 112, 179
Space-fixed axes, 125

Space wave function, 49
Spherical harmonics, 340
Spherical polar coordinates, 38, 262, 265
Spherical rotor, 127
Spin-down electron, 22
Spin–orbit coupling, 30, 56, 141, 342, 343, 344
Spin-up electron, 22
Spin wave function, 51, 147
Spontaneous emission, 328, 341, 342
Square-well model, 77
Stark effect, 330
Statistical mechanics, 87, 99
Stellar atmospheres, 142
Stimulated emission, 328, 329
Stokes radiation, 27
Sulfur hexafluoride, 127
Sutherland's model, 90
Swan bands, 142
Symmetric rotor, 127, 326
Symmetric top, 127

Taylor series, 34
Teller's theorem, 121, 295
Term values, 128, 129
Tetrahedral molecules, 127
Thermal conductivity, 89, 90, 91, 98
Thermodynamic equilibrium, 77
Thermodynamics, 15
Thomas–Fermi (TF) theory, 101, 103, 105, 113, 120, 295, 314, 322
Time average, 283
Transition moments, 128
Transition probabilities, 340
Transition state theory (absolute rate theory), 227, 356, 358–360
Translational process, 31
Trimethylene methane, 191
Triple bond, 2, 127, 175
Triplet state, 141
Two-center integrals, 46
Two-photon absorption, 33
Two-photon processes, 33, 34

Uncertainty principle, 102, 345
Ungerade state, 163
Unsaturated C compounds, 175
UPS, 145, 164, 165, 181, 193

Valence bond theory, 46, 49, 50
Valency, 2, 67
Van der Waals forces, 69, 74, 75
Variation principle, 276
Velocity, of photoelectrons, 19
Vibration–rotation band, 137
Vibrational absorption, 30
Vibrational deactivation, 29
Vibrational degrees of freedom, 123, 233
Vibrational process, 31

Vibrational spectra, 123, 133
Vibronic transition, 144
Virial coefficients, 86, 87, 98
Virial expansion, 85
Virial theorem, 103, 109, 282, 285
Viscosity, 88, 91, 98

Walsh diagram, 163, 166, 211
Walsh's rules, 107, 108, 109
Water, 91, 162, 164, 213
Wave function, 98, 261

Wave mechanics, 37, 112
Well depth, 88
WKB approximation, 377, 378

Xenon, 91
X-ray diffraction, 96
X-ray photoelectron spectroscopy (XPS), 18, 20
X-ray scattering, 119, 318

Zero differential overlap (ZDO), 205
Zero-point energy, 16, 235